CYBERARTS

2000

Cyberarts
International Compendium Prix Ars Electronica
.net, Interactive Art, Computer Animation/
Visual Effects, Digital Musics, u19-Cybergeneration

Edition 2000

Publisher: Dr. Hannes Leopoldseder,
Dr. Christine Schöpf
Editor: Christian Schrenk,
ORF, Landesstudio Oberösterreich,
Europaplatz 3, 4010 Linz
Translation: Aileen Derieg, Helmut Einfalt
Cover-Design, Layout: Arthouse, Hansi Schorn
Frontispiece: Yasuo Ohba: Zen
Coordination/German Proof-Reading:
Ingrid Fischer-Schreiber
English Proof-Reading: Aileen Derieg, Giles Tilling
Offset Reproduction, Assembly: Typeshop, Linz
Copyright 2000
by Österreichischer Rundfunk (ORF), Landesstudio
Oberösterreich

Photo Credits
Cover: Yasuo Ohba; 38, 39: Maresch / Ars Electronica
Center; 53: Maria Ziegelböck; 107: Frank Schuberth;
122 – 125: Disney/Pixar; 136: PDI; 143: sixpackfilm, Wien;
203: ludger paffrath; 207: Jon Wozencroft, Touch;
214: Mario Radinovic; 216: Dinah Frank;
219: Ute Waldhausen; 244 – 249: Fotostudio Kutzler +
Wimmer, Traun

Printing: Euroadria, Ljubljana. Printed in Slovenia

Printed on acid-free and chlorine-free bleached
paper.

Prix Ars Electronica 2000
International Competition for Computer Arts
Organizer: Österreichischer Rundfunk (ORF),
Landesstudio Oberösterreich
Idea: Dr. Hannes Leopoldseder
Conception: Dr. Christine Schöpf
Financing/Copyright: Dkfm. Augner
Liaison Office:
Prix Ars Electronica, ORF, Europaplatz 3, A-4010 Linz
Phone: 0043/732/6900-24267
Fax: 0043/732/6900-24270
Telex: (02)1616
E-mail: info@prixars.orf.at

©2000 Österreichischer Rundfunk (ORF),
Landesstudio Oberösterreich

SPIN: 10766616
ISBN 3-211-83498-2

Springer-Verlag Wien New York

CYBERARTS

HANNES LEOPOLDSEDER - CHRISTINE SCHÖPF
PRIX ARS ELECTRONICA

2000

SpringerWienNewYork

CONTENTS

COMPUTER ANIMATION / VISUAL EFFECTS

DIGITAL MUSICS

CYBERGENERATION/U 19

INDIVIDUALITY— MOBILITY—GLOBALITY

INDIVIDUALITÄT – MOBILITÄT – GLOBALITÄT

"The whole internet could re-architected by Napster-like technology", says Intel Chairman Andy Grove in "FORTUNE" about the hot idea of the year 2000. The reference is to Napster, a tiny start-up company in San Mateo, California, which was founded by 19-year-old Shawn Fanning and 20-year-old Sean Parker in late 1999. It is a new idea for the exchange of information, continued by the Gnutella technology developed by 23-year-old Gene Kann and 26-year-old Spencer Kimball, which links the PC of an individual with as many users as possible in a network of so-called "peer-to-peer" connections. Regardless of how Napster and Gnutella continue to develop, one thing is certain: the Internet is by no means at the end of its development or in a phase of maturity, but rather at the beginning. The core idea of the net, all with all communication, is only beginning to exploit its potential. As Jeff Bezos, founder of Amazon.com, drastically states it, in comparison with geological timelines, the time dimension of the Internet is currently still located in the Cambrian period, in other words about 550 years ago. Direct communication ultimately also opens up completely new perspectives for artists, especially for media artists. The process that is beginning in music distribution with Napster-like technologies, for instance, naturally goes far beyond this field alone. In fact, these technologies have the potential to turn the PC industry upside down.

Starting in the 1980's, the Prix Ars Electronica has drawn a line of development to the year 2000. It is a line along the nexus of creativity and technological development. The field of creativity ranges from music, computer graphics, animation, interactive projects all the way to the Internet. However, a reality check 2000 shows one thing: in a global culture, the separate areas are merged in the Internet; convergence results in a new cultural dimension. The Prix

„Die ganze Architektur des Internet könnte durch eine Napster-ähnliche Technologie neu strukturiert werden", sagt Intel Chairman Andy Grove in *FORTUNE* über die „heiße Idee des Jahres 2000". Gemeint ist Napster, eine winzige Startup-Firma in San Mateo, Kalifornien, die vom 19-jährigen Shawn Fanning und vom 20-jährigen Sean Parker Ende 1999 gegründet wurde. Es ist eine neue Idee des Informationsaustausches, die von der Gnutella-Technologie, entwickelt vom 23-jährigen Gene Kann und vom 26-jährigen Spencer Kimball, weitergeführt wird und den PC des Einzelnen mit so vielen Usern wie möglich in einem Netz von sogenannten Peer-to-Peer-Verbindungen zusammenschließt. Wie immer sich Napster und Gnutella entwickeln, eines ist sicher: Das Internet steht am Beginn der Entwicklung, keineswegs am Ende oder in der Reifezeit. Die Kernidee des Netzes, die Alle-mit-Allen-Kommunikation, steht noch am Beginn der Ausschöpfung des Potenzials, wie es Jeff Bezos, der Gründer von Amazon.com, drastisch ausdrückt, der die Zeitdimension des Internet, verglichen mit geologischen Zeitabläufen, derzeit noch in der kambrischen Ära, also vor etwa 550 Millionen Jahren, angesiedelt sieht. Die direkte Kommunikation untereinander ist es letztlich auch, die für die Künstler, insbesondere für die Medienkünstler, völlig neue Perspektiven eröffnet. Der Prozess, der beispielsweise derzeit mit Napster-ähnlichen Technologien in der Musikdistribution beginnt, geht selbstverständlich weit über diesen Bereich hinaus. Tatsächlich hätten diese Technologien das Potenzial, die PC-Industrie auf den Kopf zu stellen.

Der Prix Ars Electronica zieht von den 80-er Jahren des 20. Jahrhunderts eine Entwicklungslinie zum Jahr 2000. Eine Linie im Schnittpunkt von Kreativität und technologischer Entwicklung. Der Gestaltungsbereich spannt sich von der Musik, der Computergrafik, der Animation, den interaktiven Projekten bis zum Internet. Ein Reality-Check 2000 ergibt allerdings eines: In einer globalen Kultur gehen die einzelnen Bereiche im Internet auf, die Konvergenz lässt eine neue kulturelle Dimension entstehen. Der Prix Ars Electronica schafft ein Spiegelbild der Kreativität in der Ausformung der Eigengesetzlichkeiten der digitalen Medien. Im Zentrum von allem steht das Internet. Das Internet ist der Träger der neuen globalen Kultur, die jedem Einzelnen mehr als zu jeder anderen Zeit die Chancen ermöglicht, seine kreati-

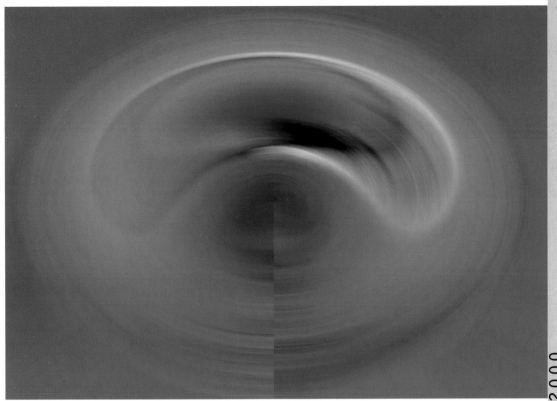

Ars Electronica provides a mirror image of creativity in the shaping of characteristic features of digital media. The focal point is primarily the Internet. The Internet is the platform of the new global culture, which provides every individual, more than ever before, with opportunities to communicate their creative ideas to other people not only in their immediate surroundings, but all around the world. Although the Internet has not created a new global culture, it has made it possible, particularly in terms of speed, but also through direct communication between the artist and the audience.

Napster and Gnutella are not quite the hottest topic of the year 2000 in Europe, but rather there is another term that seems to be the buzz word of the year 2000: mobile computing. Following e-business, e-commerce, e-life, the new magic letter is "m"—mobile. With mobile computing, e-commerce is succeeded by "m-commerce." Does this shift indicate a single fork, or does it indicate a new direction? Evidence may be found to support both views. In any case, one thing is certain: the Internet is reaching a rapidly growing number of customers with astonish-

ven Ideen anderer Menschen, in seiner Umgebung, aber auch weltweit zu vermitteln. Das Internet ist zwar nicht der Schöpfer einer neuen globalen Kultur, aber es ist der Ermöglicher, vor allem, was die Geschwindigkeit angeht, sowie die direkte Vermittlung zwischen Künstler und Publikum.

In Europa sind noch weniger Napster und Gnutella das Thema des Jahres 2000, sondern ein anderer Begriff scheint das Karrierewort des Jahres 2000 zu werden: Mobile-Computing. Nach E-Business, E-Commerce und E-Life heißt der neue Zauberbuchstabe „M" wie „mobil". Mit Mobile-Computing folgt dem E-Commerce der „M-Commerce". Bedeutet diese Verschiebung eine einzelne Verästelung oder bedeutet diese Verschiebung eine neue Richtung? Für beide Ansichten lassen sich Indizien finden. Eines allerdings steht fest: Über das Handy erreicht das Internet mit unwahrscheinlicher Geschwindigkeit eine rapid wachsende Anzahl von Kunden. Daher wird das Jahr 2000 die Wende zur Mobilität markieren, gleichzeitig aber auch den Schritt zur individuellen Unabhängigkeit von Raum und Zeit. Der individuelle mobile WAP-Zugang, in der Folge der UMTS-Standard, ist nicht nur ein Hoffnungsträger der IT-Branche, sondern legt in der digitalen Entwicklung den Fokus auf das Ich, die individuelle Unabhängigkeit, die Ortsungebundenheit, auf die Freiheit des Einzelnen. Wo immer ich bin – ich kann in nahezu allen Bereichen des Lebens, der Wirtschaft, der Unterhaltung, der Freizeit, der Kultur, in die vernetzte Welt einsteigen.

ing speed via cell phones. For this reason, the year 2000 marks a turn to mobility, but also a step toward the individual's independence from time and space. Individual mobile WAP access, consequent to the UMTS standard, is not only the great hope of the IT branch, but also focuses digital development on the individual, on individual independence of location, the freedom of every single person. No matter where I am, I can participate in nearly every area of life, business, entertainment, leisure time, culture, in the networked world.

Especially in Europe, the IT branch regards the widespread use of cell phones as a personal information and communication center as an advantage over the USA. This advantage will ultimately depend on the extent, to which customers may be provided with the appropriate contents, in particular the extent, to which it is possible to make use of the system's recognition of the location of each individual via the global medium of the Internet and provide the individual with appropriate offerings for their immediate surroundings, whether that means news, weather information, services information, community events, etc.

The most recent innovation of the Prix Ars Electronica, the Prix Ars Electronica Cyber Generation, is characteristic for the overall development that is being driven by an extremely young generation. No other technical innovation before has ever made it possible for such very young people, starting nearly at the age of ten, to take part with so much initiative and creativity in media, cultural and business projects. The Prix Ars Electronica Cyber Generation is a calling card of the generation under the age of nineteen. Quite a few of the participants have designed professional business projects, developed ideas that are far more than the mere school assignments one might perhaps expect. This generation will become the bearers of global culture, a global culture characterized by the individuality of each person and independence of time and place. The words of Anders Eriksson from the Swedish-American e-commerce company Razorfish, a consulting firm, apply to this generation: "We think and breathe digital, our soul is digital." With this qualification, it will be possible to cope with the digital transformation.

Gerade in Europa sieht die IT-Branche in der Ausbreitung des Handy als persönliches Info- und Kommunikationszentrale einen Vorsprung vor den USA – dieser Vorsprung wird letztlich davon abhängen, inwieweit die Kunden mit den entsprechenden Inhalten versorgt werden, insbesondere inwieweit es gelingt, über das globale Medium Internet die Systemkenntnis des Standortes jedes Einzelnen zu nutzen und den Einzelnen die entsprechenden Angebote seiner unmittelbaren Umwelt, seien es News, Wetter, Serviceinformationen oder Community-Angebote etc., zu vermitteln.

Die jüngste Innovation des Prix Ars Electronica, der Prix Ars Electronica Cybergeneration, ist charakteristisch für die gesamte Entwicklung, die von einer extrem jungen Generation getrieben wird. Bei keiner technischen Innovation zuvor war es möglich, dass derartig junge Menschen, nahezu von zehn Jahren an, initiativ, kreativ an medialen, kulturellen und wirtschaftlichen Projekten teilhaben können. Der Prix Ars Electronica Cybergeneration ist eine Visitenkarte der Generation unter 19. Nicht wenige unter den Teilnehmern entwerfen professionelle Wirtschaftsprojekte, entwickeln Ideen, die weit mehr als Schularbeiten sind, wie man vielleicht meinen könnte. Diese Generation wird die Trägerin der globalen Kultur sein, der globalen Kultur, die von der Individualität des Einzelnen, der Zeit- und Ortsungebundenheit geprägt sein wird. Für diese Generation gilt, was Anders Eriksson von der schwedisch-amerikanischen E-Commerce-Firma Razorfish, einer Consultantfirma, meint: „Wir denken und atmen digital, unsere Seele ist digital." Mit dieser Voraussetzung wird es möglich sein, den digitalen Wandel zu bewältigen.

Der Prix Ars Electronica liegt allerdings nicht im E-Business. Gerade in einer Zeit, in der die digitalen Medien vorwiegend von Start-ups, E-Commerce und E-Business vereinnahmt werden, kommt dem Prix Ars Electronica in seiner kulturellen Stellungnahme und Fokussierung eine wichtige Rolle zu. Denn seit seiner Gründung im Jahr 1987 war es die Absicht des Prix Ars Electronica, visionären Denkweisen und alternativen Annäherungen in Bezug auf digitale Mediengestaltung abseits industrieller Normen eine international beachtete Plattform anzubieten. In diesem Sinne konnte der Prix Ars Electronica über mehr als ein Jahrzehnt zu einer weltweiten Projektdokumentation werden, in denen sich die kulturelle und künstlerische Entwicklung im Aufbruch der digitalen Medien wiederfindet. Die kontinuierliche Dokumentation der Ergebnisse des Prix Ars Electronica in Publikationen, Videos, aber selbstverständlich auch im Internet, bedeuten eine nahezu weltweit umfassende Dokumentation, die im Gesamten global über das Internet verfügbar ist. Durch das Ars Electronica Center Linz ist es letztlich möglich geworden, von dieser Schlüsselstelle aus das Festival Ars Electronica, einschließlich des Prix Ars Electronica, aus der temporären Dauer eines Festivals in die bleibende

However, the Prix Ars Electronica is not in the e-business. Particularly at a time when the digital media are being taken over primarily by start-ups, e-commerce and e-business, the Prix Ars Electronica has an important role to play with its position and focus. The intention of the Prix Ars Electronica since its inception in 1987 has been to provide an internationally recognized platform for visionary thinking and alternative approaches to the shaping of digital media outside the realm of industrial norms. In this sense, the Prix Ars Electronica has been able to become a worldwide project documentation for over a decade, reflecting cultural and artistic development in the emergence of digital media. The continuous documentation of the results of the Prix Ars Electronica in publications, videos and, of course, the Internet signifies a virtually worldwide, comprehensive documentation that is globally accessible in its entirety via the Internet. Based on this key position, through the Ars Electronica Center it has become possible to maintain a permanent documentation of the Ars Electronica Festival, including the Prix Ars Electronica, beyond the temporality of a festival and distribution in print media, in digital media.

Yet the digital revolution is not nearly over; the Internet is still at the beginning of its development. For this reason, more than ever in the years to come, it will be the task and challenge of artists, creative minds and formative ideas to take on the peculiarities of digital media, exploit their possibilities and implement ideas accordingly. Interesting approaches are evident in every section of the competition, yet it is also clear that we find ourselves in the midst of a developmental phase. This, however, is precisely where the opportunities and obligations of forums such as the Prix Ars Electronica are to be found.

Dokumentation und über die Vermittlung im Druck hinaus in digitalen Medien erhalten zu lassen. Die digitale Revolution ist aber noch keineswegs abgeschlossen, das Internet steht nach wie vor am Beginn seiner Entwicklung. Daher wird es auch in den kommenden Jahren in besonderer Weise Aufgabe und Herausforderung der Künstler, der Kreativen, der Gestalter sein, sich mehr denn je den Eigengesetzlichkeiten der digitalen Medien anzunehmen, die Möglichkeiten auszuschöpfen und entsprechende Ideen umzusetzen. In allen Bereichen des Wettbewerbs zeigen sich interessante Ansätze, allerdings ist es auch ersichtlich, dass wir uns mitten in einer Entwicklungsstufe befinden. Aber gerade darin liegen Chance und Verpflichtung von Foren wie dem Prix Ars Electronica.

CYBERAR

The Prix Ars Electronica is the platform that cyber-artists from different disciplines all over the world have in common each year. Initiated in 1987 by the Austrian Broadcasting Corporation (ORF), Upper Austrian Regional Studio, the Prix Ars Electronica unites the artists, scientists and researchers, who reflect, discuss and comment on the digital transformation outside the mainstream and industrial norms. These are the people who represent the avant-garde in the elemental cultural upheaval of our time.

Particularly at a time, when the digital media are being taken over more and more by start-ups and e-business, the Prix Ars Electronica, as a common platform for cyberartists from various disciplines, has an important role to play with its position and focus. The intention of the Prix Ars Electronica since its inception in 1987 has been to provide an internationally recognized platform for bringing together visionary ideas and alternative approaches to digital media outside the realm of standardized patterns. With this year's results, the Prix Ars Electronica 2000 continues along this same path into the 21st century.

The Prix Ars Electronica is conceived to be not only an annual competition, but additionally also a vibrant laboratory for ideas and discussions, where new paths and trends may be examined and explored, but also where dead-ends and limitations in the world of digital media may be identified.

It is open for creations from different disciplines (in 2000 they are Digital Musics, Computer Animation and Visual Effects, Interactive Art and Internet), and over the course of the years it has shown itself to be flexible and adaptable in relation to developments in the field of digital media creativity. Like no other competition, the Prix Ars Electronica unites the otherwise separate worlds of art with those of development and research and with those of the entertainment industry. They all find a common platform in the Prix Ars Electronica.

With the additional competition category Cybergeneration—U19 Freestyle Computing, the Prix Ars Electronica once again demonstrates the creativity, freshness, vitality, but also the skill in the way that young people deal with digital media.

The presentation of the results in the form of exhibitions, artists' discussions, performances, concerts and screenings takes place not only as part of the Ars Electronica Festival, but in documentary form it also provides an ongoing survey of current positions in cyberarts.

Der Prix Ars Electronica ist die jährliche gemeinsame Plattform für Cyberkünstler unterschiedlicher Disziplinen in aller Welt.

1987 vom Österreichischen Rundfunk (ORF), Landesstudio Oberösterreich, ins Leben gerufen, vereint der Prix Ars Electronica jene Künstler, Wissenschafter und Forscher, die den digitalen Wandel abseits von Mainstream und industrieller Normierung reflektieren, diskutieren und kommentieren und somit die Avantgarde im tief greifenden kulturellen Umbruch unserer Zeit darstellen.

Gerade in einer Zeit, in der die digitalen Medien zunehmend von Start-ups und E-Business vereinnahmt werden, kommt dem Prix Ars Electronica in seiner kulturellen Stellungnahme und Fokussierung eine wichtige Rolle zu. Absicht des Prix Ars Electronica seit seiner Gründung im Jahr 1987 ist es, visionären Denkweisen und alternativen Annäherungen in Bezug auf digitale Mediengestaltung abseits normierter Muster eine international beachtete Plattform zu bieten. In seinen Ergebnissen führt der Prix Ars Electronica 2000 diesen Weg konsequent ins 21. Jahrhundert weiter.

In seiner Konzeption ist der Prix Ars Electronica nicht nur der jährliche Wettbewerb, sondern darüber hinaus ein vitales Denk- und Diskussionslabor, in dem neue Wege und Trends überprüft und ausgelotet werden, aber auch Sackgassen und Grenzen in der Welt der digitalen Medien aufgezeigt werden.

Er ist darüber hinaus offen für Kreationen unterschiedlicher Disziplinen (2000 sind das Digital Musics, Computeranimation, Visual Effects, interaktive Kunst und Internet), und er ist, über die Jahre gesehen, in Bezug auf die Entwicklungen im Bereich digitaler Mediengestaltung flexibel gestaltbar. Wie kein anderer Bewerb vereint der Prix Ars Electronica die im Regelfall getrennten Welten der Kunst mit denen der Entwicklung und Forschung ebenso wie mit denen der Unterhaltungsbranche. Sie alle finden im Prix Ars Electronica ihre gemeinsame Plattform. Mit der zusätzlichen Wettbewerbskategorie cybergeneration – u19 freestyle computing zeigt der Prix Ars Electronica einmal mehr die Kreativität, Frische, Vitalität, aber auch die Kompetenz Jugendlicher im Umgang mit den digitalen Medien auf.

Die Präsentation der Ergebnisse in Form von Ausstellungen, Künstlergesprächen, Performances, Konzerten und Screenings nicht nur innerhalb des Festivals Ars Electronica, sondern in dokumentarischer Form über den aktuellen Anlass hinaus, gewährleistet alljährlich den Überblick über aktuelle Positionen der Cyberarts.

The works presented in the book *Cyberarts 2000* were selected by five juries of international experts.

Ausgewählt wurden die im Buch *Cyberarts 2000* präsentierten Werke von fünf international besetzten Fachjuries.

A survey of the
Prix Ars Electronica 2000 prizes

Die Preise des
Prix Ars Electronica 2000 im Überblick

.net
Golden Nica (ATS 100.000/Euro 7.267/US$ 7,142):
Neal Stephenson (USA)
Two Awards of Distinction
(ATS 50.000/Euro 3.633/US$ 3,571 each):
TeleZone Group: Erich Berger, Peter Purgathofer,
Volker Christian and many more (Austria/USA)
Sharon Denning (USA)

.net
Goldene Nica (ATS 100.000/Euro 7.267/US$ 7,142):
Neal Stephenson (USA)
2 Auszeichnungen
(je ATS 50.000/Euro 3.633/US$ 3,571):
TeleZone-Gruppe: Erich Berger, Peter Purgathofer,
Volker Christian und viele andere (Österreich/USA)
Sharon Denning (USA)

Interactive Art
Golden Nica (ATS 200.000/Euro 14.534/US$ 14,286):
Rafael Lozano-Hemmer (Mexico/Canada) and Team
Two Awards of Distinction
(ATS 50.000/Euro 3.633/US$ 3,571 each):
Golan Levin (USA)
The Institute for Applied Autonomy (USA)

Interaktive Kunst
Goldene Nica (ATS 200.000/Euro 14.534/US$ 14,286):
Rafael Lozano-Hemmer (Mexiko/Kanada) und Team
Zwei Auszeichnungen
(je ATS 50.000/Euro 3.633/US$ 3,571):
Golan Levin (USA)
The Institute for Applied Autonomy (USA)

Special Prize of the Jury for Interactive Art
The RoboCup Federation:
Hiroaki Kitano, President (Japan)

Sonderpreis der Jury für Interaktive Kunst
The RoboCup Federation:
Hiroaki Kitano, Präsident (Japan)

Computer Animation
Golden Nica (ATS 200.000/Euro 14.534/US$ 14,286):
Jakub Pistecky (Canada)
Two Awards of Distinction
(ATS 50.000/Euro 3.633/US$ 3,571 each):
John Lasseter, Lee Unkrich, Ash Brannon/
PIXAR Studios (USA)
Yasuo Ohba (Japan)

Computeranimation
Goldene Nica (ATS 200.000/Euro 14.534/US$ 14,286):
Jakub Pistecky (Kanada)
Zwei Auszeichnungen
(je ATS 50.000/Euro 3.633/US$ 3,571):
John Lasseter, Lee Unkrich, Ash Brannon/
PIXAR Studios (USA)
Yasuo Ohba (Japan)

Visual Effects
Golden Nica (ATS 200.000/Euro 14.534/US$ 14,286):
Christian Volckman (France)
Two Awards of Distinction
(ATS 50.000/Euro 3.633/US$ 3,571 each):
Markus Degen (Austria)
Pierre Buffin, BUF Compagnie (France)

Visual Effects
Goldene Nica (ATS 200.000/Euro 14.534/US$ 14,286):
Christian Volckman (Frankreich)
Zwei Auszeichnungen
(je ATS 50.000/Euro 3.633/US$ 3,571):
Markus Degen (Österreich)
Pierre Buffin, BUF Compagnie (Frankreich)

Digital Musics

Golden Nica (ATS 150.000/Euro 10.901/US$ 10,715):
Carsten Nicolai (Germany)
Two Awards of Distinction
(ATS 50.000/Euro 3.633/US$ 3,571 each):
Chris Watson (Great Britain)
GESCOM / Rob Brown, Sean Booth, Russell Haswell

Cybergeneration—U19 Freestyle Computing

Golden Nica (plus free Internet connection for one year and a multimedia PC): Verena Riedl, Michaela Hermann and the project group Harvey (Waidhofen an der Thaya/Lower Austria)
Two Awards of Distinction
 (plus 1 multimedia notebook each):
Erich Hanschitz, Alexander Blaschitz, Daniel Kummer, Ronald Rabitsch, Martin Schliefnig / Project Group BG/BRG Völkermarkt (Völkermarkt/Carinthia)
Gerhard Schwoiger (Vienna)

Digital Musics

Goldene Nica (ATS 150.000/Euro 10.901/US$ 10,715):
Carsten Nicolai (Deutschland)
Zwei Auszeichnungen
(je ATS 50.000/Euro 3.633/US$ 3,571):
Chris Watson (Großbritannien)
GESCOM / Rob Brown, Sean Booth, Russell Haswell

Cybergeneration – U19 Freestyle Computing

Goldene Nica (plus kostenloser Internet-Anschluss für ein Jahr und Multimedia-PC): Verena Riedl, Michaela Hermann und die Projektgruppe Harvey (Waidhofen an der Thaya/NÖ)
Zwei Auszeichnungen
(plus jeweils ein Multimedia-Notebook):
Erich Hanschitz, Alexander Blaschitz, Daniel Kummer, Ronald Rabitsch, Martin Schliefnig / Projektgruppe BG/BRG Völkermarkt (Völkermarkt/ Kärnten)
Gerhard Schwoiger (Wien)

Sponsors

The Prix Ars Electronica 2000 is endowed with a total of ATS 1.35 million.
The prize is donated this year by the Austrian Internet provider A-Online.
The competition Cybergeneration—U19 Freestyle Computing, which is open to young people in Austria under the age of 19, is sponsored by the Austrian Postal Bank (P.S.K.) and conducted by ORF in cooperation with the Austrian Cultural Service (Ö.K.S.).
The Prix Ars Electronica can only be realized as an international competition through the financial commitment of sponsors from the business sector and support from public funding.
In this sense, the Prix Ars Electronica is made possible by Datakom Austria, VOEST-ALPINE STAHL, the city of Linz and the federal province of Upper Austria.
For additional support, the Prix Ars Electronica is grateful to Austrian Airlines, Lufthansa, Porsche Austria, Sony DADC, Courtyard by Marriott, TNT International Mail, Casinos Austria, Fujitsu Siemens and Pöstlingberg Schlössl.

Sponsoren

Der Prix Ars Electronica 2000 ist mit insgesamt ATS 1,35 Mio. dotiert.
Preisstifter ist der österreichische Internet-Provider A-Online.
Der Bewerb cybergeneration – u19 freestyle computing, der Jugendliche in Österreich unter 19 zur Teilnahme einlädt, wird von der Österreichischen Postsparkasse (P.S.K.) gesponsert und vom ORF in Zusammenarbeit mit dem Österreichischen Kulturservice (Ö.K.S.) durchgeführt.
Realisiert kann der Prix Ars Electronica als internationaler Wettbewerb nur durch das finanzielle Engagement von Sponsoren aus der Wirtschaft und Förderern der öffentlichen Hand werden.
In diesem Sinn wird der Prix Ars Electronica ermöglicht von Datakom Austria, VOEST-ALPINE STAHL, der Stadt Linz und dem Land Oberösterreich.
Für weitere Unterstützung dankt der Prix Ars Electronica Austrian Airlines, Lufthansa, Porsche Austria, Sony DADC, Courtyard by Marriott, TNT International Mail, Casinos Austria, Fujitsu Siemens und Pöstlingberg Schlössl.

.NET
INTERACTIVE ART
COMPUTER ANIMATION
VISUAL EFFECTS
DIGITAL MUSICS
U19/CYBERGENERATION

WHERE HAVE ALL THE ARTISTS GONE?
... WO SIND SIE GEBLIEBEN?

John Markoff

The Prix Ars Electronica .net category is intended to be open to a broad cross section of artistic and cultural Internet activities ranging from World Wide Web sites to new and innovative protocols that are intended to foster community and communication. This year the Prix Ars Electronica .net jury meeting began our session with a discussion of criteria to keep in mind while reviewing the entries. The .net category has a 5-year tradition of looking for work that expresses the *zeitgeist* of web and net development. The web continues to change at a radical pace, but there is an underlying essence that the jury summarized as "Web-ness". From previous years Web-ness has come to be viewed as the combination of elements that make a project inherently networked. If it can be delivered in another medium just as effectively, a project has been thought to be low in Web-ness. Some signs that Web-ness exists are thriving communities, self-replicating technologies and business models, and distributed authority. In keeping with the Prix Ars Electronica tradition, purely commercial websites were not considered in 2000.

This was an unusual year for the Prix Ars Electronica .net jury. One disappointment in this year's competition was the decline in the number of entries in our category from about 500 to 250.

The jury spent some time discussing the reason for the decline and although a number of explanations were put forward, the most likely reason appears to be the vast and unprecedented rise in e-commerce activity on the Internet. Simply put, with vast flows of venture capital available to start Internet enterprises, virtually anyone with any skill or interest in the Net has increasingly been spending his or her waking hours pursuing commercial rather than artistic endeavors.

Originality of concept was an important factor in selecting the Prix winner. As the jury reviewed the .net submissions, we came to believe that many of this year's entries were derivative of previous well-known net projects. While many of them were well-designed and more stable than their predecessors, we felt that a Golden Nica winner should reflect a vision that is at the very frontier of Internet development.

Die .net-Kategorie des Prix Ars Electronica soll einem breiten Querschnitt durch die künstlerischen und kulturellen Internet-Aktivitäten offen stehen – von World Wide Web-Sites zu neuen und innovativen Protokollen, die die Gemeinschaft und die Kommunikation vorantreiben.

In diesem Jahr begann die Sitzung der .net-Jury des Prix Ars Electronica mit einer Diskussion der Kriterien, die uns bei der Betrachtung der Einreichungen als Richtschnur dienen sollten. Die .net-Kategorie kann mittlerweile auf eine fünfjährige Tradition in der Suche nach Werken zurückblicken, die den Zeitgeist von Web- und Netz-Entwicklung darstellen sollen. Das Web ändert sich weiterhin mit rasanter Geschwindigkeit, aber dahinter liegt eine essenzielle Grundeigenschaft, die die Jury als „Web-ness" („Webgemäßheit") bezeichnete. Aus der Erfahrung vergangener Jahre ergab sich, dass „Webgemäßheit" in der Kombination jener Elemente besteht, die ein Projekt inhärent vernetzt machen. Wenn das Werk in einem anderen Medium genauso effizient dargeboten werden könnte, dann ist es als nur wenig „webgemäß" anzusehen. Und dass Webgemäßheit tatsächlich existiert, zeigen einige prosperierende Communities oder selbstreplizierende Technologien und Modelle, sie zeigt sich aber auch in einer distribuierten Autorschaft. In Übereinstimmung mit der Tradition des Prix Ars Electronica wurden rein kommerzielle Websites dieses Jahr außer Betracht gelassen.

Insgesamt war es ein ungewöhnliches Jahr für die Prix Ars Electronica .net-Jury. Enttäuschend war unter anderem der Rückgang der Einreichungen in unserer Kategorie von rund 500 auf 250.

Die Jury verbrachte einige Zeit damit, Gründe für diesen Rückgang zu finden, und obwohl sich mehrere Erklärungen anboten, scheint doch die wahrscheinlichste Ursache das beispiellose Anwachsen der E-Commerce-Aktivitäten im Internet zu sein. Einfach gesagt hat dank der enormen Menge an Risikokapital, die für die Gründung von Internet-Unternehmen zur Verfügung steht, so gut wie jeder, der nur etwas Geschick und Interesse am Netz hatte, seine Zeit eher mit der Verfolgung von kommerziellen als von künstlerischen Zielen verbracht.

Die Originalität des Konzepts war ein wichtiger Faktor bei der Auswahl des Preisträgers. Bei der Betrachtung der heurigen Einreichungen kam die Jury schnell zur Ansicht, dass viele der präsentierten Arbeiten Ableger früherer wohl bekannter Netz-Projekte sind. Und wenn auch viele davon gut designed und stabiler als ihrer Vorläufer waren, kamen wir doch zur Ansicht, dass der Gewinner der Goldenen Nica eine Vision widerspiegeln sollte, die an vorderster Front der Internet-Entwicklung steht.

In früheren Web-Jahren gab es so viele neu- und großartige Ideen, dass es für vorangegangene .net-

In earlier web-years, there were so many novel and huge ideas that it was often hard for previous .net juries to select one for the prize. This year, the commercial web appears to dominate. Thus the most creative projects on the web are increasingly found at the extremes of commercialism and anti-commercialism. It is important to note that not all of the decisions made by the jury this year were by consensus.

Golden Nica 2000

Neal Stephenson

In a postscript to his pathbreaking 1992 science fiction novel *Snow Crash* author Neal Stephenson writes about how he settled reluctantly upon writing a traditional novel when he discovered that the current interactive CD technology was not yet powerful enough to support his vision of a new kind of cyber universe.

In the text-based novel that resulted, he named this computer and network generated world the "Metaverse." The idea gave William Gibson's original notion of cyberspace a geography and although Stephenson had given up on his original plan to create an interactive computer novel, many of his readers were inspired by his vision.

The idea that there is a fabric to cyberspace and that it is somehow more than the sum of the machines that are interconnected by IP addresses is very much a reality today. VRML is a well established protocol on the web, and it has recently been complemented by a wide range of other three-dimensional navigational protocols.

Neal Stephenson is a fitting winner of the Golden Nica 2000 for more reasons than *Snow Crash*. While many writers arrive at a single vision and remain stuck there, Stephenson has continued to re-invent himself, exploring a succession of technologies and their impact on the world.

In *Diamond Age* he takes the reader on an alarming tour of the world of nanotechnology, a world which validates Arthur C. Clark's dictum: "Any sufficiently advanced technology is indistinguishable from magic."

It is particularly striking that last year Stephenson crossed back over from science fiction in his best selling novel *Cryptonomicon*. A sprawling first work of what will likely become a trilogy, *Cryptonomicon* stretches from the crucial role that cryptography played in the Second World War to an Asian Internet

Jurys oft schwer war, aus der Vielzahl der Kandidaten einen Preisträger auszuwählen. Heuer scheint das kommerzielle Web zu dominieren, und so waren die kreativsten Projekte an den Extremen von Kommerz und Anti-Kommerz zu finden.

Es soll noch festgehalten werden, dass nicht alle Entscheidungen der diesjährigen Jury einstimmig getroffen wurden.

Goldene Nica 2000

Neal Stephenson

In einem Postscriptum zu seinem bahnbrechenden Science-Fiction-Roman *Snow Crash* von 1992 beschreibt der Autor Neal Stephenson, dass er dafür nicht ganz freiwillig die Form eines traditionellen Romans gewählt hatte, doch musste er feststellen, dass die damalige interaktive CD-Technologie noch nicht leistungsfähig genug war, um seine Vision einer neuen Art von Cyber-Universum zu tragen. Im daraus resultierenden text-basierten Roman bezeichnete er diese computer- und netzwerkgenerierte Welt als „Metaverse". Diese Idee fügte William Gibsons ursprünglicher Ansicht von Cyberspace eine Geografie hinzu, und obwohl Stephenson seinen ursprünglichen Plan eines interaktiven Computerromans aufgegeben hatte, wurden doch viele seiner Leser von dieser Vision inspiriert.

Die Vorstellung, dass dem Cyberspace eine Struktur zu Eigen sei und diese mehr sei als nur die Summe der über IP-Adressen verbundenen Maschinen, entspricht durchaus der heutigen Wirklichkeit. VRML ist ein im Web wohl etabliertes Protokoll und wird mittlerweile von einem breiten Spektrum dreidimensionaler Navigationsprotokolle ergänzt.

Neal Stephenson ist ein äußerst passender Gewinner der Goldenen Nica 2000 – und nicht nur wegen *Snow Crash*. Während viele Autoren zu einer einzigen Vision gelangen und dort stecken bleiben, fährt Stephenson fort, sich selbst neu zu erfinden, eine Abfolge von Technologien und ihren Einfluss auf die Welt zu erforschen.

In *Diamond Age* führt er seine Leser auf eine erschreckende Tour durch die Welt der Nanotechnologie, eine Welt, die die Gültigkeit von Arthur C. Clarkes Diktum „Jede ausreichend fortgeschrittene Technologie ist von Zauberei nicht zu unterscheiden" unterstreicht.

Es ist besonders bemerkenswert, dass Stephenson im vergangenen Jahr in seinem Bestseller *Cryptonomicon* die Welt der Science-Fiction wieder verlassen hat. Als fesselnder erster Band einer voraussichtlichen Trilogie spinnt *Cryptonomicon* eine Geschichte, die mit der zentralen Rolle, die die Kryptografie im Zweiten Weltkrieg spielte, beginnt und bis hin zu einem asiatischen Internet-Datenspeicher in einer nicht allzu fernen Zukunft reicht. Der Protagonist des Romans, Randy Waterhouse, ein Programmierer, ist ein Unix-Hacker und der Roman berührt viele der vordringlichsten politischen, sozialen und kulturellen Fragen, die durch das globale Wachstum des Internet aufgeworfen werden.

data haven sometime in the not-to-distant future. Its protagonist, programmer Randy Waterhouse, is a Unix hacker and the novel touches on many of the most pressing political, social and cultural issues raised by the global ascendancy of the Internet.

Distinctions

Exquisite Corpse
Sharon Denning

The *Exquisite Corpse* puts an elegant interface on the surrealist writing game, in which one person begins a narrative and others add on to it. It is a well-designed collaborative storytelling tool. The unique interface allows easy additions to the non-linear story. Time and care was spent on designing the interface. As new sections are added, the interface grows to accommodate. Telling stories with the web is the idea that it explores. Text-based hyperlinking was the first big breakthrough as an alternative to the printed linear story. Interfaces like this one now explore the next generation in social storytelling.

TeleZone
TeleZone team: Erich Berger/Volker Christian/ Ken Goldberg/Peter Purgathofer

The *TeleZone* is a telerobotic art installation that creates a parallel between the physical and the virtual. The project also represents today's state of the art in virtual community. As a successor to Ken Goldberg and Joseph Santarromana's *TeleGarden*, which explored some of the same challenges in the dividing line between the physical and the virtual, a true virtual community has come to life. The goal is to permit social collaboration exclusively over the net in creating, planning, managing and growing a global community that has a physical presence. The site is designed for a city/community planner. Participants can come and view it as visitors or can become actively involved in the process of creating the community. All of the decisions are decided by the participants and it's entirely up to them to decide how to self organize. It is represented in a three dimensional VRML version online, and exists in the real world as an installation at the Ars Electronica Center in Linz. A robot at the Ars Electronica Center serves as an interface between the physical and the virtual spaces. Acting as an avatar on behalf of users in the community, it arranges and places together building

Auszeichnungen

Exquisite Corpse
Sharon Denning

Exquisite Corpse ist ein elegantes Interface für jenes surrealistische Schreibspiel, bei dem eine Person eine Erzählung startet und andere Teile hinzufügen. Es ist ein schön designetes Werkzeug zum Geschichtenerzählen. Das einzigartige Interface macht es leicht, etwas zur nicht-linearen Erzählung beizutragen, und man merkt, dass viel Zeit und Sorgfalt für dessen Entwicklung aufgewandt wurde. Text-basierte Hyperverknüpfungen waren der erste Durchbruch als Alternative zur gedruckten linearen Erzählung. Dieses Interface erforscht die sich dadurch bietenden Möglichkeiten, Geschichten mittels des Web zu erzählen, und es ist typisch für die Experimente auf der Suche nach einer neuen Generation sozialer Narration.

TeleZone
Das TeleZone -Team: Erich Berger/Volker Christian/Ken Goldberg/Peter Purgathofer

TeleZone ist eine telerobotische Installation, die eine Parallele zwischen dem Physischen und dem Virtuellen herstellt. Das Projekt ist auch repräsentativ für den gegenwärtigen technischen Stand der Dinge in Sachen virtuelle Gemeinschaften. In der Nachfolge von Ken Goldbergs und Joseph Santarromanas *TeleGarden*, der sich mit einer ähnlichen Problematik an der Grenzlinie zwischen physischer und virtueller Welt beschäftigte, ist hier eine richtige virtuelle Gemeinschaft entstanden. Ziel ist es, eine globale Gemeinschaft zu schaffen, zu planen, zu organisieren und weiterzuentwickeln, deren soziale Struktur ausschließlich über das Netz zugänglich ist, die aber dennoch eine physische Präsenz hat. Die Site ist inhaltlich auf das Thema Stadtplanung/ Raumordnung ausgelegt. Mitwirkende können entweder als Besucher oder aktiv an dem gemeinschaftlichen Prozess teilnehmen. Alle Entscheidungen werden von den Teilnehmern selbst getroffen, wobei es auch ihnen überlassen bleibt, wie sie sich organisieren. Online wird das Projekt über ein dreidimensionales VRML-Interface dargestellt, in der realen Welt existiert es als Installation im Ars Electronica Center in Linz.

Ein Roboter im Ars Electronica Center bildet die Schnittstelle zwischen Wirklichkeit und Virtualität. Als Avatar für die Benutzer der Gemeinschaft bildet er mit Bausteinen die gemeinschaftlich beschlossenen Strukturen der *TeleZone*-Website ab. Die Handlungen und Entscheidungen der Gemeinschaft werden so in der virtuellen wie in der physischen Welt sichtbar.

blocks based on design decisions made on the *Tele-Zone* web site. The actions of the community are visible both in the virtual and the physical worlds.

Honorable Mentions

@2000
Ichiro Aikawa

The jury's reaction ranged from "creative use of windows" to "pure delicious eye candy and frame play." In any case, the cascades of windows move, reframe and cascade. Sometimes they contain images and tell stories, even guide the web surfer on an short urban journey.

Intruder
Natalie Bookchin

Where is the line between video games and literature? *Intruder* takes the unique approach of reading and displaying the lines of a tale by Jorge Luis Borges to ten simple video games. It is not clear that the games add anything to the story. It is however a unique approach to the form of the story.

Reconnoitre
Tom Corby / Gavin Bailey

Corby and Baily acknowledge that *Reconnoitre* can best be described as a "dysfunctional browser." It attempts to restore serendipity to web exploration with what they call a "journey of surprises." The idea is to graze through text in a fragmentary, incomplete and aimlessly wandering fashion. They attempt to blend text with a 3D browser world, browser matches are displayed in a three-dimensional array where the "hits" are endowed with characteristics from a surreal physical world such as entropy, repulsion and attraction.

Discoder
exonemo [Kensuke Sembo, Yae Akaiwa]

What is a discoder? A *Discoder*, according to its designer, is a device "which destroys HTML informatic code and its codes of behavior, a contradiction provider for the web". In this project the user "messes" with the web's HTML structure. Or as one jury member described it, "just a small cool hack. From a programmer's perspective this has special meaning. As creators of code we strive to make our products clean, well formed and easy to read. This hack is

@2000
Ichiro Aikawa

Die Reaktion der Jury schwankte zwischen „kreativem Einsatz von Fenstern" bis „ergötzlichem Schnickschnack fürs Auge im Spiel mit Frames". Wie dem auch sei – die Kaskaden von Fenstern bewegen sich, formen sich um und bilden neue Kaskaden. Manchmal enthalten sie Bilder und erzählen Geschichten, bisweilen führen sie die Surfer sogar auf eine kurze urbane Reise.

Intruder
Natalie Bookchin

Wo liegt die Grenzlinie zwischen Videospiel und Literatur? Der einzigartige Ansatz von *Intruder* besteht darin, eine Erzählung von Jorge Luis Borges zu lesen und anzuzeigen und sie auf zehn simple Videospiele abzubilden. Es ist nicht klar geworden, ob die Spiele etwas zur Erzählung beitragen, aber das ist jedenfalls eine einzigartige Annäherung an die Form der Geschichte.

Reconnoitre
Tom Corby / Gavin Bailey

Corby und Baily geben selbst zu, dass *Reconnoitre* am besten als „dysfunktionaler Browser" beschrieben werden kann. Das Werk versucht, eine spielerische Leichtigkeit in die Erforschung des Web zu bringen, indem es eine „Reise der Überraschungen" anbietet. Der Grundgedanke ist es, auf fragmentarische, unvollständige und ziellose Weise durch Texte zu wandern. Ihre Versuche, Text mit einem 3D-Browser zu sehen und zu kombinieren, werden als dreidimensionales Array dargestellt, wobei die „Treffer" der Suchfunktion mit Eigenschaften einer surrealen physischen Welt – wie Entropie, Anziehung und Abstoßung – ausgestattet werden.

Discoder
Exonemo [Kensuke Sembo, Yae Akaiwa]

Was ist ein „Discoder"? Ein *Discoder* ist nach Ansicht seines Designers ein Gerät, „das HTML-Code und dessen Verhaltenscodes zerstört, ein Widerspruchsprovider fürs Web". Bei diesem Projekt wird mit der HTML-Struktur des Web herumgespielt, oder, wie ein Jurymitglied das ausgedrückt hat: „Ein kleiner cooler Hack. Aus Sicht eines Programmierers kommt dem besondere Bedeutung zu: Als Schöpfer von Codes bemühen wir uns, unsere Produkte klar, gut geformt und leicht leserlich zu machen. Dieser Hack tut das genaue Gegenteil: Er ist ein ganz einfacher (und klarer und gut geformter) Weg, unseren durchaus ordentlichen Code zu verwursten."

Toywar.com
Reinhold Grether

Eine Protest-Website, die gegen den Rechtsstreit um einen Internet-Domain-Namen zwischen etoy.com und eToys.com auftritt. *Toywar.com* dokumentiert

exactly the opposite, a completely simple (and clean and well formed) way to completely mess up your perfectly good code."

toywar.com
Reinhold Grether

A protest web site designed to protest the legal battle between etoy.com and eToys.com over an Internet domain name, *toywar.com* chronicles the collapse of eToys' stock price and claims credit for it. *toywar.com* describes itself as a "resistance game" which was launched to protect etoy from a hostile takeover. According to the *toywar.com* site, in its suit the toy retailer accused etoy of unfair competition, trademark delusion, security fraud, illegal stock market operation, pornographic content, offensive behavior and terrorist activity. After the resistance campaign began, eToys Inc. stock dropped from $67 to $15, a loss of $4.5 billion. *toywar.com* calls this the most expensive performance in art history: $4.5 billion in lost equity.

Silk Road
Jie Geng

Silk Road is drawn from the artist's experience a decade ago as an art student at the Fine Arts Academy of China. While there each student was required to visit and study the grottoes of the Silk Road in Dunhuang. In this site the artist attempts to present the legend of the Silk Road to the world of the Internet. It uses several navigational interfaces to traverse subjects and locations and includes documentation in an attempt to recreate the idea of a journey on the Silk Road.

Sinnzeug
Stephan Huber/Ralph Ammer/Birte Steffan

Sinnzeug uses Flash in an effort to create a dynamic search engine that plays with the idea of the spatial organization of information. Dots which represent web sites appear on the screen. After doubleclicking in the window, it is possible to enter a search word or choose a word from a pop-up menu. The dots that "feel" they have something to do with the word are attracted to it, offering visual clues.

Zeitgenossen Binary Art Site
Ursula Hentschläger/Zelko Wiener

This was one of a number of sites entered in the .net category this year that explore Macromedia's Flash

den Einbruch des Aktienpreises von eToys und heftet sich diesen auf seine Fahnen. *toywar.com* beschreibt sich selbst als „Widerstandsspiel", das gestartet wurde, um etoy vor einer feindlichen Übernahme zu schützen. Nach Angabe von *toywar.com* wirft der Spielzeughändler in seiner Klage etoy unlauteren Wettbewerb, Verstöße gegen den Schutz des Markennamens, Sicherheitsbetrug, illegale Börsenaktivitäten, Pornografie, Beleidigung und sogar terroristische Aktivitäten vor. Seit dem Beginn der Widerstandskampagne ist die Aktie von eToys Inc. von 67 USD auf 15 USD gefallen - ein Verlust von 4,5 Milliarden Dollar. *toywar.com* nennt dies die teuerste Performance in der Kunstgeschichte: Viereinhalb Milliarden verlorenes Kapital.

Silk Road
Jie Geng

Silk Road basiert auf den Erfahrungen, die die Künstlerin vor einem Jahrzehnt als Studentin an der Kunstakademie in China machte. Damals wurde jeder Student angehalten, die Grotten der Seidenstraße in Dunhuang zu besuchen und zu studieren. Auf dieser Site versucht die Autorin, die Legende von der Seidenstraße für die Welt des Internet aufzubereiten. Die Site verwendet mehrere Navigationsinterfaces, um durch Objekte und Orte zu reisen, und bietet auch eine Dokumentation an, die hilft, den Eindruck einer Reise entlang der Seidenstraße zu vermitteln.

Sinnzeug
Stephan Huber / Ralph Ammer / Birte Steffan

Sinnzeug verwendet Flash als Instrument für eine dynamische Suchmaschine, die mit der Idee der räumlichen Organisation von Information spielt. Punkte tauchen auf dem Bildschirm auf und stehen für Websites. Nach einem Doppelklick ins Fenster kann ein Suchbegriff eingegeben oder aus einem Menü ausgewählt werden. Jene Punkte, die sich vom Suchbegriff „angesprochen fühlen", werden von ihm angezogen und geben visuelle Hinweise auf ihre Relevanz.

Zeitgenossen Binary Art Site
Ursula Hentschläger / Zelko Wiener

Diese war eine von vielen heuer in der .net-Kategorie eingereichten Sites, die sich mit Macromedias Flash als künstlerischem Werkzeug beschäftigen. *Zeitgenossen* arbeitet am Zusammenfluss von Klang, Text, Farbe, künstlerischen Formen und Bewegung.

as an artistic tool. *Zeitgnossen* works at the confluence of sound, text, color, artistic forms and motion.

Grasping at Bits
Patrick Lichty

A hyper essay by artist Patrick Lichty, *Grasping at Bits* uses thebrain.com as a navigational tool. Lichty makes an effort to explore the corporate influence and control of intellectual property. He also speculates upon the consequences that will emerge from the intersection of the aesthetic and the material in the Internet age.

Electrica
Peter Mühlfriedel/Gundula Markeffsky

This interactive web site uses the notion of old electronic devices as an interface to modern music composition. This creates a nice juxtaposition. *Electrica* is an addictive set of techno instruments, beautifully designed online toys for creating ambient techno mixes. Its experimental interfaces are an intriguing pleasure to surf through.

Network Communicate Kaleidoscope
Kazushi Mukaiyama

Network Communicate Kaleidoscope is a virtual space accessible via the Internet. Users connect to a void world as a "particle" and chat and interact with others. After disconnecting, their remnant particle behaves individually with other particles' behavior reflecting activity by all who have ever connected in the virtual space. The theme is to explore and create images such as a kaleidoscope of "clustering fireflies." The jury felt it had a simple and elegant design.

0100101110101101.ORG
Art or conceptual prank? Intellectual property wars are not limited to Napster, the MP3 file trading software or to the world of video. What is authorship when anything can be copied all over the world with a single keystroke? It is something that the world has struggled over at least since the French Revolution, when the revolutionaries decried authorship. The jury felt the *0100101110101101.ORG* web site stood as a tribute in style and execution to jodi, another group of anarchists/artists. It calls into question the value of the "original" in digital media—and raises questions about the entire idea of copyright.

Grasping At Bits
Patrick Lichty

Als Hyper-Essay des Künstlers Patrick Lichty verwendet *Grasping At Bits* thebrain.com als Navigationswerkzeug. Lichty beschäftigt sich mit der Frage, welchen Einfluss und welche Kontrolle Unternehmen über geistiges Eigentum haben. Er spekuliert auch über die Konsequenzen, die aus der Durchdringung von Ästhetischem und Materiellem im Internetzeitalter folgen werden.

Electrica
Peter Mühlfriedel / Gundula Markeffsky

Diese interaktive Website verwendet die Terminologie altmodischer elektronischer Geräte als Interface zur modernen Komposition von Musik, was einen netten Kontrast darstellt. *Electrica* ist ein fast süchtig machendes Set von Techno-Instrumenten, ein Online-Spielzeug in wunderschönem Design zur Schaffung von Techno-Mixes. Seine experimentellen Interfaces faszinieren und machen Spaß.

Network Communicate Kaleidoscope
Kazushi Mukaiyama

Network Communicate Kaleidoscope ist ein über Internet zugänglicher virtueller Raum. Benutzer gelangen als „Partikel" in diesen Raum und können dort mit anderen sprechen und interagieren. Auch nachdem sich der Benutzer ausgeloggt hat, verbleibt sein Partikel im Raum und interagiert individuell mit den anderen verwaisten Partikeln, wobei er das Gesamtverhalten aller jemals eingeloggter Benutzer widerspiegelt. Aus dieser Untersuchung des Kommunikationsverhaltens entsteht eine Art „Kaleidoskop von Fliegenschwärmen". Die Jury fand sein elegantes und einfaches Design anerkennenswert.

0100101110101101.ORG
Kunst oder konzeptueller Streich? Auseinandersetzungen um geistiges Eigentum beschränken sich nicht auf Napster, MP3-Musik oder die Welt des Videos. Was bedeutet Autorschaft, wenn alles auf der ganzen Welt mit einem Tastendruck kopierbar ist? Die Frage nach dem geistigen Eigentum steht (spätestens) seit der Französischen Revolution im Raum, als die Revolutionäre die individuelle Autorschaft für abgesetzt erklärten. Die Jury war der Ansicht, dass die Website von 0100101110101101.ORG in Stil und Ausführung auch *jodi* Tribut zollt, einer anderen Gruppe von Anarchisten/Künstlern. Sie hinterfragt den Wert des „Originals" in digitalen Medien und stellt das gesamte Copyright-Konzept in Frage.

IN THE BEGINNING WAS THE COMMAND LINE (EXCERPTS)

Neal Stephenson

Computers do arithmetic on bits of information. Humans construe the bits as meaningful symbols. But this distinction is now being blurred, or at least complicated, by the advent of modern operating systems that use, and frequently abuse, the power of metaphor to make computers accessible to a larger audience ...

People who have only interacted with computers through graphical user interfaces like the Mac OS or Windows—which is to say, almost everyone who has ever used a computer—may have been startled, or at least bemused, to hear about the telegraph machine that I used to communicate with a computer in 1973. But there was, and is, a good reason for using this particular kind of technology. Human beings have various ways of communicating *to each other*, such as music, art, dance, and facial expressions, but some of these are more amenable than others to being expressed as strings of symbols. Written language is the easiest of all, because, of course, it consists of strings of symbols *to begin with*. If the symbols happen to belong to a phonetic alphabet (as opposed to, say, ideograms), converting them into bits is a trivial procedure, and one that was nailed, technologically, in the early nineteenth century, with the introduction of Morse code and other forms of telegraphy.

We had a human/computer interface a hundred years before we had computers. When computers came into being around the time of the Second World War, humans, quite naturally, communicated with them by simply grafting them on to the already-existing technologies for translating letters into bits and vice versa: teletypes and punch card machines.

When Ronald Reagan was a radio announcer, he used to call baseball games by reading the terse descriptions that trickled in over the telegraph wire and were printed out on a paper tape. He would sit there, all by himself in a padded room with a microphone, and the paper tape would eke out of the machine and crawl over the palm of his hand printed with cryptic abbreviations. If the count went to three and two, Reagan would describe the scene as he saw it in his mind's eye: "The brawny left-hander steps out of

Computer betreiben Arithmetik auf einzelnen Stückchen Information. Menschen verarbeiten diese Stückchen zu bedeutungsvollen Symbolen. Aber diese Unterscheidung scheint sich durch die Entwicklung moderner Betriebssysteme, die die Kraft der Metapher verwenden (und manchmal auch missbrauchen), um Computer für ein größeres Publikum zugänglich zu machen, zu verwischen – oder zumindest zu komplizieren.

Leute, die ausschließlich über grafische Benutzeroberflächen wie MacOS oder Windows mit Computern interagiert haben – und das ist fast jeder, der einmal einen Rechner benutzt hat –, könnten erschrecken, wenn sie von dem Telegrafenapparat erfahren, über den ich 1973 mit einem Computer kommuniziert habe. Aber es gab (und gibt) gute Gründe für die Verwendung dieser speziellen Art von Technologie. Menschen haben verschiedene Möglichkeiten, *miteinander* zu kommunizieren – Musik, Kunst, Tanz, Gesichtsausdruck – und einige davon eignen sich besser als andere, in einer Kette von Symbolen ausgedrückt zu werden. Schriftliche Sprache ist die einfachste von allen, weil sie logischerweise *von Anfang an* aus einer Kette von Symbolen besteht. Wenn das Symbol nun einem phonetischen Alphabet angehört (zum Unterschied von z. B. Ideogrammen), so ist es eine triviale Aufgabe, sie in Bits umzuwandeln, und dies wurde technologisch auch im frühen 19. Jahrhundert gelöst – eben mit der Einführung des Morse-Codes und anderer Formen der Telegrafie.

Wir hatten eine Mensch-Computer-Schnittstelle schon hundert Jahre, bevor wir Computer hatten. Und als zur Zeit des Zweiten Weltkrieges dann auch die Computer entstanden, war es nur natürlich, dass die Menschen die bisherigen Technologien zur Umwandlung von Buchstaben in Bits und zurück auf sie übertrugen: Fernschreiber und Lochkartengeräte.

Als Ronald Reagan noch Radiosprecher war, kommentierte er Baseball-Spiele, indem er die trockenen Beschreibungen las, die über den Telegrafendraht hereintröpfelten und auf einem Papierstreifen ausgedruckt wurden. Er saß da, allein mit einem Mikrofon in einem schallgedämpften Raum, und das Papierband lief aus der Maschine über seine Hand, bedruckt mit kryptischen Abkürzungen. Wenn das Ergebnis auf drei und zwei umsprang, dann beschrieb Reagan die Szene, wie er sie vor seinem geistigen Auge sah: „Der stämmige Linkshänder steigt vom Schlagmal und wischt sich den Schweiß von der Stirn. Der Schiedsrichter kratzt den Sand von der Home Plate ..." und so weiter. Wenn das Kryptogramm auf dem Papier einen Base Hit meldete, schlug Reagan wohl mit dem Bleistift auf die Tischkante – ein kleiner Soundeffekt – und beschrieb den Flug des Balls im Bogen, als könnte er ihn wirklich

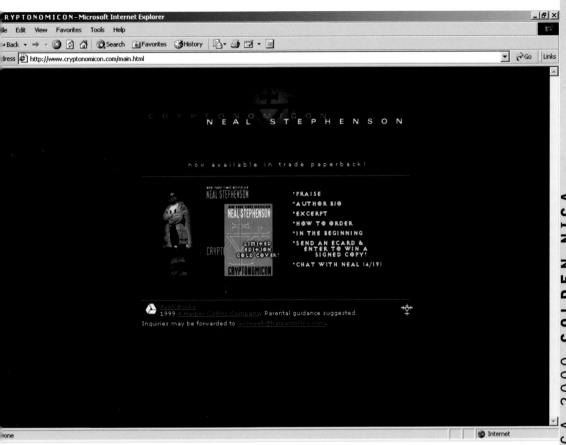

the batter's box to wipe the sweat from his brow. The umpire steps forward to sweep the dirt from home plate." and so on. When the cryptogram on the paper tape announced a base hit, he would whack the edge of the table with a pencil, creating a little sound effect, and describe the arc of the ball as if he could actually see it. His listeners, many of whom presumably thought that Reagan was actually at the ballpark watching the game, would reconstruct the scene in their minds according to his descriptions. This is exactly how the World Wide Web works: the HTML files are the pithy description on the paper tape, and your Web browser is Ronald Reagan. The same is true of Graphical User Interfaces in general. So an OS is a stack of metaphors and abstractions that stands between you and the telegrams, and embodying various tricks the programmer used to convert the information you're working with—be it images, e-mail messages, movies, or word processing documents—into the necklaces of bytes that are the only things computers know how to work with.

sehen. Seine Hörer – von denen wohl viele überzeugt waren, Reagan säße tatsächlich im Stadion und beobachte das Spiel – rekonstruierten die Szene dann in ihrer Fantasie nach Reagans Beschreibungen.
Genau so funktioniert auch das World Wide Web: Die HTML-Dateien sind die trockene Beschreibung auf dem Papierstreifen und der Webbrowser ist Ronald Reagan. Und für grafische Benutzeroberflächen gilt ganz allgemein dasselbe. Deshalb ist ein Betriebssystem ein Stapel von Metaphern und Abstraktionen, das zwischen dir und dem Telegrafen steht, und schließt verschiedene Tricks ein, die der Programmierer eingebaut hat, um die Information, mit der du arbeitest – seien das jetzt Bilder, E-Mails, Nachrichten, Filme oder Textdokumente –, in jene Perlenkette von Bytes umzuformen, die das Einzige sind, was Computer verstehen können. Als wir tatsächlich noch Telegrafen-Equipment (nämlich Fernschreiber) oder ihre etwas weiterentwickelten Substitute (wie „Glas-Fernschreiber" oder die Befehlszeile von MS-DOS) verwendet haben, waren wir dem Grund dieses Stapels noch sehr nahe. Wenn wir aber moderne Betriebssysteme einsetzen, so wird unsere Interaktion mit der Maschine stark mediatisiert. Was immer wir tun, wird interpretiert, übersetzt und wieder übersetzt, während es sich den

When we used actual telegraph equipment (tele-types) or their higher-tech substitutes ("glass tele-types," or the MS-DOS command line) to work with our computers, we were very close to the bottom of that stack. When we use most modern operating systems, though, our interaction with the machine is heavily mediated. Everything we do is interpreted and translated time and again as it works its way down through all of the metaphors and abstractions.

Hostility towards Microsoft is not difficult to find on the Net, and it blends two strains: resentful people who feel Microsoft is too powerful, and disdainful people who think it's tacky. This is all strongly remi-niscent of the heyday of Communism and Socialism, when the bourgeoisie were hated from both ends: by the proles, because they had all the money, and by the intelligentsia, because of their tendency to spend it on lawn ornaments. Microsoft is the very embodi-ment of modern high-tech prosperity—it is, in a word, bourgeois—and so it attracts all of the same gripes.

People who are inclined to feel poor and oppressed construe everything Microsoft does as some sinister Orwellian plot. People who like to think of them-selves as intelligent and informed technology users are driven crazy by the clunkiness of Windows, the exposed rivets and the leaky gaskets.

It is a bit unsettling, at first, to think of Apple as a control freak, because it is completely at odds with their corporate image. Weren't these the guys who aired the famous Super Bowl ads showing suited, blindfolded executives marching like lemmings off a cliff? Isn't this the company that even now runs ads picturing the Dalai Lama (except in Hong Kong) and Einstein and other offbeat rebels?

It is indeed the same company, and the fact that they have been able to plant this image of themselves as creative and rebellious free-thinkers in the minds of so many intelligent and media-hardened skeptics really gives one pause. It is testimony to the insidious power of expensive slick ad campaigns and, perhaps, to a certain amount of wishful thinking in the minds of people who fall for them. It also raises the ques-tion of why Microsoft is so bad at PR, when the histo-ry of Apple demonstrates that, by writing large checks to good ad agencies, you can plant a corporate

ganzen Weg hinunter durch die verschiedenen Ebe-nen von Metaphern und Abstraktionen arbeitet.

Es ist nicht schwer, im Netz Microsoft-feindliche Aus-sagen zu finden und sie vermischen zwei Grundrich-tungen: Die einen sind sauer, weil Microsoft angeb-lich zu mächtig ist, die anderen, weil sie es für schäbig halten. Das erinnert alles stark an die Blüte-zeit von Kommunismus und Sozialismus, als die Bourgeoisie von beiden Seiten her gehasst wurde: Von den *Prolos*, weil sie das ganze Geld hatte, und von den Intellektuellen wegen ihrer Tendenz, es nur für Gartenzwerge und dergleichen auszugeben. Microsoft ist der Inbegriff moderner High-Tech-Pros-perität – kurzum, es ist bourgeois – und so zieht es genau die gleichen Nörgeleien an.

Leute, die sich arm und unterdrückt fühlen, vermu-ten hinter allem, was Microsoft tut, einen hinterhäl-tigen orwellschen Plan. Leute, die sich als intelligent und als informierte Anwender von Technologie ver-stehen, reiben sich an der Schwerfälligkeit von Win-dows, an den herausstehenden Schrauben und den tropfenden Dichtungen.

Es ist auch auf den ersten Blick etwas beunruhigend, sich Apple als Überwachungsfreaks vorzustellen, weil das im Widerspruch zu Apples Firmenimage steht. Sind das nicht die Leute, die die berühmte Super-Bowl-Anzeigenkampagne gestartet haben, in der Manager im Nadelstreif mit verbundenen Augen wie die Lemminge über eine Klippe spazierten? Ist das nicht das Unternehmen, das jetzt auch Anzeigen schaltet, die den Dalai Lama (außer in Hongkong) und Einstein und andere Querdenker zeigen?

Ja, es ist dasselbe Unternehmen, und dass es ihm gelungen ist, dieses Selbstbild als kreative und rebel-lische Freidenker in die Köpfe von so vielen intelli-genten und medienerfahrenen Skeptikern einzu-pflanzen, gibt einem zu denken. Es ist der Beweis für die hinterhältige Kraft teurer, eleganter Werbekam-pagnen und bis zu einem gewissen Grad wohl auch ein wenig das Wunschdenken der Leute, die sich dafür begeistern. Es stellt sich die Frage, warum Microsoft so schlecht im PR-Bereich ist, wenn doch die Geschichte von Apple zeigt, dass man durch das Ausstellen von anständigen Schecks an gute Werbe-agenturen ein Firmenimage in die Köpfe von Leuten einpflanzen kann, das mit der Wirklichkeit nun gar nichts zu tun hat. (Für Leute, die keine Damoklesfra-gen mögen: Die Antwort lautet, dass sich Microsoft, seit es das Herz und Hirn der schweigenden Mehr-heit – der Bourgeoisie – gewonnen hat, einen Dreck um ein schickes Image kümmert, genauso wenig wie Richard Nixon das getan hat. „Ich möchte glauben", das Mantra, das Fox Mulder in *Akte X* an der Büro-wand hängen hat, passt auf unterschiedliche Weise auf beide Firmen: Mac-Anhänger wollen an das Ima-ge von Apple glauben, wie es ihnen in den Super-Bowl-Spots gezeigt wird, und daran, dass Macs sich irgendwie ganz grundlegend von anderen Compu-

image in the minds of intelligent people that is completely at odds with reality. (The answer, for people who don't like Damoclean questions, is that since Microsoft has won the hearts and minds of the silent majority—the bourgeoisie—they don't give a damn about having a slick image, any more then Dick Nixon did. "I want to believe,"—the mantra that Fox Mulder has pinned to his office wall in The X-Files—applies in different ways to these two companies; Mac partisans want to believe in the image of Apple purveyed in those Super Bowl ads, and in the notion that Macs are somehow fundamentally different from other computers, while Windows people want to believe that they are getting something for their money, engaging in a respectable business transaction).

A few years ago I walked into a grocery store somewhere and was presented with the following *tableau vivant*: near the entrance a young couple were standing in front of a large cosmetics display. The man was stolidly holding a shopping basket between his hands while his mate yanked blister-packs of makeup off the display and threw them in. Since then I've always thought of that man as the personification of an interesting human tendency: not only are we not offended to be dazzled by manufactured images, but we *like* it. We practically *insist* on it. We are eager to be complicit in our own dazzlement: to pay money for a theme park ride, vote for a guy who's obviously lying to us, or stand there holding the basket as it's filled up with cosmetics.

I was in Disney World recently, specifically the part of it called the Magic Kingdom, walking up Main Street USA. This is a perfect gingerbready Victorian small town that culminates, bizarrely, in a Disney castle. It was very crowded; we shuffled rather than walked. Directly in front of me was a man with a camcorder. It was one of the new breed of camcorders where instead of peering through a viewfinder you gaze at a flat-panel color screen about the size of a playing card, which televises live coverage of whatever the camcorder is seeing. He was holding the appliance close to his face, so that it obstructed his view. Rather than go see a real small town for free, he had paid money to see a pretend one, and rather than see it with the naked eye he was watching it on television. And rather than stay home and read a book, I was watching *him*.

If you are an intellectual type, a reader or writer of

tern unterschieden, während die Windows-Fans glauben wollen, dass sie was für ihr Geld bekommen, das sie bei einer respektablen Transaktion zahlen.)

Vor ein paar Jahren bin ich einmal in einen Laden gekommen und habe mich dort mit folgendem *Tableau vivant* konfrontiert gesehen: Nahe dem Eingang stand ein junges Paar vor einem große Kosmetik-Ständer. Der junge Mann hielt tapfer den Einkaufskorb in seinen Händen, während seine Partnerin eine Blisterpackung nach der anderen schnappte und in den Korb warf. Seit damals ist dieser Mann für mich die Personifizierung einer interessanten menschlichen Tendenz: Wir sind nicht nur *nicht* beleidigt, wenn wir künstlich fabrizierten Bildern gegenüber stehen, wir *mögen* sie sogar. Wir *bestehen* beinahe darauf. Wir sind in unserer Verblendung sogar gierig, zu Komplizen zu werden: Indem wir für eine Themenpark-Fahrt zahlen; indem wir jemanden wählen, der uns ganz offensichtlich belügt; indem wir den Korb halten, der mit jeder Art von Kosmetik gefüllt wird. Ich war kürzlich in Disney World, besonders in jenem Teil, der *Magic Kingdom* genannt wird, und bin die Main Street USA hinaufgewandert. Das ist eine lebkuchenartige viktorianische Kleinstadt, die – bizarrerweise – in einer Disney-Variante von Neuschwanstein gipfelt. Es war sehr voll, und wir wurden eher geschoben als dass wir gegangen wären. Direkt vor mir war ein Mann mit einem Camcorder. Und zwar mit einem von jener neuen Sorte, bei denen man statt durch ein Okular auf einen farbigen Flachbildschirm von der Größe einer Spielkarte schaut, der live das wiedergibt, was der Camcorder sieht. Der Mann hielt das Gerät nahe an sein Gesicht, so nahe, dass er nichts anderes im Blick hatte. Anstatt sich gratis eine echte Kleinstadt anzusehen, zahlte er gutes Geld dafür, eine vorgetäuschte zu betrachten, und anstatt sie mit seinen Augen anzusehen, betrachtet er sie im Fernsehen. Und anstatt zuhause zu bleiben und ein Buch zu lesen, habe ich *ihn* beobachtet ...

Wenn man ein intellektueller Typ ist, ein Leser oder Autor von Büchern, dann ist das Netteste, was man über Disney World sagen kann, dass die Ausführung superb ist. Aber schnell findet man das gesamte Umfeld ein wenig unheimlich, weil irgend etwas fehlt: die Umsetzung des Inhaltes in klare explizit geschriebene Worte, die Zuordnung der Ideen zu bestimmten Leuten. Es scheint, als würde da über eine Menge Dinge einfach drübergetüncht, als würde Disney World uns eins überstülpen und mit einer Menge verquerer Gedanken und versteckter Annahmen davonkommen.

Das ist genau das, was wir auch beim Übergang von der Befehlszeile zur grafischen Benutzeroberfläche verlieren.

Disney und Apple/Microsoft sind im gleichen Business tätig, nämlich anstrengende, explizite verbale Kommunikation durch aufwendig designte Interfaces kurzzuschließen. Disney ist dabei eine Art User-Interface zu sich selbst – und mehr als rein grafisch,

books, the nicest thing you can say about Disney World is that the execution is superb. But it's easy to find the whole environment a little creepy, because something is missing: the translation of all its content into clear explicit written words, the attribution of the ideas to specific people. It seems as if a hell of a lot might be being glossed over, as if Disney World might be putting one over on us, and possibly getting away with all kinds of buried assumptions and muddled thinking.

But this is precisely the same as what is lost in the transition from the command-line interface to the GUI. Disney and Apple/Microsoft are in the same business: short-circuiting laborious, explicit verbal communication with expensively designed interfaces. Disney is a sort of user interface unto itself—and more than just graphical. Let's call it a Sensorial Interface. It can be applied to anything in the world, real or imagined, albeit at staggering expense.

What is the source of our culture's rejection of explicit word-based interfaces, and the embrace of graphical or sensorial interfaces—a trend that accounts for the success of both Apple/Microsoft and Disney? Part of it is simply that the world is very complicated now—much more complicated than the hunter-gatherer world that our brains evolved to cope with—and we simply can't handle all of the details. We have to delegate. We have no choice but to trust some nameless artist at Disney or programmer at Apple/Microsoft to make a few choices for us, close off some options, and give us a conveniently packaged executive summary.

But more importantly, it comes out of the fact that, during this century, intellectualism failed, and everyone knows it. We agreed to let go of traditional folkways, mores, and religion, and let the intellectuals run with the ball, and they screwed everything up and turned the century into an abbatoir.

We Americans are the only ones who didn't get creamed at some point during all of this. We are free and prosperous because we have inherited political and values systems fabricated by a particular set of eighteenth-century intellectuals who happened to get it right. But we have lost touch with those intellectuals, and with anything like intellectualism, even to the point of not reading books any more, though we are literate. We seem much more comfortable with propagating those values to future generations nonverbally, through a process of being steeped in media. Apparently this actually works to some

man könnte es ein sensorisches Interface nennen. Das kann man es auf alles in dieser Welt anwenden, ob real oder imaginär, aber zu einem hohen Preis. Warum weist unsere Kultur explizite, auf dem Wort basierende Interfaces zurück und stürzt sich auf die grafischen oder sensorischen – ein Trend, der sowohl die Erfolge von Apple/Microsoft wie jene von Disney erklärt?

Eine Antwort ist sicherlich, dass unsere Welt nun einmal sehr kompliziert geworden ist – sehr viel komplizierter als jene des Jägers und Sammlers, mit der klarzukommen sich unser Gehirn entwickelt hat – und wir können einfach nicht mehr alle Details selbst behandeln. Wir müssen delegieren. Uns bleibt nichts anderes übrig, als irgendeinem namenlosen Künstler bei Disney oder einem Programmierer bei Apple/Microsoft zu vertrauen, der für uns einige Entscheidungen trifft, ein paar Optionen ausschließt und uns eine handlich verpackte Zusammenfassung anbietet.

Aber noch entscheidender ist die Tatsache, dass in diesem vergangenen Jahrhundert der Intellektualismus versagt hat und dass jeder es weiß. Wir haben uns darauf eingelassen, eine Menge traditioneller Volkskultur, eine Menge Brauchtum und Religion sausen zu lassen und die Intellektuellen sind mit dem Ball auf und davongegangen und haben alles durcheinandergewürfelt und aus dem Jahrhundert ein Schlachthaus gemacht.

Wir [US-]Amerikaner sind die einzigen, die sich nicht irgendwann dabei haben einseifen lassen. Wir sind frei und prosperieren, weil wir politische und Wertsysteme geerbt haben, die von einem ganz speziellen Kreis von Intellektuellen des 18. Jahrhunderts fabriziert wurden, die sie zufällig richtig hinbekommen haben. Aber wir haben den Bezug zu diesen Intellektuellen verloren – und zum Intellektualismus schlechthin, bis zu dem Punkt, dass wir keine Bücher mehr lesen, auch wenn wir keineswegs Analphabeten sind. Wir scheinen uns wohler zu fühlen, wenn wir diese Werte an spätere Generationen nonverbal weitergeben, wenn wir das harte Brot der Werte sozusagen durch Eintunken in die Medien mundgerechter machen. Dies scheint bis zu einem gewissen Grad auch zu funktionieren, denn die Polizei in etlichen Ländern beschwert sich, dass Festgenommene darauf bestehen, dass ihnen ihre Rechte vorgelesen werden, wie sie es von den Bösewichten aus den amerikanischen Fernsehkrimis kennen. Offenbar verbreiten Wiederholungen von *Starsky and Hutch* auf lange Sicht mehr Menschenrechtsbewusstsein als die Unabhängigkeitserklärung der Vereinigten Staaten.

Dass eine große, reiche, atombewaffnete Kultur ihre Grundwerte als mediale Häppchen propagiert, scheint keine allzu gute Idee. Das Risiko des Abirrens vom Weg ist offensichtlich. Das Wort ist das einzige unabänderliche Medium, das wir haben, deshalb wurde es ja auch für so extrem wichtige Konzepte wie die Zehn Gebote, den Koran und die Deklaration der Menschenrechte gewählt. Wenn die von unseren Medien transportierten Botschaften nicht irgendwie an ein fixes, geschriebenes Set von Grundregeln

degree, for police in many lands are now complaining that local arrestees are insisting on having their Miranda rights read to them as if they were perps in American TV cop shows. Starsky and Hutch reruns may turn out, in the long run, to be a greater force for human rights than the Declaration of Independence. A huge, rich, nuclear-tipped culture that propagates its core values through media steepage seems like a bad idea. There is an obvious risk of running astray here. Words are the only immutable medium we have, which is why they are the vehicle of choice for extremely important concepts like the Ten Commandments, the Koran, and the Bill of Rights. Unless the messages conveyed by our media are somehow pegged to a fixed, written set of precepts, they can wander all over the place and possibly dump loads of crap into people's minds.

Orlando used to have a military installation called McCoy Air Force Base, with long runways from which B-52s could take off and reach Cuba, or just about anywhere else, with loads of nukes. But now McCoy has been scrapped and repurposed. It has been absorbed into Orlando's civilian airport. The long runways are being used to land 747-loads of tourists from Brazil, Italy, Russia and Japan, so that they can come to Disney World and steep in our media for a while.

To traditional cultures, especially word-based ones such as Islam, this is infinitely more threatening than the B-52s ever were. It is obvious, to everyone outside of the United States, that our arch-buzzwords, multiculturalism and diversity, are false fronts that are being used (in many cases unwittingly) to conceal a global trend to eradicate cultural differences. The basic tenet of multiculturalism (or "honoring diversity" or whatever you want to call it) is that people need to stop judging each other—to stop asserting (and, eventually, to stop believing) that this is right and that is wrong, this true and that false, one thing ugly and another thing beautiful, that God exists and has this or that set of qualities.

The lesson of the Twentieth Century is that, in order for a large number of different cultures to coexist peacefully on the globe (or even in a neighborhood) it is necessary for people to suspend judgment in this way. Hence (I would argue) our suspicion of, and hostility towards, all authority figures in modern culture. As David Foster Wallace has explained in his essay "E Unibus Pluram," this is the basic message of television; it is the message that people take home, any-

gebunden sind, können sie in jede Richtung davon treiben und möglicherweise jede Menge Schrott in die Gehirne der Leute pflanzen.

Orlando hatte eine militärische Einrichtung namens McCoy Air Force Base, mit langen Rollbahnen, von denen B52-Bomber abheben und mit einer Ladung Atomwaffen nach Kuba – oder überall sonsthin – fliegen konnten. Aber McCoy ist aufgelassen und umgewidmet worden. Es ist eingebunden in den Zivilflughafen von Orlando. Die langen Rollbahnen dienen dazu, 747er-Ladungen von Touristen anzukarren, aus Brasilien, Italien, Russland, Japan, damit sie nach Disney World kommen und eine Weile in unsere Medienwelt getunkt werden.

Für die traditionellen Kulturen – insbesondere für die auf dem Wort aufbauenden wie den Islam – ist dies unendlich bedrohlicher, als es die B52 jemals waren. Jedem außerhalb der USA ist klar, dass unsere Lieblings-Modewörter „Multikulturalismus" und „Unterschiedlichkeit" eine falsche Fassade sind, hinter der sich ein (teilweise unbewusster) globaler Trend zur Ausrottung kultureller Unterschiede verbirgt. Das Credo eines wahren Multikulturalismus (oder des Respektierens von Verschiedenheiten oder wie immer man das nennen möchte) besteht darin, dass die Menschen aufhören, einander zu beurteilen – dass sie aufhören festzustellen (und irgendwann zu glauben), dass dies richtig ist und das falsch, dies wahr und das unwahr, das eine schön und das andere hässlich, dass Gott existiere und diese oder jene oder andere Eigenschaften habe.

Das 20. Jahrhunderts lehrt uns, dass eine große Zahl von Kulturen nur dann friedlich auf diesem Globus (ja, selbst auf einem klar umgrenzten Raum) zusammenleben können, wenn die Menschen eine derartige Form von Beurteilung aufgeben. Daher kommt auch (so möchte ich argumentieren) unser Vorbehalt – und unsere Abneigung – gegenüber allen Autoritätsfiguren in der modernen Kultur. Wie David Foster Wallace in seinem Essay E Unibus Pluram dargestellt hat, ist dies die Kernbotschaft des Fernsehens; es ist jedenfalls die Botschaft, die die Menschen mit nach Hause nehmen, wenn sie lange genug in unseren Medien eingeweicht wurden. Natürlich wird das nicht in so hochtrabenden Worten ausgedrückt. Es schlägt durch als die Annahme, dass alle Autoritäten – Lehrer, Generäle, Polizisten, Minister, Politiker – heuchlerische Narren sind und eine hippe elegante Coolness der einzige Weg des Seins ist.

Problematisch ist nur, dass, sobald einmal die Fähigkeit zu urteilen über Richtig und Falsch, Wahr und Unwahr und so weiter ausgeräumt ist, keine wirkliche Kultur mehr übrig bleibt. Was bleibt, ist Holzschuhtanz und Makramee. Die Fähigkeit, Urteile zu fällen, Dinge zu glauben und an Dinge zu glauben – das ist die Essenz der Kultur generell.

Durch die Verwendung von grafischen Oberflächen haben wir uns dummerweise auf einem Grundstück eingekauft, das nur wenige Leute akzeptiert hätten, wenn es ihnen unverblümt angetragen worden

way, after they have steeped in our media long enough. It's not expressed in these highfalutin terms, of course. It comes through as the presumption that all authority figures—teachers, generals, cops, ministers, politicians—are hypocritical buffoons, and that hip jaded coolness is the only way to be.

The problem is that once you have done away with the ability to make judgments as to right and wrong, true and false, etc., there's no real culture *left*. All that remains is clog dancing and macrame. The ability to make judgments, to believe things, is the entire *point* of having a culture.

By using GUIs all the time we have insensibly bought into a premise that few people would have accepted if presented to them bluntly: namely, that hard things can be made easy, and complicated things simple, by putting the right interface on them.

What made old epics like Gilgamesh so powerful and so long-lived was that they were living bodies of narrative that many people knew by heart, and told over and over again—making their own personal embellishments whenever it struck their fancy. The bad embellishments were shouted down, the good ones picked up by others, polished, improved, and, over time, incorporated into the story. Likewise, Unix is known, loved, and understood by so many hackers that it can be re-created from scratch whenever someone needs it. This is very difficult to understand for people who are accustomed to thinking of OSes as things that absolutely have to be *bought*.

But many hackers have launched more or less successful re-implementations of the Unix ideal. Each one brings in new embellishments. Some of them die out quickly, some are merged with similar, parallel innovations created by different hackers attacking the same problem, others still are embraced, and adopted into the epic. Thus Unix has slowly accreted around a simple kernel and acquired a kind of complexity and asymmetry about it that is organic, like the roots of a tree, or the branchings of a coronary artery. Understanding it is more like anatomy than physics.

Credit for Linux generally goes to its human namesake, one Linus Torvalds, a Finn who got the whole thing rolling when he used some of the GNU tools to write the beginnings of a Unix kernel that could run

wäre: nämlich dass schwierige Aufgaben einfach werden und komplizierte Dinge simpel, wenn wir ihnen nur die richtige Oberfläche draufsetzen.

Was alte Epen wie jenes von *Gilgamesch* so stark und so langlebig gemacht hat, war die Tatsache, dass sie lebende Erzählungen waren, dass viele Leute sie auswendig wussten und sie immer und immer wieder erzählt haben – mit den jeweiligen persönlichen Ausschmückungen, wo immer es dem Erzähler angebracht erschien. Schlechte Beifügungen wurden ausgepfiffen, gute hingegen von anderen aufgenommen, poliert, verbessert und im Laufe der Zeit in den Kanon der Erzählung eingegliedert. Genauso ist Unix deswegen bei so vielen Hackern bekannt, geliebt, verstanden, weil es jederzeit von Grund auf neu konstruiert werden kann, wenn jemand es braucht. Das ist schwer zu verstehen für Leute, die sich daran gewöhnt haben, Betriebssysteme als Dinge zu betrachten, die auf jeden Fall *gekauft* werden müssen. Aber zahlreiche Hacker haben mehr oder weniger erfolgreich Neu-Implementationen des Unix-Ideals auf den Weg gebracht. Jede(r) bringt neue Verschönerungen mit. Manche davon sterben schnell, manche verschmelzen mit ähnlichen, parallelen Innovationen anderer Hacker, die vor dem gleichen Problem standen, wiederum andere werden freudig begrüßt und ins Epos mit eingebaut. So hat sich Unix langsam rund um einen simplen Kernel entwickelt und eine Art von Komplexität und Asymmetrie entwickelt, die nachgerade als organisch anzusehen ist – wie das Wurzelwerk eines Baums oder die Verzweigungen eines Adernetzes. Unix zu verstehen ist eher eine Frage der Anatomie als der Physik.

Die Entwicklung von Linux wird üblicherweise seinem Namensgeber Linus Torvalds zugerechnet, einem Finnen, der das Ganze ins Rollen brachte, als er einige der GNU-Werkzeuge einsetzte, um die Grundlage einen Unix-Kernels zu schreiben, der auf PC-kompatibler Hardware lauffähig wäre. Und Torvalds verdient alle ihm dafür verliehenen Ehren und noch viel mehr. Aber er hätte dies nicht allein zustande gebracht, genauso wenig wie dies Richard Stallmann gekonnt hätte. Um überhaupt Code schreiben zu können, brauchte Torvalds billige, aber leistungsfähige Entwicklungswerkzeuge – und die kamen aus Stallmans GNU-Projekt. Und er brauchte billige Hardware, auf der er diesen Code schreiben konnte. Billige Hardware bereitzustellen ist viel schwieriger als billige Software – ein einzelner Mensch (Stallman) kann Software schreiben und sie zum kostenlosen Download ins Netz stellen, aber um billige Hardware herzustellen, bedarf es einer kompletten industriellen Infrastruktur und das ist unter keinen Umständen billig. Der einzige Weg, Hardware wirklich preisgünstig zu machen, ist, eine unglaubliche Menge an Exemplaren zu produzieren, sodass die Stückkosten irgendwann sinken. Aus den bereits erwähnten Gründen hatte Apple kein Interesse daran, dass die Hardwarepreise fallen. Der einzige

on PC-compatible hardware. And indeed Torvalds deserves all the credit he has ever gotten, and a whole lot more. But he could not have made it happen by himself, any more than Richard Stallman could have. To write code at all, Torvalds had to have cheap but powerful development tools, and these he got from Stallman's GNU project.

And he had to have cheap hardware on which to write that code. Cheap hardware is a much harder thing to arrange than cheap software; a single person (Stallman) can write software and put it up on the Net for free, but in order to make hardware it's necessary to have a whole industrial infrastructure, which is not cheap by any stretch of the imagination. Really the only way to make hardware cheap is to punch out an incredible number of copies of it, so that the unit cost eventually drops. For reasons already explained, Apple had no desire to see the cost of hardware drop. The only reason Torvalds had cheap hardware was Microsoft.

Microsoft refused to go into the hardware business, insisted on making its software run on hardware that anyone could build, and thereby created the market conditions that allowed hardware prices to plummet. In trying to understand the Linux phenomenon, then, we have to look not to a single innovator but to a sort of bizarre Trinity: Linus Torvalds, Richard Stallman, and Bill Gates. Take away any of these three and Linux would not exist.

Excerpts from: Neal Stephenson, In the Beginning ... Was the Command Line, Avon Books 1999.

Grund dafür, dass Torvalds billige Hardware bekam, war Microsoft.

Microsoft weigerte sich, ins Hardwaregeschäft einzusteigen, es bestand darauf, seine Software auf einer Plattform laufen zu lassen, die jeder bauen konnte – und so entstanden die Marktbedingungen, unter denen die Preise purzeln konnten. Um das Linux-Phänomen verstehen zu können, dürfen wir also nicht einen einzelnen Innovator betrachten, sondern wir müssen uns eine Art bizarrer Dreifaltigkeit vorstellen: Linus Torvalds, Richard Stallman und Bill Gates. Würde einer von den dreien fehlen, gäbe es kein Linux.

Auszüge aus: Neal Stephenson, In the Beginning ... Was the Command Line, Avon Books 1999.

NEAL STEPHENSON (USA) IS THE AUTHOR OF *SNOW CRASH*, *THE DIAMOND AGE*, *ZODIAC* AND *CRYPTONOMICON*. BORN ON HALLOWEEN 1959 IN FORT MEADE, MARYLAND—HOME OF THE NATIONAL SECURITY AGENCY HE GREW UP IN CHAMPAIGN-URBANA, ILLINOIS, AND AMES, IOWA, BEFORE ATTENDING COLLEGE IN BOSTON. SINCE 1984 HE HAS LIVED MOSTLY IN THE PACIFIC NORTHWEST AND HAS MADE A LIVING OUT OF WRITING NOVELS AND THE OCCASIONAL MAGAZINE ARTICLE. ■

Neal Stephenson (USA) ist Autor von *Snow Crash*, *The Diamond Age*, *Zodiac* und *Cryptonomicon*. Er wurde zu Halloween 1959 in Fort Meade, Maryland – der Heimat der National Security Agency – geboren und wuchs in in Champaign-Urbana, Illinois, und Ames, Iowa, auf, bevor er in Boston das College besuchte. Seit 1984 wohnt er an der Nordwest-Küste der USA und lebt davon, Romane und gelegentlich Artikel für Zeitschriften zu schreiben.

THE EXQUISITE CORPSE

Sharon Denning

The digital *Exquisite Corpse* studies the growth of a story, examining how stories are built collectively from individual episodes and how stories translate in the telling. It is a database of possible stories, a map of tellings and retellings, a network of routes from one beginning to many possible endings.

Like the original *Exquisite Corpse,* the digital version is made up of story sections that grow from previous contributions. It differs by collecting and presenting not just a linear continuation, but multiple variations as well. Each chapter can branch into an infinite number of stories. Users can choose to read an existing branch or to create a new one by adding their own text. It is up to the user to continue the present theme of a storyline or to send it in a different direction.

The online version takes advantage of several ways of spreading the story on the web. Though the *Exquisite Corpse* users can ask friends to continue their story and the *Exquisite Corpse* can notify them when their story is continued. Their friends ask other friends, mimicking the way stories are spread through conversation and the way jokes and stories are forwarded around the net—a viral marketing program for the *Exquisite Corpse*.

Contributions can be made through a Flash interface or HTML forms on the Web, or via email, allowing as many people as possible to contribute to the story and spreading it through several communications channels. The story, visible through several interfaces, demonstrates a flexible relationship between data and interface. Users can not only choose their stories, but choose how they view and contribute as well. The interface maps the relationships between each contribution and allows the user to navigate between them, mapping the birth and growth of a story.

The *Exquisite Corpse* has two views—a map view that looks at the whole story, and a chapter view that lets a user read the story one section at a time. The map traces the growth of the story as a whole and is

Das digitale *Exquisite Corpse* untersucht das Wachstum einer Geschichte, betrachtet, wie Geschichten aus individuellen Episoden kollektiv erstellt werden und wie sie sich beim Erzählen verändern. Es ist eine Datenbank aus möglichen Geschichten, eine Landkarte der narrativen Wege, ein Netzwerk von Routen, die von einem Anfang zu zahlreichen möglichen Enden führen.

Wie das originale *Exquisite Corpse* besteht auch die digitale Version aus Erzählabschnitten, die aus früheren Beiträgen herauswachsen. Der Unterschied zur ursprünglichen Version liegt darin, dass hier nicht nur eine lineare Fortsetzung, sondern zahlreiche Variationen gesammelt und präsentiert werden. Jedes Kapitel kann sich in eine unendliche Zahl von Geschichten verzweigen. Die Benutzer können entweder einen bestehenden Ast lesen oder durch Hinzufügung eines eigenen Texts einen neuen Zweig wachsen lassen. Es bleibt dem Anwender überlassen, ob das gegenwärtige Thema einer Erzählung beibehalten wird oder ob die Geschichte eine völlig neue Wendung nehmen soll.

Die Online-Version bedient sich der zahlreichen Möglichkeiten der Verbreitung einer Geschichte im Web. Die Benutzer von *Exquisite Corpse* können Freunde auffordern, ihre Geschichte fortzusetzen, und sich von *Exquisite Corpse* verständigen lassen, sobald eine Fortsetzung stattgefunden hat. Die Freunde wiederum fordern andere Freunde zum Weiterschreiben auf und so verbreitet sich die Geschichte sozusagen durch Konversation im Netz weiter, ähnlich, wie sich auch Witze und Anekdoten im Netz verbreiten – ein virusähnliches Marketingprogramm für *Exquisite Corpse*.

Beiträge können entweder über ein Flash-Interface, über HTML-Formulare oder per E-Mail abgeliefert werden, wodurch so viele Menschen wie möglich an der Erzählung mitwirken und sie über zahlreiche Kommunikationskanäle verbreiten können. Die Erzählung selbst ist über mehrere Interfaces einsehbar und zeigt eine flexible Beziehung zwischen Daten und Interface. Benutzer können nicht nur ihre eigenen Geschichten wählen, sondern selbst bestimmen, wie sie lesen und ihre Beiträge verfassen. Das Interface bildet die Beziehungen zwischen den einzelnen Beiträgen ab und erlaubt den Anwendern eine einfache Navigation, indem es die Geburt und das Wachstum einer Geschichte nachvollzieht.

Exquisite Corpse bietet zwei Ansichten: Eine landkartenähnliche Übersicht, die die Gesamtgeschichte anzeigt, und eine Kapitelansicht, in der die Anwender die Erzählung abschnittsweise lesen können. Die

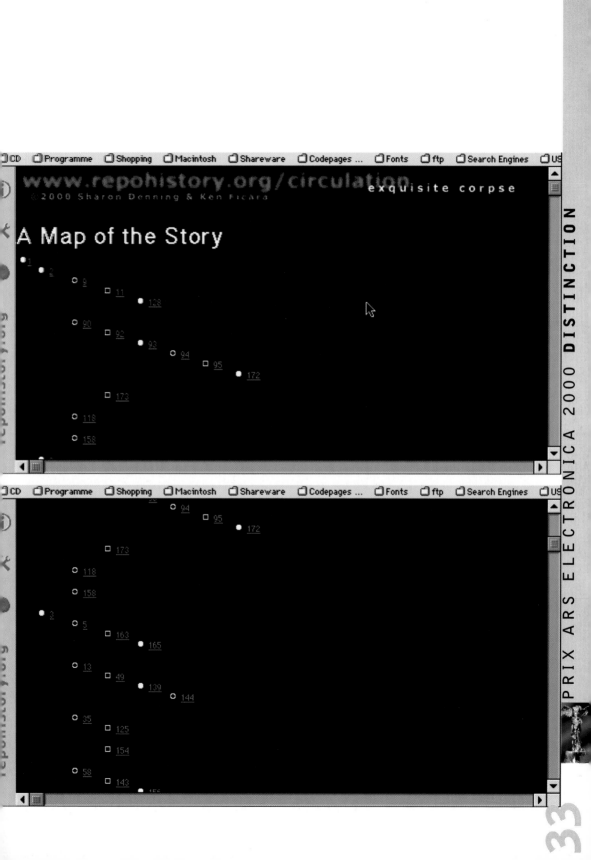

updated as chapters are added. Selecting a chapter in the map opens that section and users can then read one section and follow links to its continuations. While reading chapters users can select manual mode or automatic mode. In manual mode, users choose which branch to read. In auto mode the *Exquisite Corpse* chooses randomly from the branches and "tells" the user a story.

Under the hood, the *Exquisite Corpse* is a classical example of a tree structure that computer science students learn in their first courses on data structures. It is analogous to a family tree: a new chapter added to the story becomes a new "child" of the "parent" chapter that was continued. Others who add onto that same chapter create "siblings" to the "child". If the continuation is itself continued, "grandchildren" of the original will be created.

Email submissions to the *Exquisite Corpse* are processed by Procmail, an automated email-processing tool. Web submissions are processed by a CGI script written in Perl. Both use the same object interface to add "children" to the tree. That interface defines a "node" in the tree. Nodes have methods allowing new "children" to be created and allowing navigation of the tree by the user interface.

The web interface is written in flash. It can both open each node's file to "read in" the text, and send information to create a new node. It calls the same CGIs as the web form.

The *Exquisite Corpse* is currently included in RepoHistory's online project CIRCULATION, which explores the city as body by examining the flow of blood through the city; blood as both a physical entity and as a metaphor for identity.

Landkarte verfolgt das Wachstum der Erzählung als Ganzes und wird mit Hinzufügung eines jeden neuen Kapitels aktualisiert. Wird ein Kapitel in der Landkarte ausgewählt, öffnet sich der betreffende Abschnitt. Die Anwender können dann das Kapitel lesen und von dort über Links zu den Fortsetzungen weitergehen. Beim Lesen kann zwischen automatischem und manuellem Modus gewählt werden. Im manuellen Modus entscheidet der Anwender, welcher weiterer Zweig gelesen wird; im Auto-Modus wählt *Exquisite Corpse* nach dem Zufallsprinzip aus den Zweigen aus und „erzählt" sozusagen selbst die Geschichte.

Unter der Oberfläche ist *Exquisite Corpse* ein klassisches Beispiel einer Baumstruktur, wie sie Studenten der Informatik in den ersten Lehrveranstaltungen über Datenstrukturen erläutert wird. Es ist vergleichbar mit Familienstammbäumen: Ein neues Kapitel wird sozusagen zum „Kind" jenes „Elternkapitels", das fortgesetzt wird. Wer eine Fortsetzungsvariante hinzufügt, erzeugt sozusagen „Geschwister" der Fortsetzung; wer die Geschichte weiterspinnt, bringt „Enkel" des Originals ins Spiel.

E-Mail-Beiträge zu *Exquisite Corpse* werden mit Procmail verarbeitet, einem automatisierten E-Mail-Tool. Web-Beiträge werden über ein in Perl geschriebenes CGI-Script abgearbeitet. Beide verwenden dasselbe Objekt-Interface, um „Kinder" an den Baum anzufügen. Dieses Interface definiert einen Verzweigungsknoten am Baum und bieten die Hinzufügung weiterer Kinder sowie die Navigation durch den Baum an.

Das Web-Interface ist in Flash geschrieben. Es kann sowohl die zu jedem Knoten gehörigen Dateien öffnen, um den Text einzulesen, als auch Informationen senden, um neue Knoten zu schaffen. Es ruft die gleichen CGI-Scripts auf, wie das Web-Interface.

Exquisite Corpse ist derzeit in das *RepoHistory*-Projekt CIRCULATION integriert, das die Stadt als einen Körper darstellt, durch den Blut fließt – Blut als physische Einheit ebenso wie Blut als Metapher für Identität.

1 from "If On a Winter's Night a Traveler" by Italo Calvino

"If on a winter's night a traveler, outside the town of Malbork, leaning from the steep slope without fear of wind or vertigo, looks down in the gathering shadow in a network of lines that enlace, in a network of lines that intersect, on the carpet of leaves illuminated by the moon around an empty grave -- What story down there awaits its end? -- he asks, anxious to

3 Ken

"A story?" she said. "Well, there are
so many stories. I have happy stories,
and love stories,
and the ones that are really true, but
telling those
before bedtime, well, I'm not sure I
should."

<<

node 3
play mode manual

exquisite corpse

PRIX ARS ELECTRONICA 2000 DISTINCTION

SHARON DENNING IS A NYC BASED ARTIST WHO HAS BEEN WORKING IN INTERACTIVE MEDIA SINCE 1992. SHE RECEIVED AN MFA FROM S.V.A. IN 1998 WITH A THESIS ON INTERACTIVE NARRATIVE. HER INDIVIDUAL WORK HAS TOURED INTERNATIONALLY AND SHE IS A MEMBER OF THE ARTIST COLLECTIVE REPOHISTORY. ■■■■ Sharon Denning (USA), wohnhaft in New York, ist Künstlerin und arbeitet seit 1992 mit interaktiven Medien. Sie graduierte 1998 an der S.V.A. mit einer Arbeit über interaktive Narration zum MFA. Ihre Arbeiten wurden weltweit ausgestellt; sie ist Mitglied der Künstlerkollektivs RepoHistory.

TELEZONE
Telezone Team

An architecture for net architecture	Eine Architektur für Netz-Architektur

The *TeleZone* is a telerobotic art installation that creates a parallel between the real and the virtual. Visitors design architectural structures using a CAD interface. These structures are carefully constructed by an industrial robot arm which cements 1 cm rectangular elements within a 2 meter building area. Structures thus exist in two forms: within the virtual space of a 3D simulator, and within the real space of the Ars Electronica lobby. The result reflects both the complex social interactions of the builders and the contrast between the Cartesian world of pure design and the realities of mass, gravity, and imprecision.

Prior to the December launch, architects and theoreticians from Austria and around the world being invited to take part in an experimental design phase. The results will be published.

Communications, trade and services are getting more and more

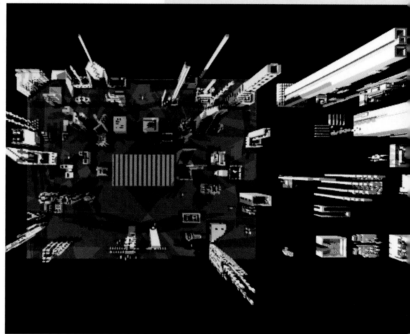

detached from urban areas and transferred to data networks. This development does not only accelerate the organization of the WorldWideWeb as a venue for social action but also affects the appearance of urban areas on the threshold of the 21st century. The *TeleZone* installation takes up these developments and intensifies them. In a process of democratic self-organization and collaborative planning, an experimental lab situation for trend-setting urban structures and architectures is created at the Ars Electronica Center and on the *TeleZone* Website.
A robot at the Ars Electronica Center acts as "interface" between real and virtual space. Representing the user, it arranges and sticks together the building

Die *TeleZone* ist eine Telerobotik-Kunst-Installation, die Parallelen zwischen dem Realen und Virtuellen zieht. Besucher konstruieren mit einem CAD-Interface architektonische Strukturen. Diese werden von einem Roboter auf einer 4 m² großen Fläche vorsichtig zusammengeklebt. So existieren die Strukturen in zwei Formen: einerseits im virtuellen Raum als 3D-Simulation und andererseits im realen Raum des Ars Electronica Center. Das Ergebnis ist eine Reflexion der komplexen sozialen Interaktionen der Erbauer und andererseits der Kontrast zwischen der karte-

sianischen Welt des puren Designs der Simulation und der Realität von Masse, Gravitation und Ungenauigkeit. Bevor die Installation den Usern des Internet übergeben wurde, waren internationale ArchitektInnen und ExpertInnen eingeladen, an der ersten Phase des *TeleZone*-Architektur-Experiments teilzunehmen. Die Ergebnisse werden veröffentlicht. Kommunikation, Handel und Dienstleistungen lösen sich zunehmend aus dem urbanen Raum und verlagern sich in die Datennetze. Diese Entwicklung beschleunigt nicht nur die Gestaltung des WorldWide Web als Raum sozialen Handelns, sondern zeigt ihre Folgen auch in den Stadtbildern des angehenden 21. Jahrhunderts. Die Installation *TeleZone* greift diese Prozesse auf und verdichtet sie. In demokratischer Selbstorganisation und kollaborativer Planung entsteht im Ars Electronica Center wie auf der *TeleZone*-Website eine experimentelle Laborsitua-

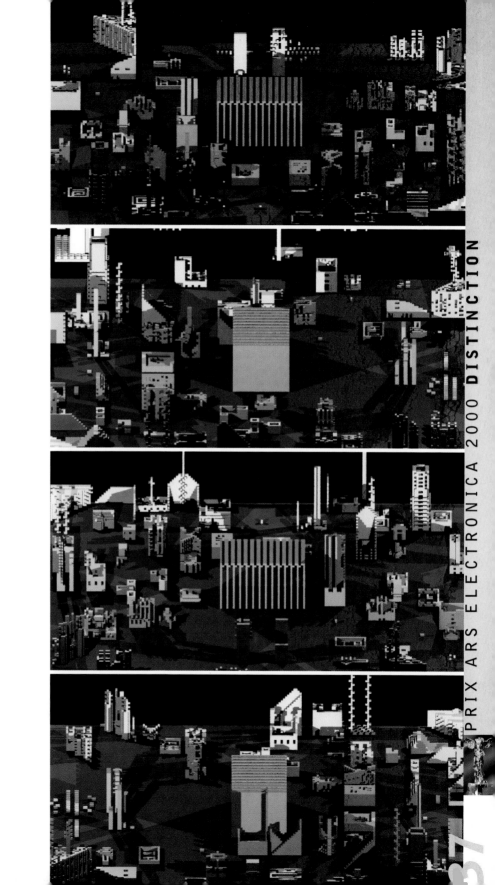

blocks on the *TeleZone* site precisely in accordance with the respective planning input. The robot makes visible the efforts undertaken by a self-organizing community. The joint actions of the community are reflected both in the real and in the virtual world; in other words, the architectural structures created in the process are not only built in reality by the robot but are also shown in a VRML world on the *TeleZone* Website.

The community is provided with a framework of rules which mainly contain "basic physical conditions." The open nature of the framework of rules complies with the target of self-organization inherent in the system, which is part of the experiment. The participants are asked regularly to decide on modifications of the rules by a democratic vote. The underlying concept of *Tele-Zone* motivates each user to interact with other participants in the project in order to expand his or her possibilities of discussion. The process of development from an empty space to a complex architectural structure reflects the degree of self-organization and activity of the participants in *TeleZone*.

tion für zukunftsweisende urbane Strukturen und Architekturen.

Ein Roboter im Ars Electronica Center bildet die Schnittstelle zwischen realem und virtuellem Raum. Als Repräsentant des „Users" platziert und verklebt er die verplanten Bausteine punktgenau auf dem *Tele-Zone*-Areal. Der Roboter ist als Kristallisationspunkt für eine sich selbst organisierende Community zu verstehen. Das gemeinsame Handeln der Community findet seine Entsprechung sowohl in der realen als auch in der virtuellen Welt. Die entstehenden Architekturen werden also nicht nur vom Roboter real gebaut, sondern auf der *TeleZone*-Website auch als VRML-Welt dargestellt.

Der Community wird ein Framework an Regeln mitgegeben, die im wesentlichen „physikalische Randbedingungen" sind. Die Offenheit des Regelwerkes kommt dem systemimmanenten Streben zur Selbstorganisation als Bestandteil des Experiments entgegen. In regelmäßigen Abständen werden die Beteiligten dazu aufgerufen, in demokratischen Abstimmungen über Modifikationen der Regeln zu entscheiden. Die Konzeption der *TeleZone* motiviert den User, mit anderen Beteiligten des Projektes zu interagieren, um den Rahmen seines Handlungsspielraumes zu erweitern. Im Entwicklungsprozess von der leeren Fläche zu einer komplexen Architekturstruktur spiegelt sich die Selbstorganisation und Aktivität der *TeleZone*-TeilnehmerInnen.

Telezone Team:

Project Direction / Projektleitung: Erich Berger/Ars Electronica FutureLab
Software Design: Mag. Volker Christian/Ars Electronica FutureLab
Interface Design: Univ.Ass. DI DR Peter Purgathofer / Technical University Vienna
Concept Consultant: Ken Goldberg/UC Berkeley USA
Programming/Graphics: Wolfgang Beer, Astrid Benzer, Ewald Dannerer, Stefan Eibelwimmer, Oliver Frommel, Tom Haberfellner, Helmut Höllerl, Thomas Martetschläger, Denis Mikan, René Pachernegg, Jörg Piringer, Nestor Pridun, Christian Retscher, Markus Seidl, Johannes Staudinger, Martin Wiesmair.
Web Design: igw in collaboration with gregerpauschitz f.o.p and Ars Electronica FutureLab.
PR and Publicity: Ursula Kürmayr (Ars Electronica Center), Mag. Romana Aumer (telekom austria), Pascal Maresch, Florian Sedmak, Martin Lengauer.

Academic Consultants/Wissenschaftliche Beratung:
Institute for Design and Effects Studies, Technical University Vienna / Institut für Gestaltungs- und Wirkungsforschung der TU Wien, Univ. Ass. DI Dr. Peter Purgathofer; Institute for Publications and Communications Studies, University of Vienna, Department of Electronic and New Media / Institut für Publizistik und Kommunikationswissenschaft der Universität Wien, Fachbereich elektronis-che und Neue Medien, Univ.Ass. Mag Gerit Götzenbrucker; Center for the Study of Aging and Social Policies Research at the St. Pölten Academy of the Federal Province of Lower Austria, Section Social Welfare and Health / Zentrum für Alternswissenschaft und Sozialpolitikforschung an der Niederösterreichischen Landesakademie St.Pölten, Bereich Soziales und Gesundheit, Mag. Bernd Löger; Study Course for Media Technique and Design at the Technical College Hagenberg / Lehrgang für Medientechnik und Design an der Fachhochschule Hagenberg, DI Dr. Wilhelm Burger; Study Course for Software Engineering at the Technical College Hagenberg / Lehrgang für Software Engineering an der Fachhochschule Hagenberg, Univ.Prof. DI Dr. Jacak Witold.

Telezone was realized at the Ars Electronica FutureLab with support from Telekom Austria and the Wittmann Company, robot production series and automatization facilities / Telezone wurde im Ars Electronica FutureLab realisiert, mit Unterstützung der Telekom Austria AG und der Fa. Wittmann, Roboterbaureihen und Automatisierungsanlagen.

@ 2000

Ichiro Aikawa

@*2000* is not the first website to move windows around, but Aikawa establishes his highly original style by controlling windows with unique graphics. *2000* is a series of simple flash animations looking toward year 2000. Two previously unseen early movies are featured. In "Message 3" the user is given the choice of terminating their browser. Different results occur based on the viewer's selection.

1999 is Aikawa's satirical look at internet technology. This wild and chaotic piece features synchronized animation in six different windows. The current version is a remixed version for Word.com (The Thing).

1998 is a revised version of Aikawa's first window moving experiment in late 1998 to early 1999. It Incorporates one expanding and four moving windows. Unfortunately, the technology was not advanced enough at that time to avoid occasional browser crashes.

206 is a fun and complete original piece using a combination of popup windows and auto-scrolling. "*9*" is a piece created for Sublimegraphics.com (Planets View). The idea of utilizing fake and real windows was a fresh idea. The animation and photo-real design combine to create a virtual walk through Aikawa's world.

4, a work in progress, is an expression of life, death and reincarnation. Aikawa's remarkable technique coordinates animation with the control of windows through viewer interaction.
After re-launching @*2000* in summer of 1999, the site was picked up by Shift.jp org's cool site reviews, Word.com, Drunktank.com, Archinect.com, Commarts.com, etc.

@*2000* ist nicht die erste Website, die Fernster umherschiebt, aber Aikawa etabliert seinen höchst originellen Stil, indem er die Fenster mit einer einzigartigen Grafik steuert.
2000 ist eine Serie einfacher Flash-Animationen im Ausblick auf das Jahr 2000, wozu auch zwei bisher noch nie gezeigte frühe Filme zählen. In „Message 3" wird dem Anwender die Möglichkeit gegeben, den Browser zu schließen. Je nach der Auswahl des Users ergeben sich unterschiedliche Resultate.
1999 ist Aikawas satirischer Blick auf die Internet-Technologie. Dieses wilde und chaotische Stück enthält synchronisierte Animationen in sechs verschiedenen Fenstern. Die derzeitige Version ist eine neu gemixte Fassung für Word.com (The Thing).
1998 ist eine überarbeitete Version von Aikawas erstem „Fenster-Schiebe-Experiment" der Jahreswende 1998/99. Es enthält ein expandierendes und vier bewegte Fenster. Leider war damals die Technologie noch nicht weit genug entwickelt, um gelegentliche Browser-Abstürze zu vermeiden.
206 ist ein amüsantes und vollkommen selbständiges Stück, das eine Kombination aus Popup-Fenstern und automatischem Scrolling einsetzt.
9 ist ein für Sublimegraphics.com (Planets View) geschaffenes Stück. Die Idee, reale und gefälschte Fenster einzusetzen, war damals neuartig, die Animation und das fotorealistische Design verbinden sich zu einem virtuellen Spaziergang durch Aikawas Welt.
4, ein Work-in-Progress, ist ein Ausdruck von Leben, Tod und Reinkarnation. Aikawas bemerkenswerte Technik koordiniert Animation mit der Steuerung der Fenster durch die Interaktion des Anwenders. Nach dem Relaunch von @*2000* im Sommer 1999 wurde die Site von Shift.jp org's „Cool Site Reviews" gelistet, ebenso wie von Word.com, Drunktank.com, Archinect.Com, Commarts.com und anderen.

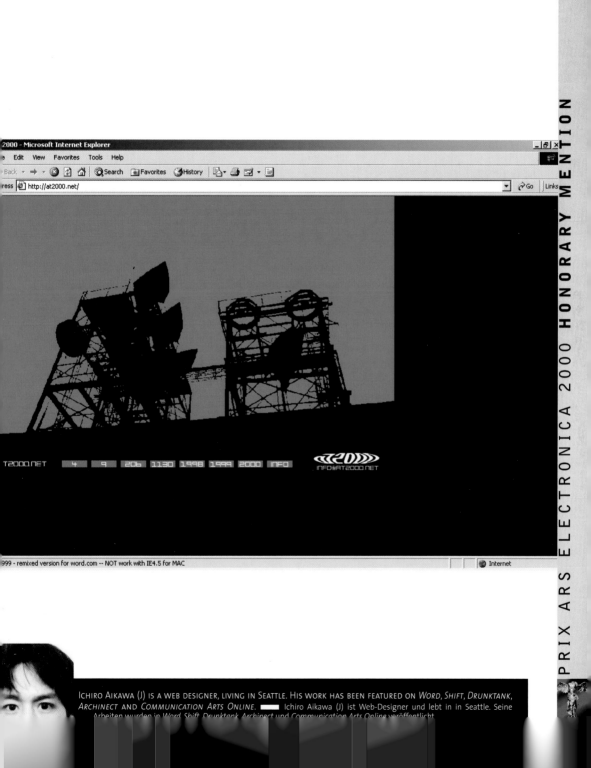

ICHIRO AIKAWA (J) IS A WEB DESIGNER, LIVING IN SEATTLE. HIS WORK HAS BEEN FEATURED ON *WORD*, *SHIFT*, *DRUNKTANK*, *ARCHINECT* AND *COMMUNICATION ARTS ONLINE*. ▬ Ichiro Aikawa (J) ist Web-Designer und lebt in in Seattle. Seine Arbeiten wurden in *Word, Shift, Drunktank, Archinect* und *Communication Arts Online* veröffentlicht.

THE INTRUDER
BETA VERSION 1.1
Natalie Bookchin

one of these mornings gonna wake up crazy, gonna take my gun, gonna shoot my baby—ain't nobody's business if i do. if i catch you with another man that's the end little girl. one of these mornings gonna wake up woozy, gonna take my gun gonna kill old suzie. i love you so much it hurts and i'd rather see you dead little girl than be with another man. i'm going out to shoot my old lady, you know i caught her messing round with another man and that ain't too cool. i killed the only woman i loved—the gun went off a root-ie-toot too-and frankie fell on the floor. it ain't nobody's business but my own.
it's an unbelievably realistic world of brutal, sense-less, gratuitous, egregious, yet strangely compelling horror.

an einem dieser morgen werd ich wütend aufste-hen, werde meine pistole nehmen, werde mein baby erschießen – geht niemanden etwas an wenn ich's tue. wenn ich dich mit einem anderen mann erwi-sche ist es das ende kleines mädchen. an einem die-ser morgen werd ich tüdelig wach werden, werde meine pistole nehmen und die alte susi erschießen. ich liebe dich so sehr dass es weh tut und ich seh dich lieber tot kleines mädel als mit einem andern mann und das ist nicht cool. ich habe die einzige geliebte frau getötet – die pistole ging los rumms-ta-bumm-bumms – und frankie fiel auf den boden. es geht niemanden was an außer mich selbst.
es ist eine unglaublich realistische welt aus bruta-lem, sinnlosem, kostenfreiem, erbaulichem und doch seltsam fesselndem horror.

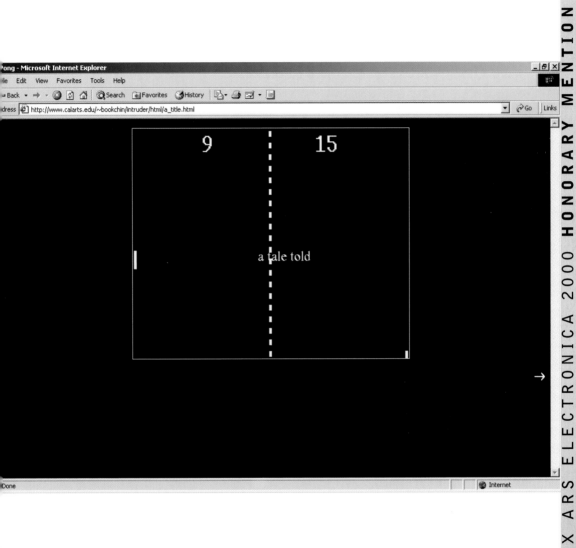

NATALIE BOOKCHIN WORKS COLLABORATIVELY AND INDEPENDENTLY ON AND OFF THE INTERNET. IN 1999-2000 SHE RECEIVED GRANTS FOR HER PROJECTS FROM THE WALKER ARTS CENTER/JEROME FOUNDATION (WITH ALEXEI SHULGIN), CREATIVE TIME AND THE DANIEL LANGLOIS FOUNDATION (WITH RTMARK) AND CREATIVE CAPITAL. SHE TEACHES AT CALIFORNIA INSTITUTE OF THE ARTS IN LOS ANGELES. ▬▬ Natalie Bookchin arbeitet allein oder mit anderen in- und außerhalb des Internet. 1999/2000 erhielt sie Stipendien für ihre Projekte vom Walker Arts Center / Jerome Fundation (zusammen mit Alexei Shulgin), von Creative Time und der Daniel Langlois Foundation (mit RTMark) und von Creative Capital. Sie unterrichtet am

RECONNOITRE

Tom Corby / Gavin Bailey

Reconnoitre is part of an ongoing series of works concerned with our experience of the network as a "bizarre_scape"—an interlocking appendage to the urban environment with a high metabolism, whose boundaries are continuously re-shaped; accreting and thickening under the influence of powerful social and commercial forces.

While *Reconnoitre* can be considered as a browser in that it allows the user to access web sites and displays the material contained on them, it is less concerned with displaying material coherently as with representing browsing as a behavioural activity. Functioning as a kind of dysfunctional browser operating in the face of the ubiquitous "add to shopping basket" button, it seeks to enunciate our access to information as a "journey of suprises" reinstating the pleasure of browsing as technologically experienced dérive (drift) in its own right—an ambient grazing of text, fragmentary, and absorbing.

How it works

The work consists of a downloadable client search engine and a 3D browser environment/program. The work may be shown as a projected installation or displayed on a standard monitor and keyboard setup.

The user is able to input search term requests directly into the browser environment by typing a request after the flashing cursor on the screen. This activates a web spider agent that hunts the Internet searching for material matches associated with the search request.

Once matches are found, the returned material is represented in the 3D environment. The user can decide to display the results as either the URL's, web page text or the "meta tags" associated with the serach "hits".

The returned search material is endowed with physical behavioural characteristics including entropy (the search results have a life span and eventually decay) and repulsion and attraction which allows emergent interaction to occur between different represented

Reconnoitre ist Teil einer Serie von Arbeiten, die sich mit der Erfahrung des Netzes als "Bizarr-Raum" auseinander setzt – ein eng verzahntes Anhängsel zur urbanen Umwelt, mit einem hohen Grad an Metabolismus, dessen Grenzen ständig neu geformt werden und das sich unter dem Einfluss starker sozialer und kommerzieller Kräfte ausdehnt und wächst. Wenn auch *Reconnoitre* als Browser angesehen werden kann, weil es den Anwendern erlaubt, Websites aufzusuchen und die dort enthaltenen Materialien anzuzeigen, so ist doch die kohärente Darstellung des Materials weniger wichtig als die Darstellung des Browsens an sich als soziale Aktivität.

Dadurch, dass es im Vergleich mit den ubiquitären „In den Warenkorb legen"-Schaltflächen wie eine Art dysfunktionaler Browser wirkt, versucht *Reconnoitre*, unseren Zugang zu Information als eine Art „Fahrt ins Blaue" zu gestalten und die Freude am Browsen als ein technologisch orientiertes Dahintreiben eigener Art darzustellen – ein fragmentarisches, fesselndes Abgrasen einer Umgebung aus Texten.

Zur Funktionsweise

Das Werk besteht aus einem herunterladbaren Search-Engine-Client und einer 3D-Browser-Umgebung (bzw. einem Programm). Das Werk kann sowohl als projizierte Installation als auch über das klassische Setup mit Standard-Monitor und Tastatur gezeigt werden.

Der User kann Suchbegriffe direkt in das Browser-Environment eingeben, indem der Begriff einfach beim blinkenden Cursor eingetippt wird. Dadurch wird ein Web-Spider-Agent aktiviert, der sich sofort im Internet auf Jagd nach materiellen Übereinstimmungen mit der Suchanfrage macht.

Sobald Treffer erzielt werden, wird das gefundene Material in einer 3D-Umgebung dargestellt. Der User kann selbst bestimmen, ob dies in der Form von URLs, als Webseitentext oder als Darstellung der mit den Treffern verbundenen Meta-Tags erfolgen soll.

Das gefundene Material wird mit physikalischen Verhaltenscharakteristika ausgestattet, darunter Entropie (die Suchergebnisse haben eine Lebensdauer und verfallen irgendwann) sowie Anziehung und Abstoßung, was in weiterer Folge zum Entstehen einer Interaktion zwischen den verschiedenen dargestellten Suchergebnissen führt. Die sich ergebenden Konstellationen sind entwicklungsfähig, fließend und unvorhersehbar.

In der virtuelle Umgebung können gleichzeitig mehrere Suchvorgänge koexistieren, indem eine räum-

search term results. The resultant constellations of form are emergent, fluid and unpredictable.

It is possible for several searches to coexist in the virtual environment simultaneously, allowing for a spatio-temporal overlaying of search term requests within the same environment. New search requests can be made by the user at any time.

Once the search results are reconstituted in the environment, the user can interact with the material by navigating the 3D environment, pausing and freezing individual search terms and deleting individual search terms.

lich-zeitliche Überlagerung von Suchanfragen innerhalb desselben Raums aufgebaut wird. Neue Suchanfragen könne vom User jederzeit gestartet werden. Sobald die Suchergebnisse im Environment neu aufgebaut sind, kann der User mit dem Material interagieren, indem er durch die 3D-Umgebung navigiert, und individuelle Suchtermini pausieren lassen, einfrieren oder löschen.

TOM CORBY IS AN ARTIST, CURRENTLY HOLDING A RESEARCH AND VISITING LECTURER POST AT CHELSEA COLLEGE OF ART AND DESIGN. HIS RESEARCH INTERESTS ARE CONCERNED WITH THE CONVERGENCE OF THE INTERNET, SOFTWARE AGENTS AND VIRTUAL ENVIRONMENT TECHNOLOGIES. GAVIN BAILEY IS INVOLVED WITH A NUMBER OF DIGITAL ART PROJECTS DEALING WITH SOFTWARE AGENTS, NEURAL NETS AND THE INTERNET. HE IS CURRENTLY WORKING AT THE ROYAL COLLEGE OF ART, LONDON ON AN A-LIFE RESEARCH PROJECT. HE STUDIED FINE ART AT THE RUSKIN SCHOOL OF DRAWING AND FINE ART AND COMPUTER SCIENCE AT UCL. ■■■ Tom Corby ist Künstler und derzeit Forscher und Gastlektor am Chelsea College of Art and Design. Sein Forschungsschwerpunkt liegt auf der Konvergenz des Internet, auf Software-Agenten und Technologien eines virtuellen Environment. Gavin Bailey arbeitet an einer Vielzahl von digitalen Kunstprojekten, die sich mit Software-Agenten, neuralen Netzwerken und dem Internet beschäftigen. Er hat Kunst an der Ruskin School of Drawing sowie Kunst und Informatik an der UCL studiert.

DISCODER
exonemo [Kensuke Sembo, Yae Akaiwa]

A "discoder" is a device which destroys HTML infor-matic *code* and its *codes* of behavior, a contradiction provider for the web. In this project the user "messes" with the web's HTML internet metastructure. The tags in HTML slip, and the integrity of HTML source code is compromised, as if eaten by bugs. The user is liberated not only from the reassuring mathematical illusions of the HTML interface (in the dismantling of the code) itself, but also from the subtle suggestions of normative behavior presented to the individual by the computer.

The *Discoder* has two modes, a "Private Mode" and an "Open Mode". In the former, the user inputs the URL of the page that he or she wishes to destroy (all pages on the WWW are *discode*able) and proceeds to wreak havoc to their little heart's content. In the lat-ter, the user can decimate a page socially, with friends. With either, the keyboard, this mundane part of our text-based lives, becomes an instrument of destruction, while typing on the keys produces sounds like a typewriter. HTML code suddenly becomes a breeding ground for inserting bugs (ASCII text, which eats HTML source code like a bug, creat-ing a great many changes in the structure of web pages) who change the face of the virtual terrain. The user is not limited to random attacks on web-sites, he can shoot at targets too. Because the bug's influence on the site can be monitored numerically in real time, the user can devise strategies for his assaults. The numbers, alphabet, symbols, backspace (delete) and other keys each carry a new function in *Discoder*, and their categories and point of introduc-tion into the HTML bear relation with previously planted bugs so that the bug's "grow" (into capital letters) and then into HTML tags in their own right, reconstructing the broken pages.

There are any number of unpredicatable effects pro-duced by the *Discoder*, especially when, in the "Open Mode", an unspecified number of users participate, and bugs engage the sedimentary layers produced by bugs that have gone before them. Then the *Discoder* produces some of its most surprising results. *Discoder*, through the invasion of unreasonable ele-ments (=bugs), and a process of destruction, becomes a mechanism for new writing. exonemo's sites aren't set on HTML. *Discoder's* cross-hairs are fixed on the vectors of mundane standardisation, and increased codification in our lives.

Ein Discoder ist ein Gerät, das HTML-Code und die dazugehörigen Verhaltenscodes zerstört – ein Widerspruchs-Provider fürs Web. In diesem Projekt spielt der Benutzer mit der HTML-Metastruktur des Internet herum. Die verschiedenen Tags in HTML ver-schieben sich und die Integrität des HTML-Quellco-des wird verändert, als würden Käfer daran nagen. Der Benutzer wird nicht nur von den Sicherheit gebenden mathematischen Illusionen des HTML-Interface befreit (durch die Demontage des Codes), sondern auch von den subtilen Suggestionen eines normativen Verhaltens, das dem Individuum vom Computer auferlegt wird.

Der *Discoder* kennt zwei Modi: einen „privaten" und einen „offenen" Modus. In Ersterem gibt der User die URL einer Website ein, die er zu zerstören wünscht (alle Seiten im WWW sind *discoder*-bar) und setzt dort sein Zerstörungswerk nach Lust und Laune fort. Im offenen Modus hingegen kann er zusammen mit Freunden eine Seite gemeinschaftlich zerlegen. In beiden Fällen dient die Tastatur, dieses allgegenwär-tige Symbol unseres auf Text basierenden Lebens, als Instrument der Zerstörung. Beim Schreiben produ-zieren die Tasten Geräusche wie eine Schreibmaschi-ne. HTML-Code wird plötzlich zur Nährlösung, in der alle Arten von Bugs wachsen, die das Antlitz des vir-tuellen Terrains verändern (HTML-Quellcode wird von ASCII-Text kahl gefressen wie Bäume von Maikä-fern, was zu den erstaunlichsten Veränderungen in der Struktur von Webseiten führt).

Der User ist nicht auf Zufallsattacken auf Websites beschränkt – auch gezielte Schüsse sind möglich. Da der Einfluss des jeweiligen Bug auf die Site in Echt-zeit numerisch überwacht werden kann, könnten Strategien für gezielte Angriffe entwickelt werden. Zahlen, Buchstaben, Symbole, Rück- und Löschtaste sowie andere Tasten haben bei *Discoder* jeweils eine neue Funktion, und diese haben zusammen mit dem jeweiligen Einsatzort im HTML-Code auch Auswir-kungen auf früher eingepflanzte Bugs, sodass der Bug „wächst" (zu Großbuchstaben und weiter zu selbstständigen HTML-Tags) und die angeknabber-ten Seiten umbaut.

Es gibt eine unendliche Anzahl unvorhersehbarer Effekte, die vom *Discoder* ausgelöst werden, beson-ders wenn im offenen Modus eine nicht näher fest-gelegte Anzahl von Usern teilnimmt und sich Bugs auf die Sedimente früherer Fraßspuren stürzen – gerade dann erzeugt *Discoder* seine allerüber-raschendsten Resultate.

Durch die Invasion von unsinnigen Elementen (= Bugs) und durch einen Prozess der Destruktion wird *Discoder* ein Mechanismus für eine neue Schrift. Die Websites von exonemo basieren übri-gens nicht auf HTML. Das Fadenkreuz von *Discoder* ist auf die Vektoren gewöhnlicher Standardisierung und auf die zunehmende Codifizierung unseres Leben gerichtet.

EXONEMO. IN 1996 KENSUKE SEMBO AND YAE AKAIWA FORMED "EXONEMO" WHEN THEY DETONAT-
ED ON IMPACT WITH THE WWW, FEELING THE INFINITELY GREAT POSSIBILITIES IN NEW SENSES OF
DISTANCE AND DYNAMICS. THEY HAVE CONTINUED TO PUBLISH THEIR ART ON THE WEB (EVER SINCE),
STILL AFTER THAT NEW FEELING. KENSUKE SEMBO, BORN 1972 IN TOKYO, JAPAN. PLANNING, PRODUC-
TION, JAVA, PERL AND HTML PROGRAMMING. YAE AKAIWA, BORN 1973 IN FUKUOKA, JAPAN. PLAN-
NING, PRODUCTION, JAVA SCRIPT, HTML PROGRAMMING, INTERFACE DESIGN. ■■■■ 1996 gründeten
Kensuke Sembo und Yae Akaiwa exonemo, das bei Berührung mit dem WWW gleich heftig
detonierte und die unendlich großen Möglichkeiten in einem neuen Gefühl von Distanz und
Dynamik fühlen ließ. Seither haben sie nicht aufgehört, ihre Kunst im Web zu publizieren, immer
auf der Jagd nach diesem Gefühl. Kensuke Sembo, geb. 1972 in Tokio (Planung, Produktion,
Programmierung in Java, Perl und HTML). Yae Akaiwa, geb. 1973 in Fukuoka (Planung, Produktion,
Javascript und HTML Programmierung, Interface Design).

TOYWAR
agent.NASDAQ
aka Reinhold Grether

The *Toywar* Platform as a
Monument to World Culture

With *Toywar*, net art has attained the power of global agency, and net culture has passed a decisive test of endurance. For this reason, the Prix Ars Electronica award anonymously honors everyone who took part in any way in the etoy campaign. Formally it recognizes the 1800 agents of the *Toywar* platform, as the competition entry consisted of nothing other than a 40-page print-out of their e-mail addresses, typographically scrambled for reasons of data protection. *Toywar* was a magnificent condensation of the feeling of life at the end of the 20th century. An abundance of scenes that had turned the Internet into a medium of world culture during
the nineties, combined with unbounded imagination, left a hostile takeover attempt of the Net by e-commerce no chance. The online toy vendor eToys lost five billion dollars worth of equity in 81 days, sponsoring the most expensive performance in cultural history to celebrate the marriage of its own network. "etoy", one of the most fascinating formations of surreal net exploration, honored as etoy.CREW with the Golden Nica in 1996 for the *Digital Hijack*, and as etoy.CORPORATION a producer of first class world cultural capital, exposed the conflict with eToys over trademarks and domain names to public attention as decisive for the future of the Net. They played cat and mouse with the immediate opponent on various levels and developed one of the most complex community platforms of the net within a remarkably brief period of time. "RTMark", the world class brokerage for anti-commercial sabotage published one of its most dreaded and most successful financial instruments with the "etoy Fund" and brought about the complete collapse of the eToys strategy with incomparable sophistication. "Electronic Civil Disobedience", spearhead of the testing and implementation of virtual forms of protest since the Chiapas uprising, created a script family in international collaboration, which was used to conduct strategic virtual sit-ins at the eToys web site repeatedly throughout the entire etoy campaign, thus revealing the opponent's infrastructural vulnerability. "Thing.net",

dryd*nk@oz.n*t - 4222 N\
ph*124 *K* m*rou*ll* l* \
G*n*v*/L*us*nn*, -, 5205
wolfg*ng@th*ng.n*t - 548
B*rb**68 *K* T*r* B*ll
N*w York, NY, 50054 ##
b*uch@schl*mp*.d* - f*rc
Roy_P*l*tus *K* ju*rg*n
gr*llp*rz*rstr*ss* 55, 46
D*v*_*mory *K* P*t*r C
58th Str**t, 84804-5005
M*ch**l Sm*th - msm*th
Ch*c*go, *L, 60640 ## Jc
bons@hom*.com - 2502 S
ypoc *K* v*ol* z*mm*rm

Die *Toywar* Plattform
als Monument der Weltkultur

Mit *Toywar* hat die Netzkunst globale Handlungs-mächtigkeit erreicht und die Netzkultur eine entscheidende Bewährungsprobe bestanden. Aus diesem Grund geht die Ehrung des Prix Ars Electronica anonym an alle, die sich in irgendeiner Form an der etoy-Kampagne beteiligten, formal an die 1800 Agenten der *Toywar*-Plattform, bestand die Einreichung für den Preis ja aus nichts anderem als einem 40-seitigen, aus Datenschutzgründen typografisch durchschossenen Ausdruck von deren E-Mail-Adressen.
Toywar war eine großartige Verdichtung des Lebensgefühls am Ende des 20. Jahrhunderts. Eine Vielzahl von Szenen, die das Internet während der Neunzigerjahre zu einem Weltkulturmedium ausgebaut hatten, ließen mit überbordender Fantasie dem feindlichen Übernahmeversuch des Netzes durch den e-Commerce keine Chance und nahmen dem Online-Spielevertreiber eToys in 81 Tagen $ 5 Mrd. an Börsenwert ab, um in der teuersten Performance der Kulturgeschichte die Hochzeit ihrer eigenen Vernetzung zu feiern.
„etoy", eine der faszinierendsten Formationen surrealer Netzerforschung, als etoy.CREW für den *Digital Hijack* 1996 mit der Goldenen Nica ausgezeichnet und als etoy.CORPORATION ein Weltkulturkapitalproduzent erster Güte, exponierte den Trademark- und Domainnamenkonflikt mit eToys als Richtungsentscheidung über die Netzzukunft, spielte mit dem direkten Gegner auf den unterschiedlichsten Ebenen Katz und Maus und entwickelte in kürzester Zeit eine der komplexesten Community-Plattformen des Netzes. „RTMark", das Weltspitzenbrokerhaus für antikommerzielle Sabotageakte, legte mit dem „etoy Fund" eines seiner gefürchtetsten und erfolgreichsten Finanzierungsinstrumente auf und sorgte mit unnachahmlicher Souveränität für den völligen

*son LN, 88484 S*lv*rd*l*, W*, 88484 ##
ll* - m*rou*ll*@hotm**l.com - -, 5205
: C*rlos *K* Wolfg*ng St**hl* -
dlow Str**t, 50005 N*w York, NY, 50005 ##
b*ll@rthl*nk.n*t - PO box 5852, 50054
mp *K* g.*. m*ll*r -
*ll*r-pl. 50, 80445 mun*ch, , 80445 ##
n - roy.p*l*tus@n*tw*y.*t -
n*rchtr*nk, upp*r*ustr**, 4654 ##
**m - mono@w*bbn*t.com - 5404 South
chmond, C*, 84804-5005 ## s*lv*r *K*
ny-d*j*.com - 5400 N W*yn*, 60640
y*ng*l *K* K*n Bouch*r -
*y St, 68005 B*ll*vu*, N*br*sk*, 68005 ##
- v*ol*@ *n*rr*s.ch - l*ngstr*ss* 62, 8004

mother of all net and media arts, achieved a master-
piece of logistics in the background and provided a
plethora of campaign elements with unlimited bandwidth. Last but not least, there were the two-thou-
sand subscribers to the leading mailing list for media
and net art, *Rhizome*, who, as an equally tenacious
and flexible base network, spread the campaign
throughout the entire net, organizing it and leading
it to victory.

The *Toywar* platform remains a monument to world
culture for all times.

Zusammenbruch von eToys' Strategie. „Electronic
Civil Disobedience", seit dem Chiapas-Aufstand
Speerspitze in der Erprobung und Durchsetzung virtueller Protestformen, erstellte in weltweiter Zusammenarbeit eine Skriptfamilie, mittels derer die
gesamte etoy-Kampagne virtuelle Sit-ins auf der
Webseite von eToys abhielt und so dem Gegner seine infrastrukturelle Verletzlichkeit vorführte.
„Thing.net", Weltkulturmutter sämtlicher Netz- und
Medienkünste, vollbrachte im Hintergrund eine logi-
stische Meisterleistung und versorgte einen Tau-
sendsassa an Kampagnenelementen mit nie versie-
gender Bandbreite. Last not least, waren es die
zweitausend Subskribenten der führenden Mailing-
liste für Medien- und Netzkunst,
Rhizome, die als so unnachgiebiges wie geschmeidi-
ges Basisnetzwerk die Kampagne über das ganze
Netz hinweg ausbreiteten, organisierten und zum
Sieg führten.
Die *Toywar*-Plattform bleibt für alle Zeiten ein
Monument der Weltkultur.

Web Addresses

Electronic Civil Disobedience
http://www.thing.net/~rdom/ecd/ecd.html
etoy http://www.etoy.com/
Rhizome http://www.rhizome.org/fresh/
RTMark http://www.rtmark.com/
Thing.net http://bbs.thing.net/login.thing
Toywar http://www.Toywar.com/

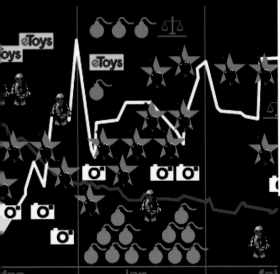

TOYWAR IS ONE OF THE FIRST SUCCESSFUL INTERACTIVE ENTER-
TAINMENT MOVEMENTS WHICH GENERATED MASSIVE IMPACT ON
THE REAL WORLD. TOYWAR IS AN EXAMPLE OF HOW DIGITAL ART
CAN TURN AGGRESSION AND GREED INTO WORLDWIDE FUN AND
CULTURAL PROFIT FOR HUNDREDS OF USERS. TODAY 1798 TOY-
WAR.soldiers (SMART PEOPLE WHO COMMITTED THEMSELVES TO
THIS ACTION BETWEEN NOVEMBER 1999 AND FEBRUARY 2000) CAN CALL THEMSELVES STOCK HOLDERS AND CO-OWNERS OF THE PROPERTY THEY
PROTECTED. ■ TOYWAR ist eine der ersten erfolgreichen interaktiven Unterhaltungs-bewegungen, die eine massive Wirkung auf die reale
Welt hatte. TOYWAR ist ein Beispiel dafür, wie digitale Kunst Aggression und Gier für Hunderte von User auf der ganzen Welt in Spaß und
kulturellen Gewinn verwandeln kann. Heute können sich 1798 „TOYWAR.soldiers" (intelligente Menschen, die sich zwischen November 1999
und Februar 2000 dieser Aktion verschrieben haben) Aktionäre und Mithesitzer an dem Eigentum nennen, das sie geschützt haben.

SILK ROAD
<u>Jie Geng</u>

As the Internet is now becoming larger and more complicated, a site map is more and more important for users to avoid getting lost in cyberspace. A good navigational site map should gracefully and interactively provide the users with the looked for details of content, while keeping them conveniently informed of where they are and provide them with an overview diagram of the information space. A site map should give a clear overall site view that would help the user to decide whether to continue going to a web site or not.

A site map can be a regional site map or a global site map, or a combination of both.

A regional site map can represent a detailed relationship between the currently viewed page and those immediately linked pages. It helps the viewer to navigate from the current page to the next page that is linked directly to the current page. The illustrated relationship is detailed but regional.

A global site map provides the overall view of the web sites that may not be directly linked. It logically links some web sites that are related in contents. It helps the users to determine whether to go into a web site or not from a higher level.

A combination of the regional map and the global map will benefit the user greatly. Imagine that a global map is always linked to a regional map, no matter which page the user looks at. With this arrangement, the user can always focus on the specific page of the current web site, and knows what other related web sites are available.

My project is a *Silk Road* web site with a dynamic regional site map and a global site map. The regional site map is a Java applet that generates a meaningful structure of the web site and visually represents it. An animated smooth movement and color change of the symbols that represent web pages are pleasing to the viewer's eyes. The user can interact with my regional site map and control and jump around my *Silk Road* web site. The global site map provides the overall view of other related and logically categorized *Silk Road* web sites authored by others on the Internet. The user can gain in-depth knowledge and understanding of Chinese history, culture, religion, custom, language etc.

Je größer und komplizierter das Internet wird, desto wichtiger wird eine Landkarte des Netzes, um zu verhindern, dass man sich in den Weiten des Cyberspace verläuft. Eine gute Navigationskarte sollte dem Benutzer elegant und interaktiv mit den gesuchten Details des Inhalts versorgen und ihn gleichzeitig mit Hilfe eines Überblicksdiagramms des Informationsraums über seine aktuelle Position informieren. Eine Site-Map sollte stets ein klares Bild einer Site bieten, damit der User entschieden kann, ob er weiter zu einer Website navigieren will oder nicht.

Eine Site-Map kann entweder regional, global oder eine Kombination aus beidem sein.

Eine regionale Site-Map kann eine detaillierte Beziehung zwischen der derzeit besuchten Seite und den unmittelbar damit verknüpften anzeigen. Sie hilft dem Betrachter, von der gegenwärtigen Seite zur nächsten direkt verknüpften zu navigieren. Die so illustrierte Beziehung ist detailliert, aber eben regional beschränkt.

Eine globale Site-Map hingegen liefert einen Überblick über Websites, die nicht notwendigerweise direkt verknüpft sind. Sie stellt Beziehungen zu Sites her, die in inhaltlichem Zusammenhang stehen, und erlaubt dem User auf einem hohen Niveau, zu entscheiden, ob eine Site eine Besuch lohnt.

Eine Kombination aus beiden Typen ist für den User von besonderem Nutzen. Man stelle sich vor, dass die globale Seite stets mit der regionalen verknüpft bleibt, egal welche Seite der User betrachtet. Bei einem solchen Arrangement kann der Anwender sich stets auf die spezifische Seite der Website konzentrieren, weiß aber trotzdem, welche verwandten Websites noch zur Verfügung stehen.

Mein Projekt ist eine Website über die Seidenstraße mit je einer dynamischen regionalen und einer globalen Site-Map. Die regionale Karte ist ein Java-Applet, das eine aussagekräftige Struktur der Website generiert und grafisch darstellt. Animierte weiche Bewegungen und Farbänderungen der für die Seiten stehenden Symbole erfreuen das Auge des Betrachters. Anwender können mit meiner regionalen Site-Map interagieren und sich frei durch meine Seidenstraßen-Website bewegen. Die globale Site-Map liefert einen Überblick über andere ähnliche und logisch kategorisierte Websites zum Thema Seidenstraße. Dem Benutzer wird umfassende Kenntnis und Verständnis für chinesische Geschichte, Kultur, Religion, Bräuche, Sprache usw. angeboten.

JIE GENG (PR CHINA) HAS ALREADY RECEIVED NUMEROUS AWARDS IN NATIONAL COMPETITIONS IN CHINA AS A PAINTER AND WORKED AS AN ART EDUCATOR IN ZHEJIANG. SHE HAS LIVED IN NEW YORK SINCE 1996 AND WORKED AS A WEBMASTER IN ADDITION TO HER STUDIES AT THE SCHOOL OF VISUAL ARTS. SHE WROTE HER DOCTORATE ON TOPOLOGY ON THE INTERNET.
Jie Geng (VR China) hat in China schon zahlreiche Preise bei nationalen Wettbewerben als Malerin erworben und als Kunsterzieherin in Zhejiang unterrichtet. Seit 1996 lebt in New York, hat als Designerin gearbeitet und zuletzt neben ihrem Studium an der School of Visual Arts als Webmaster gearbeitet. Ihre Doktorarbeit hat sie über Topologie im Internet

ZEITGENOSSEN / BINARY ART SITE

Ursula Hentschläger / Zelko Wiener

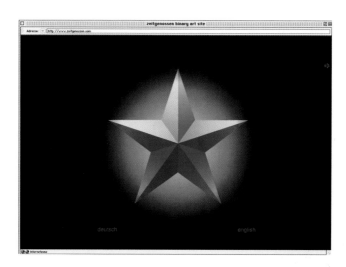

Our research and development work is based on the assumption that the technical standards of our information society include multi-media computers and broad bandwidth in communication. The Web dramaturgy which is in the foreground of the Binary Art Site of *Zeitgenossen* is the result of a radical, future-oriented concept of information architecture. As a consequence the whole navigation is reduced to one dragable moving star, which is to turn off and on. Within this meta-concept we create a three-star program: Imagination (offering space for our art projects which are specially developed for the Web), Inspiration (presenting theoretical and artistic positions on media structures) and Information (devoted to quantum communication; to be launched in the fall of 2000). Within the blue dots you will find more basic information. The whole site is programmed in Flash. Art is what we consider one of the last bastions of unfettered imagination—what we believe to be the basis of elementary processes of change. Considerations are primarily concentrated on the loaded field marked by the modern information society in its relation to individual views of the world.

Unsere Forschungs- und Entwicklungsarbeit basiert auf der Annahme, dass der technische Standard unserer Informationsgesellschaft Multimedia-Computer und große Bandbreite in der Kommunikation umfasst. Die Web-Dramaturgie, die bei der Binären Kunst-Site *Zeitgenossen* im Vordergrund steht, ist das Ergebnis eines radikalen, zukunftsorientierten Konzepts der Informationsarchitektur, das darin resultiert, dass die gesamte Navigation auf einen ziehbaren Stern reduziert wird, der ein- und ausgeschaltet werden kann. Innerhalb dieses Metakonzepts schaffen wir ein Drei-Sterne-Programm: *Imagination* (indem wir Raum für speziell für das Web entwickelte Kunstprojekte zur Verfügung stellen), *Inspiration* (indem wir theoretische und künstlerische Positionen zu Medienstrukturen veröffentlichen) und *Information* (weil wir der Quantenkommunikation verpflichtet sind, die im Herbst 2000 eingeführt wird). Unter den blauen Punkten finden Sie weitere Basisinformationen. Alles ist in Flash programmiert. Wir halten die Kunst für eine der letzten Bastionen einer entfesselten Vorstellungskraft – die wir als Ausgangspunkt elementarer Veränderungsprozesse ansehen. Unsere Überlegungen konzentrieren sich vor allem auf das von der modernen Informationsgesellschaft abgesteckte Feld und seiner Beziehung zu einer individuellen Sicht der Welt.

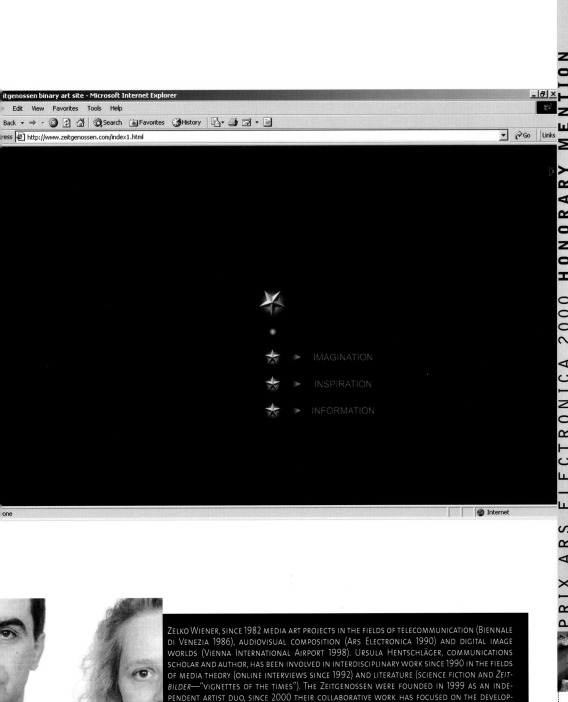

itgenossen binary art site - Microsoft Internet Explorer

Edit   View   Favorites   Tools   Help

Back   ▾   ⇨   ▾   ⊗   ⌂   ⌂ | ⊘Search   ⌂Favorites   ⊙History | ⧉▾   ⊜   ⌂   ▾   ⊟

ress ⊠ http://www.zeitgenossen.com/index1.html   ▾   ⌀Go | Links

IMAGINATION

INSPIRATION

INFORMATION

one   ◉ Internet

ZELKO WIENER, SINCE 1982 MEDIA ART PROJECTS IN THE FIELDS OF TELECOMMUNICATION (BIENNALE DI VENEZIA 1986), AUDIOVISUAL COMPOSITION (ARS ELECTRONICA 1990) AND DIGITAL IMAGE WORLDS (VIENNA INTERNATIONAL AIRPORT 1998). URSULA HENTSCHLÄGER, COMMUNICATIONS SCHOLAR AND AUTHOR, HAS BEEN INVOLVED IN INTERDISCIPLINARY WORK SINCE 1990 IN THE FIELDS OF MEDIA THEORY (ONLINE INTERVIEWS SINCE 1992) AND LITERATURE (SCIENCE FICTION AND *ZEIT-BILDER*—"VIGNETTES OF THE TIMES"). THE ZEITGENOSSEN WERE FOUNDED IN 1999 AS AN INDE-PENDENT ARTIST DUO, SINCE 2000 THEIR COLLABORATIVE WORK HAS FOCUSED ON THE DEVELOP-MENT OF WEB PROJECTS. ▬▬▬ Zelko Wiener, seit 1982 Medienkunstprojekte in den Bereichen Telekommunikation (Biennale di Venezia 1986), audio-visuelle Komposition (Ars Electronica 1990) und digitale Bildwelten (Vienna International Airport 1998). Ursula Hentschläger, Kommu-nikationswissenschaftlerin und Autorin, arbeitet seit 1990 interdisziplinär in den Bereichen Medientheorie (Online-Interviews seit 1992) und Literatur (Science-Fiction und Zeitbilder). Die ZEITGENOSSEN wurden 1999 als unabhängiges Künstlerduo gegründet, seit 2000 liegt der

SINNZEUG
Stefan Huber / Ralph Ammer / Birte Steffan

Sinnzeug is a novel dynamic search-engine for links to intelligent websites we think are worth spending time on. All the represented websites are special for some reason. There is a variety ranging from beautiful programming ideas to up-to-date online magazines or experimental art sites.

Each dot on the screen represents a website. After double-clicking anywhere in the window you can enter a searchword or chose one from the pop-up-menu. You can start your individual search for the websites you are interested in by placing an index of catchwords in the window and arranging them.

The websites, i.e. the dots that feel they have something to do with this word, will be attracted by it. To remove a searchword just drag it out of the window. There will be lines drawn between the visited sites so you can easily find your way "back" to websites you may have enjoyed during your last use of *Sinnzeug*. *Sinnzeug* is shockwave-application programmed with macromedia director 7.

Sinnzeug ist eine neuartige dynamische Suchmaschine für Links zu intelligenten Websites, die es unserer Ansicht nach verdienen, dass man ihnen Zeit widmet. Alle dargestellten Websites haben irgendwelche Besonderheiten – die Bandbreite reicht von schönen Programmier-Ideen zu modernen Online-Magazinen und experimentellen Kunst-Sites.

Jeder Punkt auf dem Bildschirm stellte eine Website dar. Wenn man irgendwo im Fenster doppelklickt, kann man einen Suchbegriff eingeben oder aus einem Auswahlmenü wählen. Die Suche nach interessanten Websites kann je nach Geschmack individualisiert werden, indem ein Katalog von Kernbegriffen ins Fenster eingegeben und dort geordnet wird. Webssites – genauer gesagt die sie repräsentierenden Punkte –, die sich von den eingegebenen Schlüsselbegriffen angesprochen fühlen, werden von diesen angezogen. Um einen Schlüsselbegriff zu löschen, wird er einfach aus dem Fenster gezogen. Zwischen den besuchten Sites werden Linien gezogen, sodass man jederzeit seinen Weg „zurück" zu den in der letzten Sitzung mit *Sinnzeug* besuchten Sites finden kann.

Sinnzeug ist eine Shockwave-Applikation, programmiert mit Macromedia Director 7.

CAT SCAN CONTEST

Websites are a pain in the ass.
Avoid having one if you can.
http://www.cat-scan.com/

Done   🌐 Internet

PRIX ARS ELECTRONICA 2000 HONORARY MENTION

STEFAN HUBER (D), 25 YEARS OLD, 5TH SEMESTER VISUAL COMMUNICATION AT THE HOCHSCHULE DER KÜNSTE, BERLIN, WITH PROF. SAUTER. BIRTE STEFFÀN (D), 27 YEARS OLD, 10TH SEMESTER VISUAL COMMUNICATION AT THE HOCHSCHULE DER KÜNSTE, BERLIN, IN THE CLASS FOR DIGITAL MEDIA WITH PROF. JOACHIM SAUTER. RALPH AMMER (D), BORN 1973. 1995—2000 VISUAL COMMUNICATION AT THE HOCHSCHULE DER KÜNSTE, BERLIN. ▬ Stefan Huber (D), 25 Jahre, 5. Semester Visuelle Kommunikation an der Hochschule der Künste Berlin bei Prof. Sauter. Birte Steffan (D), 27 Jahre, 10. Semester Visuelle Kommunikation an der Hochschule der Künste Berlin in der Klasse für Digitale Medien bei Prof. Joachim Sauter Ralph Ammer (D), geb. 1973. 1995–2000 Visuelle Kommunikation an der Hochschule der Künste Berlin.

GRASPING AT BITS

Patrick Lichty

Art and Intellectual Control in the Digital Age

At the turn of the millennium, the international art community has begun to recognize the significance of the Internet as a milieu for expression and critical inquiry of issues such as the globalization of capitalist culture. The increasingly Blade Runner-esque role of corporate culture and "big money" in global society, and cyberspace in particular raises questions *vis-à-vis* freedom of expression and the controlling influence on intellectual property by multinational corporations. Artists who critique the expanding role of corporate power make visible the cultural terrain of this power relation, frequently through the subsequent litigation by those very same institutions under scrutiny. In addition, events such as the Leonardo and etoy controversies have brought to light corporations wishing to enforce their brand identity over artistic groups that predate them through the exertion of legal force.

This essay is a continuation of my six years of research into the use of aesthetic metaphors and spatial narrative in critical discourse. *Grasping at Bits* uses a mindmap interface metaphor powered by Natrificial's SiteBrain to create a dynamic associative map of relations between topics. In addition, the entire essay employs embedded annotations, rich cross-linking and a system of relative color coding to one of four dominant themes, from which emerges a rich information architecture. In this way the essay exhibits no less than five layers of organization, and so seeks to maximize the interactive, non-linear potential of new media discursive structures.

The site requires no more than a Netscape or Microsoft 4+ browser.

Kunst und intellektuelle Überwachung im digitalen Zeitalter

An der Wende zum neuen Jahrtausend hat die internationale Kunst-Community begonnen, die Bedeutung des Internet als Milieu für die Äußerung und kritische Untersuchung von Anliegen wie z. B. der Globalisierung der kapitalistischen Kultur zu erkennen. Die zunehmend *Blade-Runner*-ähnliche Rolle der Unternehmenskultur und des „Großen Geldes" in der Gesellschaft – und besonders im Cyberspace – wirft Fragen nach der Freiheit des Ausdrucks und dem überwachenden und steuernden Einfluss auf, den multinationale Unternehmen auf geistiges Eigentum ausüben. Künstler, die die Expansion der Unternehmensmacht hinterfragen, machen das kulturelle Terrain dieser Machtbeziehungen sichtbar, und nicht selten nimmt diese Auseinandersetzung die Form eines Rechtsstreits mit den so in Frage gestellten Unternehmen an. Darüber hinaus haben Ereignisse wie die Kontroversen um *Leonardo* und *etoy* aufgezeigt, wie Unternehmen versuchen, Künstlergruppen gerichtlich zu belangen, um ihre Markennamensprüche durchzusetzen, auch wenn diese Gruppen den Namen schon wesentlich länger führen.

Diese Arbeit ist die Fortsetzung meiner sechsjährigen Forschungen im Reich der ästhetischen Metaphern und räumlichen Narration im kritischen Diskurs. *Grasping at Bits* verwendet eine Mindmap-Metapher, die mit SiteBrain von Natrificial arbeitet, um eine dynamische assoziative Landkarte der Beziehungen zwischen einzelnen Themen zu gestalten. Zusätzlich verwendet der gesamte Essay eingebettete Anmerkungen, jede Menge Kreuzverknüpfungen und ein System von Farbkodierungen, um die Zugehörigkeit zu einem von vier dominanten Themenkreisen zu illustrieren, wodurch sich eine reiche Informationsarchitektur ergibt. Nicht weniger als fünf Organisationsebenen werden so konstruiert, um das interaktive und nonlineare Potenzial der diskursiven Strukturen zu maximieren.

Zur Betrachtung der Website ist lediglich ein Netscape- oder Microsoft-Browser (4.0 oder höher) erforderlich.

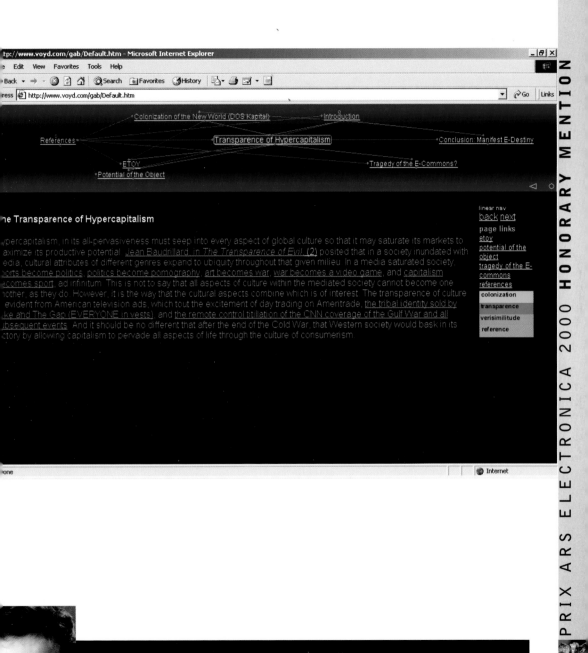

tp://www.voyd.com/gab/Default.htm - Microsoft Internet Explorer

e Edit View Favorites Tools Help

Back ▾ → ▾ ◎ ◻ ⌂ ◎Search ⊡Favorites ◎History ◻▾ ◎ ◻ ▾ ◻

ress ◻ http://www.voyd.com/gab/Default.htm ▾ ⌐Go | Links

- Colonization of the New World (DOS Kapital) · Introduction

References · · Transparence of Hypercapitalism · · Conclusion: Manifest E-Destiny

· ETOY · Tragedy of the E-Commons?

· Potential of the Object

he Transparence of Hypercapitalism

ypercapitalism, in its all-pervasiveness must seep into every aspect of global culture so that it may saturate its markets to aximize its productive potential. Jean Baudrillard, in *The Transparence of Evil*, (2) posited that in a society inundated with edia, cultural attributes of different genres expand to ubiquity throughout that given milieu. In a media saturated society, ports become politics, politics become pornography, art becomes war, war becomes a video game, and capitalism ecomes sport, ad infinitum. This is not to say that all aspects of culture within the mediated society cannot become one nother, as they do. However, it is the way that the cultural aspects combine which is of interest. The transparence of culture ; evident from American television ads, which tout the excitement of day trading on Ameritrade, the tribal identity sold by ke and The Gap (EVERYONE in vests), and the remote control titillation of the CNN coverage of the Gulf War and all ibsequent events. And it should be no different that after the end of the Cold War, that Western society would bask in its ctory by allowing capitalism to pervade all aspects of life through the culture of consumerism.

linear nav
back next
page links
etoy
potential of the
object
tragedy of the E-
commons
references
colonization
transparence
verisimilitude
reference

one ● Internet

PATRICK LICHTY IS A CONCEPTUAL MEDIA ARTIST AND CULTURAL THEORIST. HIS WORKS INCLUDE A VARIETY OF MEDIA, INCLUD-ING PRINTMAKING, KINETICS, VIDEO, GENERATIVE MUSIC, ONLINE TECHNOLOGIES, AND NEON. COLLABORATIONS INCLUDE WORK WITH SITO.ORG AND RTMARK, AND EVENTS INCLUDE THE NEW YORK DIGITAL SALON, AND ARS ELECTRONICA. AS A SCHOLAR, HE HAS BEEN PUBLISHED WIDELY IN ACADEMIC AND CYBERJOURNALS SUCH AS *CTHEORY*, *FRAME*, *LEONARDO*, AND *SOCIOLOGICAL SPECTRUM*. ▬ Patrick Lichty ist konzeptueller Medienkünstler und Kulturtheoretiker. Seine Arbeiten umfassen eine Vielzahl von Medien: Druck, Kinetik, Video, Generative Musik ebenso wie Online-Technologie und Neon. Kooperationsprojekte unter anderem mit SITO.org und RTMark, Events beim New York Digital Salon und Ars Electronica. Seine wissenschaftlichen Arbeiten wurden in vielen akademischen und Cyber-Zeitschriften publiziert, darunter *CTHEORY*

ELECTRICA

Peter Mühlfriedel / Gundula Markeffsky /
Leonard Schaumann

*"It's like standing in a power station on acid. If you just stand in
the middle of a really massive one so you get a really weird
presence and you've got that hum. You feel electricity around
you. That's totally dreamlike for me."*   *Richard James*

Electrica is an interactive sound web site, an experimental sound installation investigating the possibilities of controlling sound in real time.
The project is a collaboration between Gundula
Markeffsky and Peter Mühlfriedel (skop) with
Leonard Schaumann (Leonid) and can be accessed on
the Internet at <http://www.electrica.leonid.de>.
An important component of this audio-visual experiment is its great potential for interaction.
The visitor/listener/player exploring the web site
moves through user interfaces that are unusual for
the web. Partly intuitively, partly assimilatively, they
find their own strategies for use, manipulating the
sound cosmos consciously and unconsciously.
The point of departure for the project is a transformer station in Jena that was built in the 30's by
Bauhaus architects and is now a historical monument. This location is reminiscent of the vibrating
atmosphere of a mad scientist's laboratory from 50's
movies. The visual and acoustic material recorded in
the transformer station and the circuitry aesthetic of
the electricity became the source of both inspiration
and material for the project.
Interwoven elements also include samples from old
science fiction movies like "Forbidden Planet" and
original recordings of electrical experimentation
facilities in technical museums.
The web site consists of five different audio environments.
Until now, the bandwidth problem of the Internet
has presented an almost insurmountable obstacle to
the integration of sound in web sites. Sound files are
generally very large or become distorted through
compression.
With *Electrica* this reciprocal relation between the
amount of data and the sound quality has been
solved through the use of a new sound format from
the company Beatnik (www.beatnik.com), the so-
called "Rich Music Format" (RMF). By way of comparison: a 5-minute RMF file with a size of only 5 KB in 16-

„*It's like standing in a power station on acid. If you just stand
in middle of a really massive one so you get a really weird
presence and you've got that hum. You feel electricity around
you. That's totally dreamlike for me.*"   *Richard James*

Electrica ist eine interaktive Sound-Website, eine
experimentelle Klanginstallation, welche die derzei-
tigen Möglichkeiten der Echtzeitsteuerung von
Sound im Netz untersucht.
Das Projekt ist eine Zusammenarbeit von Gundula
Markeffsky und Peter Mühlfriedel (Skop) mit Leonard
Schaumann (Leonid) und im Internet unter der
Adresse <http:///www.electrica.leonid.de> erreich-
bar.
Eine wichtige Komponente dieses audiovisuellen
Experiments ist sein großes Interaktionspotenzial.
Beim Erforschen der Website bewegt sich der Be-
trachter/Hörer/Spieler in für das Web ungewöhnlichen User-Interfaces. Teils intuitiv, teils assimilato-
risch erschließt er sich eigene Benutzungsstrategien
und manipuliert dabei den Klangkosmos in bewuss-
ter wie unbewusster Weise.
Ausgangspunkt des Projektes ist ein in den 30er-Jah-
ren von Bauhausarchitekten errichtetes, jetzt denk-
malgeschütztes Umspannwerk in Jena.
Dieser Ort erinnert an die vibrierende Atmosphäre
der Mad-Scientist-Labors in Filmen aus den 50er-Jah-
ren. Das im Umspannwerk aufgenommene visuelle
und akustische Material sowie die Schaltplan-Ästhe-
tik der Elektrik wurden zur Inspirations- und Mate-
rialquelle des Projekts.
Eingeflossen sind unter anderen Samples aus alten
Science-Fiction-Filmen wie *Forbidden Planet* und Ori-
ginalaufnahmen elektrischer Versuchsanlagen aus
Technikmuseen.
Bisher stellte das Bandbreitenproblem des Internet
ein kaum zu überwindendes Hinternis für die Integration von Sound in Websites dar. Sounddateien
sind in der Regel sehr groß und werden durch Kompression entstellt.
Bei *Electrica* ist dieses reziproke Verhältnis von
Datenmengen und Soundqualität durch den Einsatz
eines von der Firma Beatnik (www.beatnik.com) entwickelten neuen Soundformats – das sogenannte
„Rich Music Format" (RMF) – gelöst.

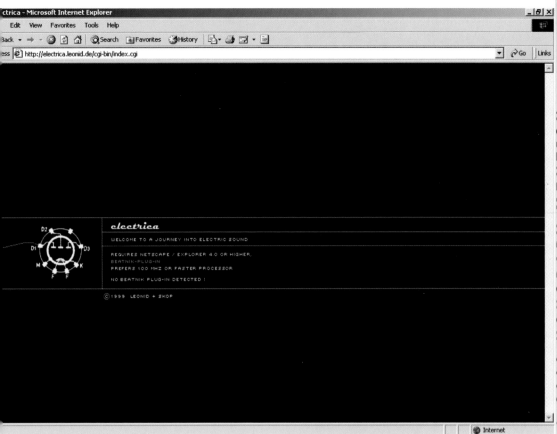

bit stereo 44 Khz would need 50 MB of memory in a conventional CD format.

The RMF format makes it possible to use sounds in CD quality and yet still keep downloading time low. An interface of the plug-in Java, the programming language of the WWW, creates a previously unknown potential for the user to manipulate sound in real time.

The Beatnik plug-in and the new possibilities of JavaScript 1.2 form the technical basis of *Electrica*.

(Zur Veranschaulichung: Ein 5 Minuten spielendes RMF-Files mit nur 5 KB Größe in 16-bit Stereo 44kHz würde in derselben Qualität im konventionellen CD-Format 50 MB Speicherplatz benötigen.) Das RMF-Format ermöglicht Sounds in CD-Qualität zu verwenden und gleichzeitig die Downloadzeiten der Seite gering zu halten. Eine Schnittstelle des Plug-In zu JavaScript, der Programmiersprache des WWW, schafft ein bisher nicht gekanntes Potential zur Manipulation von Sound in Echtzeit durch den User. Das Beatnik-Plug-In und die neuen Möglichkeiten von JavaScript 1.2 bilden die technische Basis von *Electrica*.

PETER MÜHLFRIEDEL, BORN 1964 IN FREIBERG/SACHSEN; 1993—1999 STUDIED AT THE HOCHSCHULE DER KÜNSTE, BERLIN, GRADUATED WITH A DIPLOMA IN DESIGN; 1996/97 STUDIED AT CALIFORNIA INSTITUTE OF THE ARTS, FINISHED WITH BA. GUNDULA MARKEFFSKY, BORN 1970 IN BERLIN; 1993—1999 STUDIED AT THE HOCHSCHULE DER KÜNSTE, BERLIN, GRADUATED WITH A DIPLOMA IN DESIGN. ■■■■   Peter Mühlfriedel (D), geb. 1964. 1993–1999 Studium an der Hochschule der Künste, Berlin, Abschluss als Diplom-Designer. 1996–1997 Studium am California Institute of the Arts, Abschluss als BA. Gundula Markeffsky (D), geb. 1970. 1993–1999 Studium an der Hochschule der Künste, Berlin, Abschluss als Diplom-Designerin.

NETWORK COMMUNICATE KALEIDOSCOPE

Kazushi Mukaiyama

To explore the substance of beauty

I have a simple question: What is beauty, if fine art is the representation of beauty? Since this idea is confused by multiple points of view, I want to promote the beauty of nature.

A spectacular sunset or a fascinating coastline seems to hypnotize people for a long time, so people can feel the beauty of nature. What do we perceive as beauty in nature? If this were a work of fine art, it would be the artist's personality, but this beauty is not made by anyone. Why do we feel nature's beauty, if this is not its purpose?

I have an answer to this: It is relation and reflection. Objects in nature are not always beautiful, there are many dirty objects in the environment as well. Yet all of nature (even the dirty object) is very important to beauty. Take a sunset scene like in La Jolla, for example. The elements of a beautiful sunset include the sun, sea, sky, etc. If the sun is beautiful, we should notice that it is always beautiful, but in fact, we don't think so. At noon we wear sunglasses to protect our eyes from the sunshine. So the sun is not what makes a sunset beautiful. It is the same with the sea and everything else. What makes a sunset beautiful, then? It is not beautiful because the scene contains beautiful objects, but because all the elements relate to and reflect on one another. That color of the clouds was made by sunshine filtered through the air, reflecting on the cloud and the surface of the sea. Thus the beauty that we feel is the complex relation and reflection of each simple object. This is the substance of the beauty of nature. We call it "harmony." Remarkably, this is composed simply of refrains of objects. This is what leads me to the basic structure of beauty.

Die Substanz der Schönheit erforschen

Wenn die Kunst die Darstellung des Schönen ist, dann habe ich eine einfache Frage: „Was ist Schönheit?" Zahlreiche Gesichtspunkte beeinflussen diesen Begriff; ich möchte die Schönheit der Natur allen anderen voranstellen.

Ein spektakulärer Sonnenuntergang oder eine faszinierende Küstenlandschaft scheinen Menschen über lange Zeit zu hypnotisieren – sie fühlen die Schönheit der Natur. Aber was erkennen wir in der Natur als schön? Wäre dies ein Kunstwerk, dann wäre es die Persönlichkeit des Künstlers. Die Natur aber wird von niemandem gemacht. Warum also empfinden wir die Schönheit der Natur, wenn sie nicht zu genau diesem Zweck entstanden wäre?

Ich habe auch eine Antwort darauf: Sie lautet „Beziehung und Reflexion". Objekte der Natur sind nicht notwendigerweise schön – es gibt genug „schmutzige" Dinge in der Natur. Aber alle gemeinsam (auch die unangenehmen) sind für den Begriff der Schönheit wichtig. Betrachten wir einen Sonnenuntergang in La Jolla: Die Gegenstände seiner Schönheit sind die Sonne, das Meer, der Himmel usw. Wenn die Sonne schön ist, dann ist sie es immer. Aber wir Menschen sehen das keineswegs so: Tragen wir nicht mittags Sonnenbrillen, um unsere Augen vor ihr zu schützen? Also ist es nicht die Sonne, die den Sonnenuntergang „schön" macht. Das Gleiche gilt für das Meer und alles andere. Was also macht den Sonnenuntergang „schön"? Wirkt er Wunder? Keineswegs. Er ist nicht deshalb schön, weil seine Szenerie aus „schönen" Objekten besteht, sondern weil jedes Objekt mit jedem anderen in einer faszinierenden Beziehung steht. Die Farbe der Wolke entsteht, weil der Sonnenschein, gefiltert durch die Luft, auf die Wolken und die Meeresoberfläche reflektiert wird. Wir müssen erkennen, dass Schönheit ein komplexes Gefüge aus Beziehungen und Reflexionen aller Objekte untereinander ist.

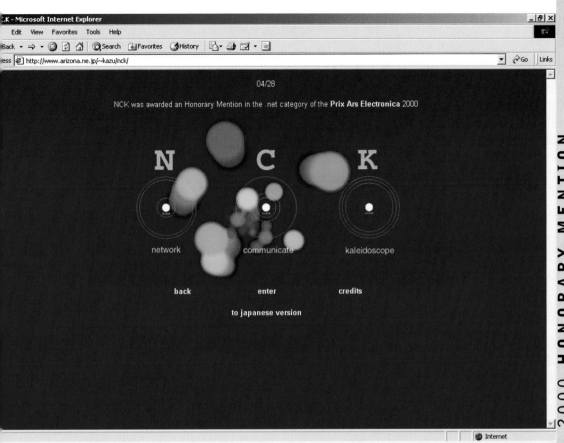

04/28

NCK was awarded an Honorary Mention in the .net category of the **Prix Ars Electronica** 2000

N C K

network   communicate   kaleidoscope

back   enter   credits

to japanese version

Network Communicate Kaleidoscope is a virtual space accessible via the Internet.

People connect to a void world as "particles", where they enjoy chatting and interacting with others. After they disconnect, their remnant particle interacts individually with other particles' behavior, reflecting the activity of all those who have ever connected with the virtual space. The theme of the project is to explore and create beautiful images like a kaleidoscope of "clustering fireflies." Yet the images can be made by anonymous personalities beyond the artist's. In this context, the user/joiner contributes to and knows the substance of the art.

Dies ist die Substanz der Schönheit der Natur. Wir nennen es „Harmonie" – und daraus kann ich die Grundstruktur der Schönheit ableiten.
Network Communicate Kaleidoscope ist ein über das Internet zugänglicher virtueller Raum. Besucher treten mit einer leeren Welt in Verbindung und erschienen dort als „Partikel": Sie unterhalten sich und interagieren mit anderen. Wenn sie sich wieder ausloggen, bleiben ihre „Partikel" erhalten und interagieren individuell mit den anderen. Das Verhalten der Partikel reflektiert die Aktivität aller User, die jemals den virtuellen Raum besucht haben. Das Thema des Projekts ist es, schöne Bilder zu generieren und zu erforschen – eine Art Kaleidoskop voller spielender Schmetterlinge, die von anonymen Personen geschaffen werden und die sich einer Kontrolle durch den Künstler entziehen. In diesem Sinne trägt jeder Besucher dazu bei und erkennt die Substanz der Kunst.

KAZUSHI MUKAIYAMA (J) GRADUATED IN CONCEPT AND MEDIA ART FROM THE UNIVERSITY OF KYOTO. HE IS CURRENTLY ARTIST IN RESIDENCE AT THE "CENTER FOR RESEARCH IN COMPUTING AND THE ARTS", UCSD, USA. EXHIBITIONS AND PERFORMANCES IN JAPAN, THE USA AND EUROPE. ■■■ Kazushi Mukaiyama (J) graduierte in Konzept- und Medienkunst an der Universität von Kyoto. Er ist gegenwärtig Artist in Residence am „Center for Research in Computing and the Arts", UCSD, USA.

0100101110101101.ORG

The difference between net.art and every other form of art seems to be "interactivity", at least this is what we become accustomed to hearing. Well: "interactivity", as it's usually intended, is a delusion, pure falsehood. When people reach a site (net.art or not), by their mouse clicks, they choose one of the routes fixed by the author, they only decide what to see before and what after; this is not interactivity. It would be the same as stating that an exposition in a museum is interactive because you can choose from which room to start, which works to see before and which one after.

But recently something's changed. In fact, 0100101110101101.ORG is trying to show that art on the web can really become "interactive": the public must use it interactively, we must use an artwork in an unpredictable way, one that the author didn't foresee, to rescue it from its normal routine (studio/gallery/museum or homepage/hell.com/ Moma) and re-use it in a different and novel way. The first files that appear in 0100101110101101.ORG are what we will call "hybrids": pages by other net.artists all mixed in a random way. This section of the site is centered around a random concept, so that the interface changes every time you visit it. The toolbar becomves useless, the "back" command loses its logical function: every page is set in the unpredictable sequence of chance. 0100101110101101.ORG downloads the websites of the most popular net.artists and then s/he/it/ manipulates them as "it" wants, using them in an interactive way.

The night of 9th June, was the turn of "Art.Teleportacia". "Art.Teleportacia" is the first net.art gallery to have appeared on the web, and also the first attempt to sell works of net.art. The exhibition we are talking about was "Miniatures of the Heroic Period", and consisted of some pages by five of the most well known net.artists in the world for sale at $ 2000 each. 0100101110101101.ORG downloaded the gallery, manipulated the contents, and uploaded it in a new "anticopyright" version, obviously without asking permission from anyone and violating the copyright of the original site. The exhibition changed its name into "Hybrids of the Heroic Period", and the five "original" works were replaced with as many "hybrids", files obtained by mixing pages by net.artists with some trash of the web.

Der Unterschied zwischen net.art (Netz-Kunst) und jeder anderen Kunstform scheint in der „Interaktivität" zu bestehen, zumindest bekommt man diese Definition immer wieder zu hören. Nun, „Interaktivität" wie sie normalerweise verstanden wird, ist eine Enttäuschung, ein Falsifikat. Wenn Benutzer eine Website erreichen (egal ob es sich um net.art handelt oder nicht), wählen sie mit ihren Mausklicks eine der vom Künstler vorgegebenen Routen aus – sie entscheiden lediglich, was sie zuerst ansehen und was später, und das ist noch lange keine Interaktivität. Genauso gut könnte man eine Ausstellung in einem Museum als „interaktiv" bezeichnen, bloß weil man sich aussuchen kann, in welchem Raum man beginnt und welche Werke man in welcher Reihenfolge wie intensiv betrachtet.

Allerdings hat sich in letzter Zeit etwas verändert und 0100101110101101.ORG versucht zu zeigen, dass Kunst im Web tatsächlich „interaktiv" werden kann: Das Publikum muss Kunst interaktiv benützen; wir müssen ein Kunstwerk auf unvorhersehbare Weise verwenden – auf eine, die der Künstler nicht vor-(her)gesehen hat –, um es aus seiner normalen Routine zu erretten (sei es Studio / Galerie / Museum oder Homepage / hell.com /Moma) und um es auf neuartige, andere Weise zu nutzen.

Die ersten bei 0100101110101101.ORG aufgetauchten Dateien waren sozusagen Hybride – Seiten anderer Netzkünstler, die auf zufällige Weise gemixt wurden. Dieser Teil unserer Website dreht sich um ein Konzept des Zufalls, sodass das Interface sich bei jedem Besuch ändert. Die Werkzeugleiste wird unnütz, die Rücktaste verliert ihre logische Funktion: Jede Seite steht in der unvorhersehbaren Sequenz des Zufalls. 0100101110101101.ORG lädt die Websites der bekanntesten Netzkünstler herunter und manipuliert sie nach Lust und Laune – eine interaktive Anwendung. In der Nacht zum 9. Juni war Art.Teleportacia an der Reihe: die erste Netzwerkgalerie, die im Web aufgetaucht ist, und der erste Versuch, Netzkunstwerke zu verkaufen. Die Ausstellung hieß „Miniatures of the heroic period" und bestand aus Seiten mit Arbeiten von fünf der bekanntesten Netzwerkkünstler der Welt, die um jeweils 2000 Dollar zu haben waren. 0100101110101101.ORG hat die Galerie heruntergela-

http://plagiarist.org/www.0100101110101101.ORG
http://plagiarist.org/www.0100101110101101.ORG
http://plagiarist.org/www.0100101110101101.ORG
http://plagiarist.org/www.0100101110101101.ORG
http://plagiarist.org/www.0100101110101101.ORG
http://plagiarist.org/www.0100101110101101.ORG
http://plagiarist.org/www.0100101110101101.ORG
http://plagiarist.org/www.0100101110101101.ORG
http://plagiarist.org/www.0100101110101101.ORG
http://plagiarist.org/www.0100101110101101.ORG
http://plagiarist.org/www.0100101110101101.ORG
http://plagiarist.org/www.0100101110101101.ORG
http://plagiarist.org/www.0100101110101101.ORG
http://plagiarist.org/www.0100101110101101.ORG

Art.Teleportacia is based mainly on three theoretical pillars: 1) A work of net.art can be sold as well as any other work of art. 2) Each net.art work must be covered by copyright and nobody, except the artist, can download it or even link to it without the permission of the author. 3) the "sign" of a net.art work is in the "location bar", so the URL is the only guarantee of originality.

Duplicating Art.Teleportacia *0100101110101101.ORG* brought down all the presuppositions of the gallery; the contradictions which this way of thinking runs into became evident. Technically, whoever visits a site automatically downloads into the cache [kash], all the files he sees. In fact s/he already owns them, therefore it is nonsense to sell pages already on the hard disks of millions of people—it would be more useful to tell the public the fastest way to download the whole website.

We wish to see hundreds of *0100101110101101.ORG* repeating sites of net.artists endlessly, so that nobody realizes which was the "original" one; we would like to see hundreds of Jodi and hell.com, all different, all original, and nobody filing lawsuits for copyright infringement. If there are no more originals to preserve, will "WebDevil" be the brush of a new generation of artists?

From: Rhizome, June 99
http://www.rhizome.org/cgi/to.cgi?t=1486

den, den Inhalt manipuliert und in einer „Anticopyright"-Version wieder hochgeladen – offensichtlich ohne jemanden zu fragen und unter flagranter Verletzung des Copyrights der Ursprungssite. Der Ausstellungsname änderte sich zu „Hybrids of the heroic period" und die fünf „Original"-Werke wurden durch ebenso viele Hybriden ersetzt, die durch die Mischung von Seiten der Webkünstler mit zufälligem Schrott aus dem Netz entstanden sind.

Art.Teleportacia geht im Wesentlichen von drei Prinzipien aus: 1. Ein Netz-Kunstwerk kann verkauft werden wie jedes andere auch. – 2. Jedes Netzkunstwerk wird durch Copyright geschützt und niemand außer dem Künstler selbst kann es ohne Zustimmung des Autors herunterladen oder sich dort hinlinken. – 3. Die „Signatur" eines Netzkunstwerks liegt einzig im Location Bar, d. h. die URL ist die einzige Garantie für Originalität.

Durch die Duplizierung von *Art.Teleportacia* hat *0100101110101101.ORG* all die Annahmen der Galerie widerlegt und die darin liegenden Widersprüche offensichtlich gemacht: Technisch gesehen lädt jeder Besucher mit einer Site ganz automatisch den gesamten Inhalt, den er sieht, in den Cache-Speicher herunter. Kurzum, er hat ihn schon in seinem Besitz, und deshalb ist es unsinnig, Seiten zu verkaufen, die bereits auf der Festplatte von Millionen Leuten gespeichert sind – vernünftiger wäre es, dem Publikum den schnellsten Weg zum Downloaden zu zeigen.

Aus: Rhizome, Juni 99
http://www.rhizome.org/cgi/to.cgi?t=1486

PRIX ARS ELECTRONICA 2000

.NET
INTERACTIVE ART
COMPUTER ANIMATION
VISUAL EFFECTS
DIGITAL MUSICS
U19/CYBERGENERATION

Joachim Sauter / Jon Snoddy

THE YEAR OF THE ROVING ART PIECE DAS JAHR DES FREI WANDERNDEN KUNSTWERKS

The essential objective of our jury has been to examine interactive digital art generated by extraordinary multi-cultural/cross-disciplined teams and individuals. The selection of the most significant pieces must ultimately stimulate the imaginative potential of an equally diverse audience. Charged with this responsibility our jury demonstrated considerable trust and mutual respect to fulfill the highly respected cultural initiatives of Ars Electronica. It is in this spirit of open mindedness that we appeal to all participants and supporters of Prix Ars Electronica to rise above any particular political activity attempting to deflect our liberty to celebrate the collective cultural aspirations. The reputation of this institution has been hard earned, and now, more than ever we need to protect its status within the international community of art and science.

Background

This category always tracks technology development somewhat and the maturity of the PC is beginning to make itself evident. Gone are the days of science as art. We remember just a few years ago, that nervous anticipation as the engineers worked feverishly just to get an image to appear on a screen. Gone is the image of the artist wrestling with technology forcing an idea into existence.

The PC is has become the generic interactive device. As art installations become more sophisticated, the equipment lists get shorter and more manageable.

Hauptaufgabe unserer Jury war es, die von sehr unterschiedlichen multikulturellen und/oder disziplinenübergreifenden Teams und Individuen geschaffene interaktive digitale Kunst zu bewerten, und zwar auch unter dem Gesichtspunkt, dass die von uns getroffene Auswahl der signifikantesten Werke letztlich auch das imaginative Potenzial eines nicht minder breit gefächerten Publikums anregen soll. Wir waren und sind uns unserer Verantwortung gegenüber den kulturellen Anliegen von Ars Electronica bewusst. Aus dieser Verantwortung und dem in der Jury herrschenden Geist von Offenheit und Toleranz heraus fordern wir alle Teilnehmer und Freunde des Prix Ars Electronica auf, politischen Ambitionen, die auf eine Einschränkung der von uns allen gefeierten künstlerischen Freiheit abzielen, entschieden entgegenzutreten. Der gute Ruf des Prix Ars Electronica ist hart erarbeitet und wohl begründet und mehr denn je bedarf seine Stellung innerhalb der internationalen Gemeinschaft von Kunst und Wissenschaft unseres Schutzes und unserer Unterstützung.

Hintergrund

Diese Kategorie verfolgt immer ein wenig die Entwicklung der Technologie und das Heranreifen des PCs beginnt augenfällig zu werden. Vorbei sind die Tage der Wissenschaft als Kunst. Wir erinnern uns an die nervöse Erwartung, als noch vor wenigen Jahren die Ingenieure fieberhaft arbeiteten, bloß um ein Bild auf einen Bildschirm zu bekommen. Vorbei ist das Bild des Künstlers, der mit der Technik ringt, um einer Idee zur Existenz zu verhelfen.

Der PC ist zum gängigen interaktiven Gerät geworden. In dem Maße, in dem Kunstinstallationen immer ausgefeilter werden, werden die Gerätelisten immer kürzer und handlicher. Klang und Video kommen heute fast immer aus dem Computer und nicht

Sound and video almost always come from computers now rather than separate components. There is still room for the highly customized installation, but thankfully, that is the artist's choice, not imperative allowing more energy and time to be focused on the idea. The area in which we saw artists pushing the technology was in taking the hardware out of the boxes. This trend will likely accelerate.

Each year a particular technology seems to be in vogue. Circular screens last year. Projections and sounds that viewers triggered, the year before. This could turn out to be the year of the roving art piece, heralding what might turn out to be a new era in which the art approaches the viewer on its own terms rather than the other way around. *RoboCup, Free Range Appliances in a Light Dill Sauce*, and *GraffitiWriter* all explored different aspects of mobile intelligent objects.

Embedding intelligence in mobile machines is a timely subject as we are entering a new era. The news is filled with examples. Indeed the *RoboCup* project sets as a goal for itself, the winning of the World Cup soccer event by a robot team. This week the news is filled with stories of laser systems that shoot down missiles fired from many miles away. Robots as avatars warring against one another. Virtual Reality indeed.

The lack of development in the CAVE is an interesting trend. The system's breathtaking ability to transport must be balanced with the complexity of development. While there were a few entries, none were awarded prizes. One must wonder about the future of such large expensive installations. Maybe it is time to abandon the current hardware in favor of lower cost video game systems.

Vectorial Elevation, Relational Architecture #4
Raffael Lozano-Hemmer

This piece is interesting as a work of public art, as a sophisticated but simple user interface, and as a highly successful collaborative work. It was also interesting in its reappropriation of fascist lighting themes in a celebration of egalitarian access to public artistic expression.

von separaten Geräten. Sicherlich ist noch immer Raum für die ganz individuell entwickelte Installation, aber dankenswerterweise bleibt das jetzt der Wahl des Künstlers überlassen und ist nicht mehr zwingend, lässt also mehr Raum für eine Konzentration der Zeit und Energie auf die Ideen. Wo wir einen durch Künstler ausgelösten technologischen Fortschritt gesehen haben, entstand er, weil die Hardware aus den Boxen genommen wurde, und dieser Trend wird sich wahrscheinlich noch beschleunigen. Jedes Jahr scheint jeweils eine spezielle Technologie in Mode zu sein. Vergangenes Jahr waren es runde Schirme, im Jahr davor Projektionen und Klänge, die von den Besuchern ausgelöst wurden. Dieses Jahr könnte das Jahr des frei wandernden Kunstwerks sein und eine neue Ära einläuten, in der sich das Kunstwerk einen Zugang zum Betrachter sucht anstatt umgekehrt. *RoboCup, Free Range Appliances in a Light Dill Sauce* und *GraffitiWriter* untersuchen jeweils unterschiedliche Aspekte mobiler intelligenter Objekte.

Der Einbau von Intelligenz in mobile Maschine ist ein sehr zeitgemäßes Thema am Beginn dieser neuen Epoche. Die Nachrichten sind voll mit Beispielen: Das *RoboCup*-Projekt setzt sich zum Ziel, dass eines Tages ein Roboterteam die Fußballweltmeisterschaft gewinnt. Diese Woche sind die Nachrichten voll von Geschichten über ein Lasersystem, das Raketen abschießt, die viele Meilen entfernt abgefeuert wurden. Roboter als Avatare, die miteinander Krieg führen – das ist virtuelle Realität!

Ein interessanter Trend ist das Fehlen einer Weiterentwicklung im CAVE. Die atemberaubenden Vermittlungsfähigkeiten des Systems müssen sich die Waage halten mit der Komplexität der Entwicklung. Auch wenn einige wenige Werke eingereicht wurden, so erhielt doch keines einen Preis. Man muss sich die Frage nach der Zukunft solch großer teurer Installationen stellen. Vielleicht ist es an der Zeit, die gegenwärtige Hardware zugunsten billigerer Videospiel-Systeme einzumotten.

Vectorial Elevation, Relational Architecture #4
Raffael Lozano-Hemmer

Dieses Stück ist als öffentliches Kunstwerk interessant, als ausgefeiltes und dennoch simples User-Interface und als höchst erfolgreiches Beispiel für Kollaboration. Interessant war auch, wie es sich die faschistischen Lichtspiel-Themen aneignete und sie in einen für alle zugänglichen öffentlichen künstlerischen Ausdruck umsetzte.

Das Werk ist eine Apotheose an jenen Platz, der seit Jahrhunderten das Zentrum des öffentlichen Lebens der Bewohner des heutigen Mexico City ist. Starke ferngesteuerte Scheinwerfer wurden auf den Dächern der Gebäude rund um den Platz montiert.

The work celebrates the city square that has been the center of public life for the people of what is now Mexico City, for hundreds of years. Powerful, remote controlled spot lights were placed on top of the buildings that surround the square. A 3-D model of the square was created and a client applet was written to run on the user's machine. Using this applet and a normal web browser, users can very easily arrange the lights over the virtual square and view it from any vantage point in the area. The design is submitted to the site and it is added to the queue. A web page for the design is automatically created showing the virtual design and after the design is loaded into the lighting control system, actual photographs of the work are added to the web page for the design. Control of the lighting was offered to and accepted by people from all walks of life throughout the world, but appropriately, most of the participants were from Mexico. The people enjoying the work helped create it.

GrafittiWriter
Institute for Applied Autonomy

Entries like *GrafittiWriter* are seductive on their entertainment value alone but this one uses its clever presentation to force a deeper examination of our relationship to society, law and self expression. The system consists of a remote controlled battery powered car. On its rear bumper is a line of spray paint cans with electronically actuated valves. A message is loaded into the car's memory and as the car moves forward, a message can be written on the ground by pulsing the paint sprayers on and off like an ink jet printer.

As the anonymity of the internet enables a separation of the body and the personality we are forced to rethink many concepts that were previously assumed to be settled. Music companies worry openly about the difficulty of enforcing copyright laws in an anonymous society. Discussions are turning up on the web about the difference between anonymity and privacy with some industry leaders openly calling for an end to anonymity.

While few of us paint our political positions on the sides of buildings in public, if those acts could be done with complete anonymity the numbers would surely be different.

Ein 3D-Modell des Platzes wurde geschaffen und eine kleine Client-Anwendung geschrieben, die auf dem Rechner des Anwenders läuft. Mit Hilfe dieses Applets und des normalen Web-Browsers können die Benutzer auf einfache Weise die Lichter über dem virtuellen Platz anordnen und aus jedem beliebigen Standpunkt im Areal betrachten. Das Design wird auf die Website übertragen und in die Warteschlange eingereiht. Für jedes Design wird eine Webseite angelegt, auf der die virtuelle Darstellung sichtbar ist, und nach der Übernahme des Designs in die Steuerung der Installation werden auch Fotos der Realisierung vor Ort auf die Website eingespielt. Die Steuerung der Lichtinstallation wurde allen Menschen weltweit angeboten und von ihnen auch angenommen, aber klarerweise stammte ein Großteil der Mitwirkenden aus Mexiko. Die Leute, die das Werk genossen, haben auch mitgeholfen, es zu schaffen.

GraffitiWriter
Institute for Applied Autonomy

Einreichungen wie *GraffitiWriter* sind schon allein wegen ihres Unterhaltungswerts verführerisch, aber diese hier verwendet ihre clevere Präsentation, um uns zu einer tiefer gehenden Untersuchung unserer Beziehung zu Gesellschaft, Gesetz und Selbstdarstellung zu zwingen.

Das System besteht aus einem ferngesteuerten batteriebetriebenen Auto. An dessen hinterer Stoßstange ist eine Reihe von Sprühdosen montiert, deren Ventile elektronisch gesteuert werden. Eine Botschaft wird in den Speicher des Autos geladen und kann bei fahrendem Auto auf den Boden geschrieben werden, indem die einzelnen Sprühdosen wie ein Tintenstrahldrucker ein- und ausgeschaltet werden.

Je stärker die Anonymität des Internet eine Trennung von Körper und Persönlichkeit erlaubt, desto mehr sind wir gezwungen, viele jener Konzepte zu überdenken, die wir bisher als gegeben angenommen haben. Musikfirmen denken öffentlich über die Schwierigkeit nach, Urheberrechtsbestimmungen in einer Gesellschaft der Anonymen durchzusetzen. Im Web tauchen Diskussionen über den Unterschied zwischen Anonymität und Privatsphäre auf und einige Industrieführer rufen schon öffentlich nach dem Ende der Anonymität. Wenn auch nur wenige von uns ihre politischen Positionen öffentlich an Gebäudewände schreiben, würde diese Zahl doch sehr viel höher sein, wenn dies unter Wahrung der vollständigen Anonymität geschehen könnte.

Audiovisual Environment Suite
Golan Levin

Kunst mit Interaktivität oder Interaktivität als Kunst. Die Jurysitzung begann mit den mahnenden Worten, dass dem Prix Ars Electronica Letzteres wichtig sei, nicht Ersteres, und die Diskussion entwickelte sich weiter rund um die Reife des Mediums. Immer wieder tauchen ähnliche Konzepte auf, wenn sie auch immer geschliffener und interessanter in der Durchführung sind. Die interaktive Kunst nähert sich

Die Website bestand aus einem 3D-Java-Interface, das es den Mitwirkenden ermöglichte, ein vektorielles Design über die Innenstadt zu legen und es aus fast jedem Blickwinkel zu betrachten. Sobald der Projektserver in Mexico einen Beitrag empfing, wurde dieser nummeriert und in eine Warteschlange eingereiht. Alle sechs Sekunden richteten sich die Scheinwerfer automatisch aus und mittels dreier Webcams wurden Aufnahmen zur Dokumentation des Designs der einzelnen Mitwirkenden geschossen. Für jeden Teilnehmer wurde eine Archivseite mit Kommentaren, Informationen und elektronisch gekennzeichneten Fotos des jeweiligen Designs angelegt. War diese aktive Webseite fertig gestellt, wurde der Teilnehmer automatisch über E-Mail benachrichtigt.

Vectorial Elevation received participants from over 50 countries and all the regions of Mexico. To facilitate access, free terminals were also set up in public libraries and museums across the country.

The Zócalo's monumental size makes the human scale seem insignificant, an observation that has been noted by some Mexican scholars as an emblem of a rigid, monolithic and homogenizing environment. Searchlights themselves have been associated with authoritarian regimes, in part due to the military precedent of anti-aircraft surveillance. Indeed, the Internet itself is the legacy of a military desire for distributed operations control. By ensuring that participants were an integral part of the artwork, *Vectorial Elevation* attempted to establish new creative relationships between control technologies, ominous urban landscapes and a local and remote public. It was intended to interface the post-geographical space of the Internet with the specific urban reality of the world's most populous city.

Vectorial Elevation wurde von Teilnehmern aus über 50 Ländern und aus allen Regionen Mexikos besucht. Um den Zugang im Land selbst zu erleichtern, wurden kostenlose Terminals in öffentlichen Bibliotheken und Museen im ganzen Land aufgestellt.

Die monumentale Größe des Zócalo lässt die Dimension des Menschen unbedeutend werden – eine Beobachtung, die mexikanische Wissenschaftler als Ausdruck einer rigiden, monolithischen und gleichmacherischen Umgebung interpretierten. Auch die Suchscheinwerfer als solche wurden mit autoritären Regimes in Zusammenhang gebracht – zum Teil deswegen, weil sie ihre Vorgänger in der militärischen Luftabwehr haben. Und nicht zuletzt ist ja auch das Internet selbst ein Erbe des Wunsches des Militärs nach Steuerung und Überwachung dezentralisierter Operationen. Dadurch, dass *Vectorial Elevation* die Mitwirkenden zu einem integrierten Bestandteil des Kunstwerkes machte, sollte eine neue kreative Beziehung zwischen Steuerungs- und Überwachungstechnologie, zwischen urbanen Landschaften und dem lokalen wie dem remoten Publikum hergestellt werden – ein Versuch, den post-geografischen Raum des Internet mit der spezifischen städtischen Wirklichkeit der bevölkerungsreichsten Stadt der Welt zu verknüpfen.

Credits

Rafael Lozano-Hemmer --Concept, direction, interface.

Relational Art (E/ CDN)
Will Bauer--Project manager.
Conroy Badger--Lead programmer, Java engine, DMX control, Shout3D applet.
Crystal Jorundson---Webcam watermarking, differential GPS, vrml jpeg generation.
Dragos Ruiu---Server, security and streaming video programming.
Kimihiko Sato --Streaming video client, servlets.
Emilio López-Galiacho---3D Modelling, vrml export.
Paul Pelletier ---Case and MSD programming
Kelly Myers, Ana Parga, Susie Ramsay, Therese Gaetz, Rob Lake, Greg Bodnar --------------------Production assistance

Conaculta (Mexico)
Rafael Tovar ---President
Ignacio Toscano---Coordinator for the Millennium celebrations
Juan Ramón Ayala ---Production manager
Alicia Martínez, Lilia Vera, Fernanda Garfias, Lourdes Melgoza,
Guillermina Ochoa, Ernesto Betancourt, Felipe Leal, Jorge Bracho, Ariel Rojo ----------------------Production assistants

Rac Producciones, Grupo CIE (Mexico)
Guadalupe de Anda ---Production Manager
Luis J. Vargas, Mario Torres, José Antonio Barona, Luis Pérez, Miguel Angel Villa, Armando Sánchez,
Arturo Mendoza, Alejandro Echenique ---Production team

Syncrolite (USA)
Jorge Gallegos ---Lead Technician
Alberto Meza, Renato del Castillo, Jeff Moss, Sergio Martínez, Mauricio Martínez,
Omar Rivas, Raúl Rios, Roberto Diaz --- Technicians
Jerry Woods, Harold McLallen --Clifford Power

Telmex, Centec (Mexico)
José Manuel Cortés---Director
Ricardo Medina --Development Manager
Jorge Huesca Salas, Alejandro Fuentes, Héctor León ---Centec engineering
Ricardo Rodriguez Aguilar--Sistemas support

RAFAEL LOZANO-HEMMER (MEX/CDN), CURRENTLY LIVES IN MADRID. HIS INSTALLATION AND PERFORMANCE WORK INVOLVES TELEPRESENCE, NETWORKS, AND LARGE-SCALE INTERVENTIONS IN PUBLIC SPACES. HIS PIECES HAVE BEEN SHOWN IN OVER A DOZEN COUNTRIES, MOST RECENTLY IN AUSTRIA, MEXICO, GERMANY AND SWEDEN. ▬▬ Rafael Lozano-Hemmer (MEX/CDN) lebt zurzeit in Spanien. Seine Installations- und Performancewerke umfassen Telepräsenz, Netzwerke sowie Großereignisse im öffentlichen Raum und waren in über einem Dutzend Länder zu sehen, zuletzt in Österreich, Mexiko, Deutschland und

GRAFFITI WRITER
Institute for Applied Autonomy

A Tactical Platform for
Remote Content Deployment

The advent of next generation military/police technologies for urban use has made engaging in active
social insurgency an increasingly risky venture. Real-
time video surveillance systems[1], networked databases, urban infiltration robots[2], and a flurry of "nonvio-
lent" restraint and subjugation technologies
threaten to have a chilling effect on traditional methods of cultural resistance, particularly the creation
and dissemination of subversive texts. The Robotic
GraffitiWriter (GW) was developed in response to the

need for a high
speed, teleoper-
ated, portable
platform that
operates
beyond the line
of sight (BLOS)
to disseminate
unsanctioned
content in the
dynamic adver-
sarial urban
environment. In
repeated test-

ing, this system has proven its effectiveness on such
high risk/high profile targets as the U.S. Capital
Building as well as numerous urban commercial and
municipal spaces in the US and abroad.
Following its first full year of active service, an in-
depth technological assessment was performed on

Eine taktische Plattform
für ferngesteuerte Entladung von Inhalt

Das Heranwachsen einer nächsten Generation von
militärisch/polizeilicher Technologie für urbane
Anwendung lässt es immer riskanter werden, sich in
aktivem sozialem Aufruhr zu engagieren. Echtzeit-
Videoüberwachungssysteme[1], vernetzte Datenbanken, urbane Infiltrationsroboter[2] und eine Vielzahl
von „gewaltlosen" Einschränkungs- und Unterwer-
fungstechnologien drohen einen einfrierenden
Effekt auf die traditionellen Methoden kulturellen
Widerstandes auszuüben, besonders auf die Schaffung und Verbreitung von subversiven Texten. Der
Robotic GraffitiWriter (GW) wurde als Antwort auf
den Bedarf nach einer schnellen, ferngesteuerten,
tragbaren Plattform entwickelt, die außerhalb des
Sichtbereichs (beyond the line of sight – BLOS) operiert, um unsanktionierten Inhalt in das dynamische
gegnerische urbane Environment zu verteilen. In
wiederholten Testanwendungen hat dieses System
seine Effizienz an so hochriskanten wie hochprofi-
lierten Zielen wie dem US Capital Building und zahlreichen urbanen kommerziellen und öffentlichen
Räumen in den USA und im Ausland bewiesen.
Im Anschluss an das erste Jahr in aktivem Dienst
wurde eine umfassende technologische Bewertung
von GraffitiWriter vorgenommen. Während dieser
Phase wurden signifikante Verbesserungen an GraffitiWriter vorgenommen, darunter eine komplette
Überholung der mechanischen und elektronischen

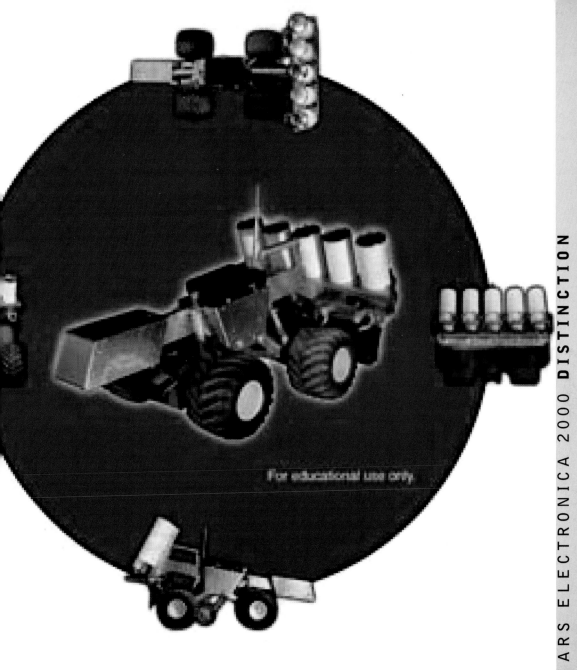

For educational use only.

GraffitiWriter
Institute for Applied Autonomy

For educational use only.

GraffitiWriter. During this time several significant upgrades were made to *GraffitiWriter* including a full mechanical and electronic sub-system overhaul. With these improvements, GW now meets the requirements of strategic transportability, operating with extreme confidence in standard threat scenarios including public parks, federal buildings, and shopping malls.

References
[1] Kanade, Collins and Lipton. "Advances in Cooperative Multi-Sensor Video Surveillance". Robotics Institute, Carnegie Mellon University, 1998.
[2] US Army Research Laboratories, "Pandora: A Robotic System for Operations in Urban Environments - Final Design Document", official contract report submission, March 1998.

Subsysteme. Mit diesen Verbesserungen erfüllt *GraffitiWriter* nunmehr die Anforderungen an strategische Transportierbarkeit und operiert mit extremer Zuverlässigkeit in Standard-Bedrohungsszenarien wie öffentlichen Parks, Bundesgebäuden und Einkaufszentren.

Anmerkungen
[1] Kanade, Collins and Lipton: „Advances in Cooperative Multi-Sensor Video Surveillance". Robotics Institute, Carnegie Mellon University, 1998.
[2] US Army Research Laboratories: „Pandora: A Robotic System for Operations in Urban Environments - Final Design Document", Offizieller Vertragsprüfungsbericht, März 1998.

THE INSTITUTE FOR APPLIED AUTONOMY (IAA), WAS FOUNDED IN 1998 AS A TECHNOLOGICAL RESEARCH AND DEVELOPMENT ORGANIZATION CONCERNED WITH INDIVIDUAL AND COLLECTIVE SELF-DETERMINATION. THE IAA IS COMPRISED OF ENGINEERS, COMPUTER SCIENTISTS, ARTISTS, DESIGNERS AND ACTIVISTS WHO CHOOSE TO WORK COLLECTIVELY WHILE REMAINING ANONYMOUS AS INDIVIDUALS. THE IAA HAS EXHIBITED AS TECHNOLOGISTS AT THE INTERNATIONAL CONFERENCE ON ROBOTICS AND AUTOMATION; AS ARTISTS AT THE ZKM IN KARLSRUHE, GERMANY; AND AS ACTIVISTS IN NUMEROUS ILLEGAL ACTIONS. ▬

Das Institute for Applied Autonomy (IAA) wurde 1998 als technologische Forschungs- und Entwicklungsorganisation mit dem Untersuchungsschwerpunkt individuelle und kollektive Selbstbestimmung gegründet. Das IAA besteht aus Ingenieuren, Computerwissenschaftlern, Künstlern, Designern und Aktivisten, die gerne im Kollektiv arbeiten, dabei aber als Individuen anonym bleiben wollen. Das IAA hat im Technologiebereich bei der International Conference on Robotics and Automation ausgestellt, im künstlerischen Bereich am ZKM in Karlsruhe und als Aktivist bei zahlreichen illegalen Aktionen

AUDIOVISUAL ENVIRONMENT SUITE

Golan Levin

The *Audiovisual Environment Suite* is a collection of five interactive artworks which allow people to create and perform dynamic animation and sound in ways which are direct, engaging, gestural and communicative.

While the computer is an excellent starting point for abstract visual experimentation, sound is often overlooked as an expressive element beyond music and literally recorded sounds. In this work I have attempted to bond visual and aural abstractions that exhibit a unique property of unification in expressive voice. This work is important as it represents a vision for creative activity on the computer, in which uniquely ephemeral dynamic media blossom from the expressive "voice" of a human user.

As artworks, the pieces of the *Audiovisual Environment Suite* extend an established twentieth century tradition, originated by such pioneers as Marcel Duchamp, Sol Le Witt, and Myron Krueger, in which artworks are themselves generative systems for other media; in Marshall McLuhan's terms, such systems are characterized by an "outer medium" (in this case, gestural performance and interaction) whose forms make possible the articulation of yet other expressions in an "inner medium" (for this work, synthetic sound and image). Distinguishing such meta-artworks from the kinds of artifacts we conventionally call "tools" or "instruments" is largely a question of semantics and context; my intent in creating the software sytems in the *Audiovisual Environment Suite* was as a set of vehicles through which I could explore and present a strictly personal audiovisual vocabulary, and suggest provocative new possibilities for human-machine interaction.

Die *Audiovisual Environment Suite* ist eine Sammlung von fünf interaktiven Kunstwerken, die es ermöglichen, dynamische Animation und Klänge auf direkte, herausfordernde, gestische und kommunikative Weise zu schaffen und aufzuführen. Wenn auch der Computer ein ausgezeichneter Ausgangspunkt für abstrakte visuelle Experimente ist, so wird doch häufig die Bedeutung des Klanges als expressives Element jenseits der Musik und der eigentlichen Tonaufzeichnung übersehen. In dieser Arbeit habe ich versucht, visuelle und auditive Abstraktionen so zu verbinden, dass in einer expressiven Stimme eine einzigartige Form der Vereinigung entstehen konnte. Diese Arbeit ist insofern bedeutend, als sie die Vision einer kreativen Arbeit am Computer darstellt, in der ephemere dynamische Medien aus der expressiven „Stimme" eines menschlichen Benutzers „herauswachsen".

Vom künstlerischen Gesichtspunkt gesehen bewegen sich die Stücke der *Audiovisual Environment Suite* in der Tradition des 20. Jahrhunderts, wie sie von Pionieren wie Marcel Duchamp, Sol Le Witt und Myron Krueger etabliert wurde, deren Kunstwerke selbst wiederum generative Systeme für andere Medien werden. In Marshall McLuhans Terminologie ausgedrückt, werden solche Systeme charakterisiert durch das „äußere Medium" (in diesem Falle die gestische Performance und die Interaktion), dessen Formen erst die Artikulation anderer Ausdrucksformen in einem „inneren Medium" ermöglichen (in diesem Werk eben synthetische Klänge und Bilder): Solche Meta-Kunstwerke von jenen Artefakten zu unterscheiden, die wir gemeinhin „Werkzeuge" oder „Instrumente" nennen, ist weitgehend eine Frage der Semantik und des Zusammenhangs; meine Absicht bei der Schaffung der Software-Systeme für die *Audiovisual Environment Suite* war, ein Set von Vehikeln zur Erforschung und Präsentation eines strikt persönlichen audiovisuellen Vokabulars zu schaffen und gleichzeitig neue provokante Möglichkeiten der Mensch-Maschine-Interaktion vorzuschlagen.

BORDERLAND

JULIEN ALMA (F) ATTAINED A DEGREE IN SCULPTURE AND THEN A MASTERS DEGREE FOR ELECTRONIC IMAGE PROCESSING AND INTERACTIVE IMAGES. EXHIBITIONS AT GRAND PRIX OF MULTIMEDIA/ TOKYO AND AT YOUNG TALENTS PAVILLON / CANNES. LAURENT HART (F), STUDIED SCULPTURE IN POITIERS, MASTERS DEGREE WITH DISTINCTION IN ART AND NEW MEDIA. SEVERAL VIDEO GAMES, INCLUDING SOME WITH THE CHARACTER "CARTONMAN", WHICH HE DEVELOPED. ▬

Julien Alma (F) hat eine Ausbildung in Bildhauerei abgeschlossen und anschließend einen Masters Degree über elektronische Bildbearbeitung und interaktive Bilder erworben. Ausstellungen beim Grand Prix of Multimedia/Tokio und im Young Talents Pavillon / Cannes. Laurent Hart (F), Ausbildung zum Bildhauer in Poitiers, Masters Degree mit Auszeichnung in Kunst und Neue Medien. Mehrere Videospiele, u. a. mit dem von ihm entwickelten Charakter „Cartonman".

ZIMMER MIT AUSSICHT

Michael Bielicky / Bernd Lintermann

In the land of Bohemia, with its capital city of Prague, the perception of time and space has always worked differently than in the rest of Europe. This phenomenon has been manifested in the most various realms of Bohemian existence. People from very different circles and disciplines have attained astonishing achievements in science and also in culture. In the strange atmosphere of this land, there have been alchemists, astronomers, musicians, scientists, artists, architects, engineers and others, who have attempted again and again to create "the impossible." *Das Zimmer mit Aussicht* ("Room with a View") is a metaphor, and it represents the diversity of the Bohemian "alchemist's kitchen." In a cathedral with a diameter of 12 meters, visitors are able to decide for themselves with the help of the "navigator" how they want to arrange "their view."

The navigator is an "interface" installed in the middle of the "room with a view." It consists of a revolving column with an integrated flat screen. This screen displays the picture segment that is captured by a video camera installed on the column. This means that each time the navigator is turned, a different section of the filmed canvas is displayed. In addition, touch-sensitive icons repeatedly appear on the picture. By touching these icons, the visitor can trigger different visual and acoustic events within the separate worlds. As the visitor "navigates" through the various "worlds", he is not merely a passive observer of the show, but an active co-designer of the worlds that he experiences in the "room with a view." The navigator becomes a catalyst that can set human powers of creativity in motion.

In the worlds confronting the visitor, "cultural patterns" from the land of Bohemia repeatedly surface. The project *Das Zimmer mit Aussicht* was developed as a research prototype at the ZKM in Karlsruhe. It was developed for the so-called Autostadt in Wolfsburg. The program used here is called MTK (Mapping Toolkit) and it has a unique ability to generate the most diverse dynamic graphic processes in real time.

Im Land Böhmen, mit seiner Hauptstadt Prag, herrschten schon immer etwas andere Gesetze, was die Zeit- und Raumwahrnehmung anging, als im übrigen Europa. Dieses Phänomen hatte sich in den unterschiedlichsten Bereichen des Bohemia-Daseins manifestiert. Es waren Personen aus verschiedensten Kreisen und Disziplinen, die erstaunliche Leistungen in der Wissenschaft, aber auch in der Kultur erbrachten. Es waren Alchimisten, Astronomen, Musiker, Wissenschaftler, Künstler, Baumeister, Ingenieure und andere, die immer wieder den Versuch unternahmen, in der eigenartigen Atmosphäre dieses Landes das „Unmögliche" zu kreieren. *Das Zimmer mit Aussicht* ist eine Metapher und stellt die Vielfalt der böhmischen „Alchimistenküche" dar. In einem Dom mit einem Durchmesser von zwölf Metern wird es den Besuchern möglich sein, mit Hilfe des „Navigators" selbst zu entscheiden, wie sie „ihre Aussicht" gestalten.

Der Navigator ist ein „Interface", welches in der Mitte des Zimmers mit Aussicht installiert ist. Es besteht aus einer sich drehenden Säule, in der ein Flachbildschirm integriert ist. Dieser Bildschirm zeigt den Bildausschnitt, der von der an der Säule installierten Videokamera aufgenommen wird. Das bedeutet, dass sich jedes Mal, wenn der Navigator gedreht wird, ein anderer Ausschnitt der aufgenommenen Leinwand zeigt. Zusätzlich erscheinen immer wieder Icons über dem Bild, die berührungssensitiv sind. Der Besucher kann diese Icons berühren und innerhalb der einzelnen Welten unterschiedliche visuelle und akustische Ereignisse auslösen. Der Besucher „navigiert" durch die verschiedenen „Welten" und ist so nicht nur passiver Betrachter der Show, sondern aktiver Mitgestalter der Welten, die er im Zimmer mit Aussicht erlebt. Der „Navigator" wird zu einem Katalysator, mit dem die menschliche Schöpfungskraft in Gang gesetzt werden kann.

In den Welten, mit denen der Besucher konfrontiert wird, tauchen immer wieder „kulturelle Muster" aus dem Land Böhmen auf.

Das Projekt *Zimmer mit Aussicht* wurde als ein Forschungs-Prototyp am ZKM in Karlsruhe entwickelt. Es wurde für *Autostadt* in Wolfsburg entwickelt. Das Programm, welches hier eingesetzt wird, heißt MTK (Mapping Toolkit) und hat die einmalige Eigenschaft, verschiedenste dynamische Echtzeitgrafikvorgänge zu erzeugen.

Artistic concept, visual design:
Michael Bielicky, Bernd Lintermann
Graphics, artistic advisor:
Francis Wittenberger, Jeffrey Shaw
Graphics software: Bernd Lintermann
Audio software, sound design:
Torsten Belschner
Overall technical concept and production:
ZKM/Karlsruhe, Institut für Bildmedien
User interface conception and production:
Jeffrey Shaw, ZKM Karlsruhe,
Nelissen Dekorbouw, Amsterdam

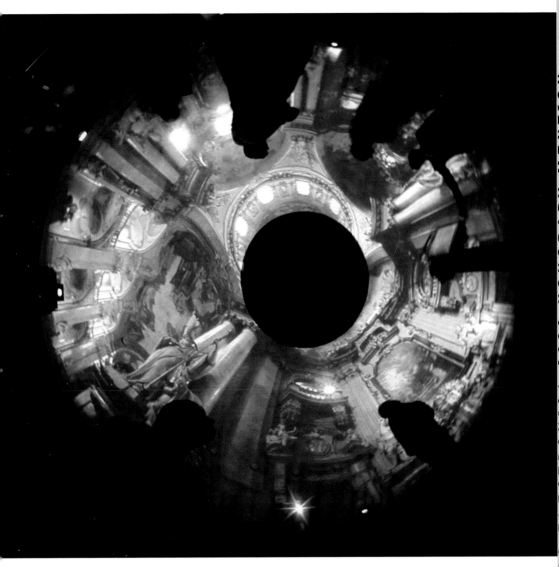

MICHAEL BIELICKY (CZ/D) WAS BORN IN GRAZ AND EMIGRATED TO GERMANY. HE PHOTOGRAPHED FOR *MONOCHROME-MAGAZIN* AND STUDIED WITH NAM JUNE PAIK IN DÜSSELDORF. HE IS CURRENT-LY DEVELOPING A 360° VIRTUAL ENVIRONMENT AT ZKM IN KARLSRUHE. LIVES IN DÜSSELDORF AND IN PRAGUE. BERND LINTERMANN (D) STUDIED COMPUTER SCIENCE IN KARLSRUHE AND HAS BEEN GUEST ARTIST AND SCIENTIST AT THE ZKM INSTITUTE FOR IMAGE MEDIA SINCE 1996. SINCE 1996 INTERACTIVE INSTALLATIONS WITH JEFFREY SHAW, AGNES HEGEDUES, TORSTEN BELSCHNER AND OTHERS. ▬▬▬ Michael Bielicky (CZ/D) wurde in Graz geboren und emigrierte nach Deutschland. Er fotografierte für das *Monochrome-Magazin* und studierte bei Nam June Paik in Düsseldorf. Derzeit Entwicklung einer 360° virtuellen Umgebung am ZKM in Karlsruhe. Lebt in Düsseldorf und in Prag. Bernd Lintermann (D) studierte Informatik in Karlsruhe und ist seit 1996 Gastkünstler und Wissenschaftler am ZKM Institut für Bildmedien. Seit 1996 interaktive

EXPERIMENTS IN TOUCHING COLOR

Jim Campbell

Experiments in Touching Color consists of a small black pedestal in a small dark room. The pedestal has a glass screen in the top surface and is hollow with a video projector inside. An image is projected onto the glass rear screen from below. When the viewers first walk into the room they see an image on top of the pedestal in a silent room. When a viewer puts their hand on the glass screen two things happen at the same time. A sound from the image fades up in the space and the image dissolves to a single color. As long as the hand stays on the glass screen, the sound remains and the image stays a single but changing color. As the hand moves around the image, the color field emanating from the pedestal goes through a sequence of colors. This color is based upon where specifically the viewer is touching. For example, if the image is of a talking head, as the viewer moves his fingers from the lips up to the eyebrow, the emanating color from the screen changes from pale pink to varying flesh tones to dark brown. The color seen at any one time is based upon the color of the pixel that the viewer's middle fingertip is touching in the now unseen image that was there. The viewer's memory of the image that was there, along with the exploration of the colors at the various points in the now hidden image, define a new image that is part memory and part detail. And the image is moving. So after the image fades out and the color field fades in, the image continues to change. This is why the sound is important. The sound which contains the movement of the event helps the viewer to imagine what the image is now doing.

Experiments in Touching Color besteht aus einem kleinen schwarzen Podest in einem kleinen dunklen Raum. In die Oberfläche des Podests ist ein Glasschirm eingelassen, auf den aus dem hohlen Podest ein Videobild projiziert wird. Wenn Betrachter der Raum betreten, sehen sie zunächst ein Bild auf dem Podest in einem stillen Raum. Sobald ein Betrachter die Hand auf den Glasschirm legt, passieren gleichzeitig zwei Dinge: Ein Klang steigt aus dem Bild in den Raum und das Bild löst sich in eine einzige Farbe auf. Solange die Hand auf dem Glasschirm liegen bleibt, bleiben auch der Klang und die Einzelfarbe bestehen, wobei sich Letztere verändert. Wenn sich die Hand über den Glasschirm bewegt, durchläuft die Bildfarbe eine ganze Farbsequenz. Dieses Farbfeld hängt unmittelbar vom jeweiligen Berührungsort ab. Wenn das Bild beispielsweise ein sprechendes Gesicht darstellt, so verändert sich die darauf folgende Farbfläche von hellrosa über verschiedene Fleischfarben zu dunklem Braun, sobald der Betrachter seine Hand von der Position der Lippen zu den Augenbrauen bewegt. Die dargestellte Farbe basiert auf den Grundfarben jenes Pixels des ursprünglichen Bildes, das von der Spitze des Mittelfingers berührt wird, auch wenn das Bild selbst nicht mehr sichtbar ist. Die Erinnerung des Betrachters an das ursprüngliche Bild und seine Erforschung der Farbe an verschiedenen Stellen des mittlerweile verschwundenen Bildes bestimmen eine neue visuelle Erscheinung, die teils Erinnerung, teils Detail ist. Und dieses Bild bewegt sich, sodass nach dem Ausblenden des Bildes und dem Einblenden der Farbe stets Bewegung auf dem Schirm stattfindet. Deshalb ist auch der Klang so wichtig – er enthält und repräsentiert die Bewegung und hilft dem Betrachter, sich vorzustellen, was das Bild gerade tut.

Experiments in Touching Color

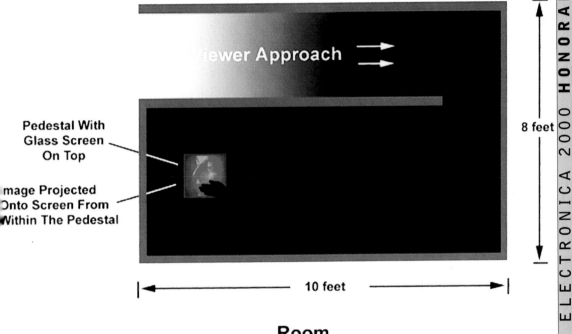

Viewer Approach →
→

8 feet

Pedestal With
Glass Screen
On Top

Image Projected
Onto Screen From
Within The Pedestal

10 feet

Room
Top View

JIM CAMPBELL (USA) WAS BORN IN CHICAGO IN 1956 AND NOW LIVES IN SAN FRANCISCO. HE RECEIVED 2 BACHELOR OF SCIENCE DEGREES IN MATHEMATICS AND ENGINEERING. HE HAS SHOWN INTERNATIONALLY AND THROUGHOUT NORTH AMERICA. IN 1992 HE CREATED ONE OF THE FIRST PERMANENT PUBLIC INTERACTIVE VIDEO ARTWORKS IN THE U.S. IN PHOENIX, ARIZONA. HE RECENTLY RECEIVED A ROCKEFELLER GRANT AND A EUREKA FELLOWSHIP AWARD. AS AN ENGINEER HE HOLDS MORE THAN A DOZEN PATENTS IN THE FIELD OF IMAGE PROCESSING AND IS EMPLOYED AT FAROUDJA LABORATORIES IN SUNNYVALE WORKING ON HDTV RELATED PRODUCTS. ■ John Campbell (USA), geb. 1956, lebt jetzt in San Francisco. Er erwarb zwei Bachelors of Science aus Mathematik und Ingenieurwesen. Er hat in den USA und international ausgestellt und erhielt kürzlich ein Rockefeller-Stipendium und einen Eureka Fellowship Award. Er ist Inhaber von mehr als einem Dutzend Patente zur Bildbearbeitung und ist bei den Faroudja Laboratories in Sunnyvale angestellt, wo er an Produkten im

FREE RANGE APPLIANCES IN A LIGHT DILL SAUCE

Rania Ho

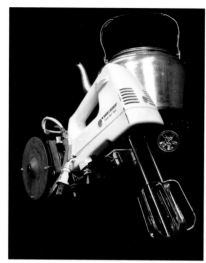

Free Range Appliances in a Light Dill Sauce is an exploration of anthropomorphic qualities inherent in household appliances and an irreverent look at the meaning of "smart" appliances. Kitchen gadgets are liberated from their mundane existences and taught motor skills, enabling them to fully realize their suppressed ambulatory desires.

Each of the appliances have been retrofitted with "state of the art" mechanics and sensors. Each machine has been carefully and lovingly enhanced in accordance with its individual personality. The work is done the old-fashioned way: handcrafted by skilled artisans living in the depths of the remote island Manhattan. This timeworn process is the key to our success. Most of our products are photosensitive and react to the bright lights. Of course, all of them are friendly and enjoy the company of humans. Shining an ordinary household flashlight directly on the body of any one of the appliances will cause it to move towards the light. Removing the flashlight will cause it to seek light and heat elsewhere. Some days, if the machines are in a good mood, they may break out in a celebratory dance. We hope you enjoy our *Free Range Appliances in a Light Dill Sauce*. If you're happy, they're happy.

Battery replacement and periodic battery charging is needed.

Free Range Appliances in a Light Dill Sauce ist eine Untersuchung der anthropomorphen Qualitäten, die in Haushaltsgeräten stecken, und gleichzeitig ein respektloser Blick auf die Bedeutung von so genannten „intelligenten" Haushaltsprodukten. Küchenutensilien werden von ihrer beschränkten alltäglichen Existenz befreit, es werden ihnen motorische Fähigkeiten beigebracht, die ihnen erlauben, ihre bislang stets unterdrückten Wandereigenschaften endlich auszuleben.

Jedes der Haushaltsgeräte wurde im Nachhinein mit der dem Stand der Technik entsprechenden Mechanik und Sensorik ausgestattet. Jede einzelne Maschine wurde liebevoll und sorgfältig in Übereinstimmung mit ihrer Persönlichkeit behandelt. Die technische Arbeit erfolgte auf traditionelle Weise in Handarbeit durch geschickte Kunsthandwerker, die in den Tiefen der fernen Insel Manhattan hausen. Dieser althergebrachte Arbeitsprozess ist der Schlüssel zu unserem Erfolg. Die meisten unserer Produkte sind lichtempfindlich und reagieren auf große Helligkeit und natürlich sind alle freundlich und genießen es, in Gesellschaft menschlicher Wesen zu sein. Leuchtet man den Körper eines unserer Haushaltsgeräte mit einer gewöhnlichen Taschenlampe an, so wird es sich auf das Licht zu bewegen. Entfernt man die Taschenlampe wieder, so wird es dazu veranlasst, Licht und Wärme anderswo zu suchen. An manchen Tagen, wenn die Geräte besonders guter Laune sind, kommt es vor, dass sie alle gemeinsam in einen Freudentanz ausbrechen. Wir hoffen, Sie haben viel Freude an unseren *Free Range Appliances in a Light Dill Sauce*. Wenn Sie glücklich sind, sind auch sie glücklich.

Regelmäßiger Austausch beziehungsweise Nachladen von Batterien ist erforderlich.

TOSHIBA

1 ~ 2	HOT DOG
1 ~ 2	FRENCH BREAD
1 ~ 2	HAMBURGER
2 ~ 3	TOASTING BREAD
5 ~ 6	GRATIN
8 ~ 9	CUP CAKE

TO SET THE DIAL KNOB AT DIAL NUMBER
5 OR BELOW, TURN THE DIAL KNOB
BEYOND DIAL NUMBER 5 FIRST, AND THEN
RETURN IT TO THE POSITION DESIRED.

HTR-861A

RANIA HO RECEIVED HER MASTER'S DEGREE FROM NEW YORK UNIVERSITY'S INTERACTIVE TELECOMMUNICATIONS PROGRAM AND HER B.A. IN THEATER FROM THE UNIVERSITY OF CALIFORNIA, LOS ANGELES. SHE IS CURRENTLY AN INTERVAL RESEARCH FELLOW AT ITP AND IS A MEMBER IN ABSENTIA OF "THE WORLD'S MOST PSYCHOTIC ASIAN AMERICAN SKETCH COMEDY GROUP", THE 18 MIGHTY MOUNTAIN WARRIORS. ▬▬ Rania Ho graduierte im Zuge des Interactive Telecommunications Program der New York University zum Master. Ihr BA aus Theaterwissenschaften erwarb sie an der University of California, Los Angeles. Derzeit ist sie als Research Fellow beim ITP beschäftigt und ist in absentia Mitglied der „psychotischsten ... 8 Mighty Mountain Warriors.

INTERCOURSE— THE FILE CABINET PROJECT

Istvan Kantor (Monty Cantsin)

The File Cabinet Project investigates the sculptural system and kinesonic mechanism of information storage furniture.

The digital age didn't kill the old hardware of information technology, and the metal office furniture is now part of the new electronic communication network. There are file cabinets in every office all around the world.

While in the old times all data was stored in the metal hardware of this specific furniture, today file cabinets are extended with computers and are interconnected through the electronic system of communication. Information kept in the cabinet drawers is transferred to the computer and distributed through cyberspace. Files are accessible through the opening and closing of the cabinet drawers. The kinetic movements of the drawers produce mechanical sounds.

The single monolithic file cabinets are linked together by computers and integrated into a giant network that functions as both kinesonic mega-machine and world wide monument serving the entire planet with information.

In January 1993 I opened my new office, Puppet Government, located at 372 Richmond Street West, in Toronto. My interest in file cabinets developed at around this time, and, as it happened on so many previous occasions, it was induced by chance. I found a four drawer lateral cabinet at the freight elevator waiting to be thrown in the garbage. I moved it immediately into my office and at that moment I knew it was the object I was waiting for. For the first time in my life I looked at a file cabinet as someone would look at an alien device with astonishing interest. I started pulling, pushing the drawers in and out and I was fully amazed by the noise they made.

I have been using scrapmetal junk for performances and installations for many years collecting them from junk yards and from the streets, but, for some reason, I've never paid enough attention to the file cabinet. It took a special moment to discover the nature of this perfect noise instrument and potential monument.

Das *File Cabinet Project* untersucht das skulpturale System und den kinesonischen Mechanismus von Informationsspeicherungsmöbeln.

Das digitale Zeitalter hat der alten Hardware zur Informationsspeicherung keineswegs den Garaus gemacht und die metallenen Büromöbel sind auch weiterhin Teil des neuen elektronischen Kommunikationsnetzwerks. Und Aktenschränke finden sich noch immer in jedem Büro der Welt.

Während früher alle Daten in der metallenen Hardware dieser speziellen Möbel gespeichert wurden, werden die Datenspeicher von einst heute durch Computer erweitert und mittels elektronischer Kommunikation miteinander vernetzt. Die Information aus den Aktenschränken wird auf den Computer übertragen und über den Cyberspace hinweg verteilt.

Die Akten sind durch das Öffnen und Schließen der Schubladen des Aktenschranks zugänglich. Die kinetische Bewegung dieser Laden erzeugt mechanische Klänge.

Die einzelnen monolithischen Aktenschränke werden über Computer vernetzt und in ein gigantisches Netzwerk eingebunden, das sowohl als kinematisch-klangliche Mega-Maschine funktioniert als auch als weltweites Monument, das den Planeten mit Information versorgt.

Im Jänner 1993 habe ich mein neues Büro eröffnet – Puppet Government, in der Richmond Street West 273 in Toronto. Mein Interesse an Aktenschränken ist um diese Zeit herum erwacht und wurde, wie das auch früher schon häufiger der Fall war, durch einen Zufall geweckt: Ich fand neben dem Lastenaufzug einen Vier-Schubladen-Mappenschrank, der für den Sperrmüll vorgesehen war. Sofort verfrachtete ich ihn in mein Büro und erkannte augenblicklich, dass ich genau auf dieses Objekt gewartet hatte. Zum ersten Mal in meinem Leben betrachtete ich einen Aktenschrank, wie jemand ein fremdartiges Gerät mit Erstaunen und Interesse betrachten würde. Ich begann, die Schubladen auf- und zuzuziehen und war völlig fasziniert von dem Geräusch, das sie machten.

Seit Jahren verwende ich Metallschrott, den ich auf Schrottplätzen und auf der Straße einsammle, für Installationen und Performances, aber aus irgendwelchen Gründen habe ich Aktenschränke nie beachtet. Es bedurfte eines besonderen Moments, um die Natur dieses perfekten Klanginstruments und potenziellen Denkmals zu entdecken.

ISTVAN KANTOR, BEST KNOW AS MONTY CANTSIN, FOUNDER OF NEOISM?!, IS A MEDIA ARTIST/PRODUCER, ACTIVE IN MANY FIELDS, PERFORMANCE, ROBOTICS, INSTALLATION, SOUND, MUSIC, VIDEO AND NEW MEDIA. HIS WORK HAS BEEN SHOWN AT MANY MAJOR NEW MEDIA AND PERFORMANCE FESTIVALS THROUGHOUT NORTH AMERICA AND EUROPE. HE HAS PERFORMED AT CENTRE POMPIDOU IN PARIS (1989), AND AT DOCUMENTA 8, TECHNOLOGY/MEDIA SECTION, IN KASSEL (1997). HIS MAIN SUBJECTS ARE THE DECAY OF TECHNOLOGY AND THE STRUGGLE OF THE INDIVIDUAL IN TECHNOLOGICAL SOCIETY. KANTOR HAS LIVED IN BUDAPEST, PARIS, MONTREAL, NEW YORK AND PRESENTLY IS A RESIDENT OF TORONTO. ■■■ Istvan Kantor, besser bekannt als Monty Cantsin, Gründer des Neoism?!, ist als Medienkünstler und Produzent in vielen Bereichen aktiv – Performance, Robotik, Klang, Musik, Video und Neue Medien. Seine Arbeiten waren und sind bei vielen bedeutenden Festivals in Nordamerika und Europa zu sehen. Er ist unter anderem am Centre Pompidou aufgetreten (1989) und bei der Technologie/Medien-Sektion der documenta 8 in Kassel (1987). Seine Hauptthemen sind der Verfall der Technologie und der Kampf des Individuums in einer technologische Gesellschaft. Kantor hat in Budapest, Paris, Montreal und New York

AS MUCH AS YOU LOVE ME

Orit Kruglanski

As Much As You Love Me is an interactive poetry project that consists of a force feedback mouse and an interactive poem. The project deals with guilt and tries to convey its message using the special properties of the physical and graphic interface.

The text of *As Much As You Love Me* is a series of non-apologies, which all start with the words "don't forgive me" (e.g. "Don't forgive me for the things I've said"; "don't forgive me for the mistakes I've made"). In order to hear the text the user must collect objects on the screen. Each object collected will sound a non-apology and increase the physical difficulty of moving the mouse. The collected objects cling to the cursor and appear to be tied to the edges of the screen. The combination of the graphic representation and the actual physical effort required to move the mouse as more objects are collected reinforces the attempt of the speaker in the poem to relieve herself from guilt. As moving becomes harder and the burden of guilt heavier, the non-apologies become harsher and almost accusing as the speaker demands "don't forgive me for my father's lies" and "don't forgive me for saving myself."

But facing the guilt and carrying the burden can be avoided: on the main screen of *As Much As You Love Me*, which is gray, there is a white circle. If the user enters this area the force feedback will turn off and movement will become easy again. The user may move within the circle with complete freedom, but the guilt-objects collected will be waiting outside. So eventually, the user must leave the circle (which I call denial), face the guilt and keep collecting it, until with the last piece of guilt comes the final release, and final rejection. The last piece of guilt will sound the phrase "forgive me this: I can't remember loving you", the force feedback will shut off, and all the pieces collected will retreat, returning the user to the opening screen of the piece.

The force feedback is achieved by two electromagnets attached to the bottom of the mouse. The electromagnets are controlled by a Pic chip microcontroller that is connected serially to the computer. Every object collected on screen will send a command to the microcontroller to increase the power of the magnets. The force feedback mouse sits on a large metal plate, so the user will never have to lift the mouse and break the force feedback.

As Much As You Love Me ist ein interaktives Poesie-Projekt, das aus einer Force-Feedback-Maus und einem interaktiven Gedicht besteht. Das Projekt beschäftigt sich mit der Frage nach Schuld und versucht, seinen Inhalt mit Hilfe der speziellen Eigenschaften des physischen und grafischen Interface zu transportieren.

Der Text von *As Much As You Love Me* ist eine Serie von Nicht-Entschuldigungen, die alle mit den Worten „Vergib mir nicht" beginnen (z. B. „Vergib mir nicht die Dinge, die ich gesagt habe"; „Vergib mir nicht die Fehler, die ich gemacht habe"). Um den Text zu hören, muss der Benutzer Objekte am Bildschirm sammeln. Jedes aufgelesene Objekt spielt dann eine Nicht-Entschuldigung ab und erhöht gleichzeitig den physischen Widerstand der Maus. Die eingesammelten Objekte scheinen einerseits am Mauszeiger zu kleben, andererseits hängen sie am Bildrand fest. Die Kombination aus grafischer Darstellung und physischer Anstrengung, die zur Bewegung der Maus erforderlich ist und mit steigender Objektzahl wächst, verstärkt den Wunsch der Sprecherin nach einer Befreiung von Schuld. Im gleichen Maß, in dem die Bewegung schwieriger wird und das angesammelte Schuldgefühl wächst, werden auch die Nicht-Entschuldigungen immer harscher und fast schon anklagend, wenn die Sprechstimme fordert „Vergib mir nicht die Schuld meines Vaters" oder „Vergib mir nicht, dass ich mich gerettet habe".

Aber die Last der eigenen Schuld kann vermieden werden: Auf dem ansonsten grauen Hauptschirm von *As Much As You Love Me* befindet sich ein weißer Kreis. Wenn der Benutzer sich in diesen Kreis hineinbegibt, wird das Force Feedback abgeschaltet und die volle Bewegungsfreiheit wiederhergestellt. Die angesammelten Schuldobjekte jedoch warten weiter draußen und wenn der Benutzer früher oder später den Kreis (den ich „Verneinung" nenne) verlässt, sieht er sich wieder mit der Schuld konfrontiert und muss sich weiter Schuld aufladen, bis mit dem letzten Schuldobjekt die Erlösung und gleichzeitig endgültige Abweisung kommt: Das letzte Objekt spricht „Vergib mir dies: Ich kann mich nicht erinnern, dich zu lieben". Jetzt schaltet sich das Force Feedback ab, die gesammelten Objekte verschwinden und man findet sich in der Ausgangssituation wieder.

Das Force Feedback wird durch zwei an der Unterseite der Maus angebrachte Elektromagneten ausgelöst, die von einem über eine serielle Leitung mit dem Computer verbundenen Pic-Chip gesteuert werden. Jedes auf dem Bildschirm aufgelesene Objekt sendet dem Mikrocontroller einen Impuls, durch den die Magnetkraft erhöht wird. Die Force-Feedback-Maus sitzt auf einer großen Metallplatte, sodass der Benutzer sie niemals anzuheben braucht, wodurch er das Feedback unterbrechen würde.

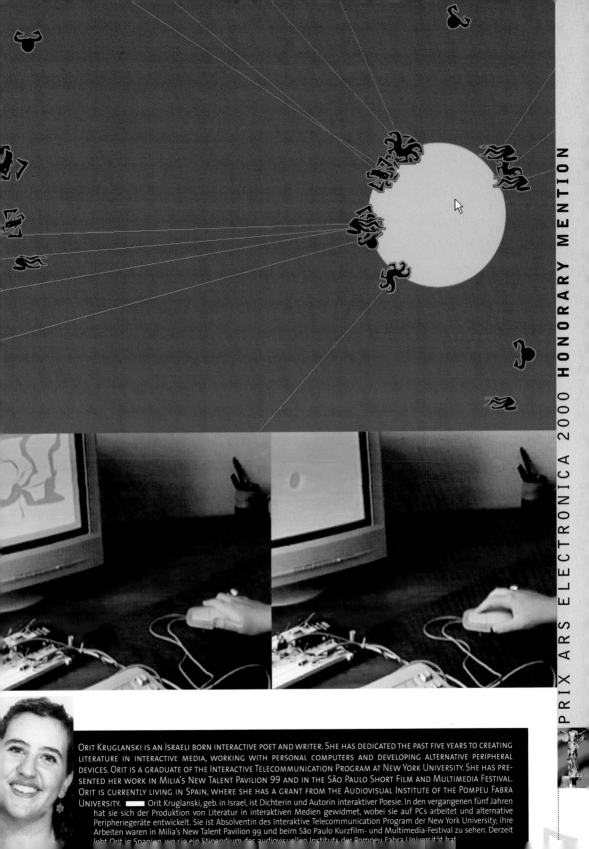

ORIT KRUGLANSKI IS AN ISRAELI BORN INTERACTIVE POET AND WRITER. SHE HAS DEDICATED THE PAST FIVE YEARS TO CREATING LITERATURE IN INTERACTIVE MEDIA, WORKING WITH PERSONAL COMPUTERS AND DEVELOPING ALTERNATIVE PERIPHERAL DEVICES. ORIT IS A GRADUATE OF THE INTERACTIVE TELECOMMUNICATION PROGRAM AT NEW YORK UNIVERSITY. SHE HAS PRESENTED HER WORK IN MILIA'S NEW TALENT PAVILION 99 AND IN THE SÃO PAULO SHORT FILM AND MULTIMEDIA FESTIVAL. ORIT IS CURRENTLY LIVING IN SPAIN, WHERE SHE HAS A GRANT FROM THE AUDIOVISUAL INSTITUTE OF THE POMPEU FABRA UNIVERSITY. ■■■■ Orit Kruglanski, geb. in Israel, ist Dichterin und Autorin interaktiver Poesie. In den vergangenen fünf Jahren hat sie sich der Produktion von Literatur in interaktiven Medien gewidmet, wobei sie auf PCs arbeitet und alternative Peripheriegeräte entwickelt. Sie ist Absolventin des Interaktive Telecommunication Program der New York University; ihre Arbeiten waren in Milia's New Talent Pavilion 99 und beim São Paulo Kurzfilm- und Multimedia-Festival zu sehen. Derzeit lebt Orit in Spanien, wo sie ein Stipendium des audiovisuellen Instituts der Pompeu Fabra Universität hat.

THE ACTIVE TEXT PROJECT

Jason E. Lewis / Alex Weyers

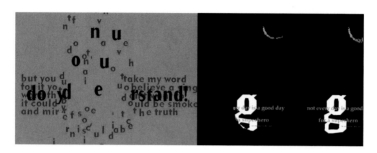

Overview

The Concrete Poets of the '60s, and their Dada and Futurist forebears, treated the visual appearance of text as a principal participant in the production of meaning. Their interests in such experimentation grew out of a deeply held belief that traditional forms of written communication, with its clean spacing, rectilinear layout, and sober letterforms, no longer could speak to the cultural schizophrenia of the modern age. Working within the constraints of traditional letterpress, these artists and poets managed to explore text's visual presence in ways that seem fresh even to the MTV-jaded eyes of today. As concrete poets practicing at the beginning of a new century, we have embarked upon the *Active Text Project*, an on-going experiment in radical ways of treating and interacting with the visual appearance of text in ways which reflect our post-millennial data devotion. Users can set glyphs, words, or entire passages in motion, pull them apart, blow them up, infect them with dynamic behaviors and even reconstitute them, in an attempt to deconstruct standard notions of text presentation and reception.

It's Alive!

The first major application of our project is *It's Alive!* This program allows the user to author dynamic and interactive texts in a fluid environment. Its functionality is best summarized as the mutant offspring of a text editor and Adobe AfterEffects. The user can enter, edit and lay out text as usual. Unlike a text editor, however, which requires her to constantly move from text to pull-down menu and back to adjust basic visual characteristics such as font, size and col-

Übersicht

Die Konkreten Dichter der 6oer-Jahre und ihre Vorläufer im Dada und Futurismus haben die visuelle Erscheinung des Textes als ein Hauptelement in der Erzeugung von Bedeutung angesehen. Ihr Interesse an solchen Experimenten erwuchs aus der festen Überzeugung, dass die traditionellen Formen schriftlicher Kommunikation mit ihren klinisch reinen Abständen, ihrem rechtwinkeligen Layout und den nüchternen Letternformen nicht mehr der kulturellen Schizophrenie der modernen Zeit entsprächen. Obwohl sie im traditionellen Raum der Druckerpresse arbeiteten, gelang es diesen Künstlern und Poeten dennoch, die visuelle Präsenz des Textes auf eine Weise zu erforschen, die selbst für unsere MTV-verklärten Augen noch immer frisch erscheint. Als konkrete Dichter am Eingang eines neuen Jahrhunderts haben wir das *Active Text Project* in Angriff genommen, ein kontinuierliches Experiment zur radikalen Gestaltung von und Interaktion mit der visuellen Erscheinung von Texten auf eine Weise, die unsere post-millenniale Hingabe an die Daten reflektiert. Die Benutzer können Glyphen, Wörter oder ganze Passagen in Bewegung setzen, sie auseinander ziehen, sie aufblasen, sie mit dynamischen Verhaltensweisen infizieren und sie neu zusammensetzen – alles im Versuch, die gängigen Begriffe von Textpräsentation und -rezeption zu dekonstruieren.

It's Alive!

Die erste größere Applikation in unserem Projekt ist *It's Alive!*. Dieses Programm erlaubt dem Anwender, als Autor dynamischer und interaktiver Texte in einer fließenden Umgebung aufzutreten. Die Funktionalität des Programms lässt sich am besten zusammenfassend als mutierter Nachkomme von Adobe AfterEffects und einer Textverarbeitung beschreiben. Der User kann wie üblich Texte eingeben, editieren und layouten. Im Unterschied zur Textverarbeitung jedoch – bei der man zur Bearbeitung selbst elementarster visueller Parameter wie Schriftart, -größe und -farbe ständig zwischen Text und Pulldown-Menüs hin und her wechseln muss – gibt *It's*

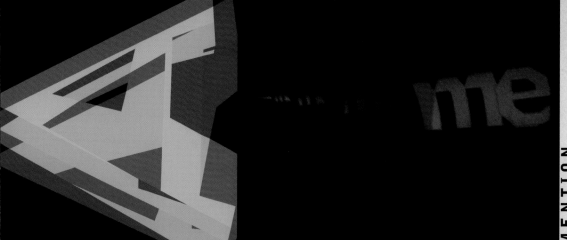

or, *It's Alive!* gives the user continuously variable controls which provide continuous feedback.

TextOrgan

Equal parts digital graffiti and digital concrete poetry, *TextOrgan* is the rave cousin of *It's Alive! TextOrgan* takes the *It's Alive!* environment, adds a MIDI keyboard for quick access to all the functionality, and allows the user to either input text directly or to select among a set of prepared texts that are then streamed onto the screen. She can quickly build up an immensely rich collage of performance-specific poetry and found texts. *TextOrgan* was developed for use in conjunction with a DJ at musical events. By collaborating beforehand, the DJ and the TextOrganist can choose a selection of texts and discuss what sort of mood or themes they want develop in the set. The TextOrganist can also extract simple signal information such as pitch and intensity from either the ambient environment or directly from the DJ's decks. This provides a mean of automating some of the text dynamics to be in synch with the beat, or to ebb and swell as the intensity of the music decreases and increases.

Alive! dem Anwender kontinuierlich variable Steuerungsmöglichkeiten, die ebenso kontinuierlich ein Feedback liefern.

TextOrgan

TextOrgan – zu gleichen Teilen digitales Graffiti und digitale konkrete Poesie – ist sozusagen der Rave-Ableger von *It's Alive!*. *TextOrgan* verwendet das Environment von *It's Alive!*, fügt ein MIDI-Keyboard für den schnellen Zugang zur gesamten Funktionalität hinzu und erlaubt dem Benutzer, den Text entweder direkt einzugeben oder aus einem Set von vorgefertigten Textpassagen auszuwählen, die auf den Bildschirm fließen. So kann schnell eine immens reiche Collage aus performance-spezifischer Poesie und gefundenen Texten aufgebaut werden.
TextOrgan wurde zur Anwendung in Verbindung mit einem Diskjockey bei Musikereignissen entwickelt. Bereits vor dem Event können der DJ und der *TextOrgan*ist eine Selektion von Texten treffen und diskutieren, welche Art von Stimmung oder Themen sie letztlich entwickeln wollen.
Der *TextOrgan*ist kann auch einfache Signalinformationen wie Tonhöhe und Intensität entweder aus den Umweltgeräuschen oder direkt aus den Decks des DJs übernehmen. Dies ermöglicht es, einige Teile der Textdynamik automatisch an den Beat anzupassen oder mit der Musikintensität an- und abschwellen zu lassen.

JASON E. LEWIS (USA) IS A DIGITAL ARTIST AND TECHNOLOGIST INTERESTED IN CREATING NEW FORMS OF INTERACTIVE EXPERIENCE. HE HAS RECEIVED COMMISSIONS FROM LAURIE ANDERSON & BRIAN ENO AND THE ENGLISH ARTS COUNCIL, AND BEEN FEATURED IN *I.D. MAGAZINE*, *UPPER & LOWER CASE* AND *CTHEORY MULTIMEDIA*, AMONG OTHERS. ALEX WEYERS (USA). THE ENIGMA KNOWN AS ALEX WEYERS FIRST EMERGED FROM THE ALEXANDRIA, VIRGINIA MUSIC SCENE AS A FOUNDING MEMBER OF THE EXPERIMENTAL, ELECTRONIC COLLECTIVE BUS PEOPLE. HE SERVES AS THE PRIMARY PROGRAMMER AT THE ARTS ALLIANCE LABORATORY. ■■■■ Jason E. Lewis (USA) ist digitaler Medienkünstler und Techniker, der sich für neue Formen interaktiver Erfahrung interessiert. Er hat Aufträge von Laurie Anderson, Brian Eno und vom English Arts Council bekommen; sein Werk wurde unter anderem von *I.D. Magazine*, *Upper & Lower Case* und *CTheory Multimedia* präsentiert. Alex Weyers (USA). Das als Alex Weyers bekannte Rätsel tauchte zuerst in der Musikszene von Alexandria, Virginia, als Gründungsmitglied des experimentellen elektronischen Kollektivs BuS PeOpLe auf. Er arbeitet als Hauptprogrammierer beim Arts Alliance Laboratory.

ASYMPTOTE
Douglas Edric Stanley

In both its mathematical and figurative uses, the term "asymptote" describes a paradox: two elements, by nature incompatible, extend themselves nevertheless towards an infinite point at which their impossible contact can take place.

The installation *Asymptote* takes its point of departure from the narrative-essay *On the Marionette Theater*, written by Heinrich von Kleist in 1810. Although Kleist makes direct reference to the asymptote, his central allegory revolves around the similarly paradoxical relationship of the marionette and the marionettist, in which an object devoid of gravitational forces takes on the emotional weight of its operator, thus effectuating a transfer or translation between matter and the disjunctive imagination of the human subject. The operator's gravity-bound gestures are uplifted into a state of grace, described by Kleist as a sort of spiritual prosthesis. The central conduit of this transfer takes place through the wire that both connects and divides the two worlds.

In the center of a darkened room: a small cylindrical pedestal. A series of thin wires connected to pulleys sit atop the cylinder, which in turn connect up to three computers lying in the cylinder chamber, and which are visible through the transparent glass of the top surface. The wires are tightly stretched and offer a force-feedback resistance when pulled. When the museum visitors pull these strings, a series of interactive video images projected onto the walls of the installation react to these interventions: appearing, disappearing and evolving according to the rhythm of the interactor's gesture. By pulling on one of the strings violently, for example, the hand of a marionette appears, spinning in circles frenetically, hitting itself with each rotation. With another, softer, pull of the string, another marionette's hand gently caresses a third marionette's hair. Slowly, as the visitors pull and release the wires, an interactive narrative develops, based on stories from Kleist's

Im mathematischen wie im übertragenen Gebrauch beschreibt der Begriff „Asymptote" ein Paradoxon: Zwei Elemente, von Natur aus inkompatibel, erstrecken sich bis zu einem Punkt im Unendlichen, an dem ihre eigentlich unmögliche Berührung dennoch stattfindet.

Die Installation *Asymptote* geht von Heinrich von Kleists Essay *Über das Marionettentheater* aus dem Jahr 1810 aus. Obwohl Kleist darin wörtlich auf die Asymptote anspielt, dreht sich die zentrale Allegorie doch um die ähnlich paradoxe Beziehung zwischen der Marionette und dem Puppenspieler, die als ein von Kräften der Gravitation befreites Objekt das emotionale Gewicht ihres Operateurs übernimmt und so einen Transfer oder besser eine Translation zwischen der Materie und der disjunkten Imagination des menschlichen Subjekts vollzieht. Die Schwerkraft-gebundenen Gesten des Puppenspielers werden in einen Zustand der Grazie erhoben, was Kleist als eine Art geistiger Prothese beschreibt. Die Übertragung dieses Transfers erfolgt mit Hilfe der Drähte, die beide Welten zugleich verbinden und trennen.

Im Zentrum eines verdunkelten Raums steht ein kleines zylindrisches Podest. Eine Reihe von dünnen Drähten geht von der Oberfläche des Zylinders aus und ist über Rollen mit drei Computern in der Zylinderkammer verbunden, die durch die transparente Grundfläche sichtbar sind. Die Drähte sind straff gespannt und bieten einen Widerstand, wenn man an ihnen zieht. Wird an den Drähten gezogen, so reagieren interaktive Videobilder, die auf die Wände der Installation projiziert werden, mit verschiedenen Reaktionen: Bilder erscheinen, verschwinden oder bewegen sich als Reaktion auf den Rhythmus der Gesten des Interagierenden. So löst beispielsweise ein heftiger Ruck an einem bestimmten Draht das Erscheinen einer Marionettenhand aus, die frenetisch im Kreis wirbelt und sich selbst bei jeder Umdrehung schlägt. Ein sanfter Zug an einem anderen Draht wiederum veranlasst eine Marionettenhand, zart das Haar einer dritten Marionette zu streicheln. Wenn die Besucher an den Drähten ziehen, entwickelt sich also langsam eine interaktive Erzählung, die auf Geschichten aus Kleists Text basiert. Diese Geschichten werden – wie bei Kleist – durch einen Dialog zusammengehalten, den der Erzähler mit einem zufällig im Park getroffenen renommierten Balletttänzer führt. In dem Maße, in dem der oder die Interagierenden das Draht-Interface bearbeiten, entwickelt sich dieser Dialog, nimmt unerwartete Wendungen oder gabelt sich irgendwann, um den Inkonsistenzen und Abweichungen in der Interaktion zu folgen.

text. These stories are related, as in the text, through a dialogue that takes place between the narrator and a renowned ballet dancer the narrator happens to meet in a public park. As the interactor(s) activate the wire interface, this dialogue develops, takes unexpected turns, and eventually bifurcates to follow the digressions of the interactors interventions.

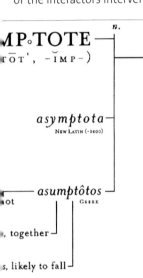

IP∘TOTE —— *n.*
ōt', -ĭmp-)

—— *Math. A line considered a limit to a curve in the sense that the perpendicular distance from a moving point on the curve to the line approaches zero as the point moves an infinite distance from the origin. Fig. Two or more elements streching towards each other, achieving contact only at infinity.*

asymptota —
New Latin (~1600)

—— *asumptôtos* —
ot   Greek

, together

s, likely to fall

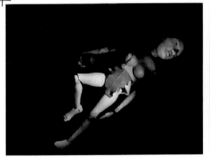

└ INTERACTIVE FORCE-FEEDBACK INSTALLATION

patati, to fly, to hurry, to fall
SANSKRIT

petesthai, to fly
INDO-EUROPEAN

DOUGLAS EDRIC STANLEY WAS BORN IN THE HEART OF SILICON VALLEY IN 1969. BORED OF SILICON VALLEY AND THE UNITED STATES, HE MOVED TO FRANCE WHERE HE HAS BEEN A RESIDENT SINCE 1991. HE IS CURRENTLY ARTIST-IN-RESIDENCE AT THE VILLA ARSON, FRANCE. SINCE 1998 HE HAS BEEN PROFESSOR OF DIGITAL ARTS AT THE AIX-EN-PROVENCE SCHOOL OF FINE ARTS. HE HOLDS BACHELORS' DEGREES IN LITERATURE AND CINEMA FROM SAN FRANCISCO STATE UNIVERSITY, AND WROTE HIS MASTER'S THESIS AT THE UNIVERSITY OF PARIS. HE IS CURRENTLY PREPARING HIS THESIS AT JEAN-LOUIS BOISSIER'S LABORATOIRE ESTHÉTIQUE DE L'INTERACTIVITÉ, AT PARIS 8. ▬▬ Douglas Edric Stanley wurde 1969 im Herzen des Silicon Valley geboren. Gelangweilt vom Silicon Valley und den USA übersiedelte er 1991 nach Frankreich, wo er derzeit als Artist-in-Residence an der Villa Arson lebt. Seit 1998 ist er Professor für Digitale Kunst an der Kunsthochschule von Aix-en-Provence. Er hat einen BA aus Literatur und Film (San Francisco State University) und hat seine Magisterarbeit an der Universität Paris geschrieben. Derzeit schreibt er an einer weiteren Diplomarbeit an Jean-Louis Boissiers Laboratoire

UNCONSCIOUS FLOW
Naoko Tosa / Sony-Kihara Research Center

In face-to-face communication, the occasional need for intentional lies is something with which everyone can identify. For example, when we get mad, circumstances may force us to put on a big smile instead of expressing our anger; when we feel miserable, good manners may dictate that we greet others warmly. In short, to abide by social norms, we consciously lie. On the other hand, if we consider the signs that our bodies express as communication (body language), we can say that the body

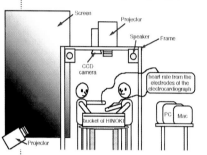

does not lie even while the mind does. Considering this phenomenon, we propose "touching the heart" in a somewhat Japanese way by measuring the heartbeat of the "honest" body and using other technologies to reveal a new code of non-verbal communication from a hidden dimension in society. We call this "techno-healing art."

Two computer-generated mermaids function as individual agents for two viewers. Each mermaid agent moves in sync with the heart rate detected by an electrode attached to the collarbone of its viewer. Then, using a synchronization interaction model that calculates the mutual heart rate on a personal computer, the two mermaids express hidden non-verbal communication. The data of relax-strain calculated from the heart rate and the interest calculated from the variation in the heart rate are mapped on the mode. The synchronization interaction model reveals the communication codes in the hidden dimension that do not appear in our superficial communication. Then, using a camera to pick up hand gestures and a personal computer to analyze the images, the synchronization interaction model is applied to determine the mermaid's behavior. For a high degree of synchronisation, the agents mimic the hand gestures of their subjects, but for a low degree of synchronisation, the agents run away. In the event that one mermaid agent touches the other, a pseudo-touch can be felt through the use of a vibration device. As for background sound, the heart sound of the subjects are picked up by an electronic stethoscope and processed for output on a personal computer.

Jeder von uns weiß, dass es in der Kommunikation von Angesicht zu Angesicht manchmal notwendig ist, bewusst zu lügen. Es kann zum Beispiel vorkommen, dass wir vor Ärger fast platzen, die Umstände uns aber zwingen, freundlich zu lächeln, statt unserem Ärger Ausdruck zu verleihen; obwohl wir uns miserabel fühlen, können es uns die üblichen „guten Manieren" vorschreiben, jemanden freundlich zu grüßen ... kurzum: Um den sozialen Normen zu entsprechen, sind wir ständig am Lügen. Betrachten wir andererseits die Signale, die unser Körper als Teil der Kommunikation aussendet (die Körpersprache), so kann man sagen, dass unser Körper nicht lügt, selbst wenn unser Geist dies ganz bewusst tut. Ausgehend von diesem Phänomen stellen wir ein Werkzeug vor, das „das Herz zu rühren" vermag – und das auf eine irgendwie japanische Art: Wir messen den Herzschlag des „ehrlichen" Körpers und bedienen uns anderer Technologien, um in einer versteckten Dimension der Gesellschaft einen neuen Code non-verbaler Kommunikation zu entdecken. Wir nennen dies „techno-heilende Kunst". Zwei computergenerierte Meerjungfrauen dienen als individuelle Repräsentanten für zwei Betrachter. Jede Nixe bewegt sich synchron zum Herzschlag, der über eine am Schlüsselbein des Betrachters angebrachte Elektrode gemessen wird. Mit einem Synchronisations-Interaktions-Modell wird der beidseitige Puls auf einem PC errechnet und die beiden Nixen bringen dann die non-verbale Kommunikation optisch zum Ausdruck, indem die Daten von An- und Entspannung und Interesse aus dem Herzschlag bzw. dessen Veränderungen errechnet und auf das Modell übertragen werden. Dieses Synchronisationsmodell kann die Codes in den verborgenen Dimensionen offen legen, die in unserer oberflächlichen Kommunikation unbeachtet bleiben. Eine Kamera erfasst die Gesten der Hand, die von einem PC analysiert und ebenfalls auf die Nixen übertragen werden, wobei der Synchronizitätsgrad zwischen Herzschlag und Gestik deren Verhalten bestimmt. Bei hoher Übereinstimmung imitieren die Nixen die Handgesten ihrer Subjekte, bei geringer bewegen sie sich weg. Sollte eine der Nixen eine andere berühren, so kann dies über ein Vibrationsgerät auch für die Mitspieler fühlbar gemacht werden. Der Hintergrundklang des Herzschlags wird über ein elektronischen Stethoskop aufgenommen und durch Bearbeitung auf dem PC dargestellt.

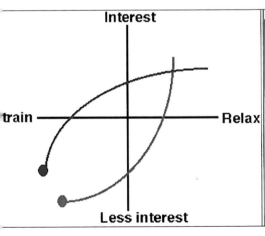

Naoko Tosa (J) is a Media Artist & Researcher in the ATR Media Integration & Communications Research Laboratories. She is also a guest professor at Kobe University and a visiting lecturer at Musashino Art University. She received a Ph.D. in engineering for technology research from the University of Tokyo. In particular, she focused on the topic of communication and used computers, video and electronics to design art work that relates to the intelligence of emotions, consciousness and unconsciousness. She specializes in the creation of experimental film, video art, computer graphics animation, and interactive arts. Her work has been exhibited at the Museum of Modern Art New York, the New York Metropolitan Art Museum, SIGGRAPH, Ars Electronica, the Long Beach Museum, and other locations worldwide. ▬ Naoko Tosa (J) ist Medienkünstlerin und Forscherin am ATR Media Integration & Communications Research Laboratories. Sie ist Gastprofessorin an der Universität Kobe und Gastlektorin an der Musashino Art University. Sie promovierte an der University of Tokyo in Kunst- und Technologieforschung und konzentriert sich vorwiegend auf den Bereich der Kommunikation. Sie verwendet Computer, Video und Elektronik zur Gestaltung von Kunstwerken, die sich mit Aspekten der Intelligenz von Emotionen, mit dem Bewusstsein und Unbewusstem beschäftigen. Schwerpunkt ihrer Arbeit sind Experimentalfilme, Videokunst, Computeranimationen und interaktive Kunstwerke. Ihre Arbeiten wurden am Museum of Modern Art, New York, im New York Metropolitan Art Museum, bei der SIGGRAPH, im Long Beach Museum, bei Ars Electronica und anderen Ereignissen auf der

WATASHI-CHAN
Tomoko Ueyama

Watashi-chan is clothing that makes it possible to see invisible sounds in a space. Balloons are attached to the clothing, which inflate when there is a sound in the space. If there is any sound at all in the space, even if the person wearing *Watashi-chan* does not hear it, the balloons attached to the clothing inflate. In other words, I can visually recognize what kind of sound exists around you, if you are wearing *Watashi-chan*.

"I" means "oneself" in Japanese, "chan" is added to a noun to represent a person, and it is a word used for a friendly feeling and calling a person. So *Watashi-chan* recognizes the sound that you do not consciously perceive, do not recognize, and it tells you, instead of you telling yourself. It is like having another sense organ. Therefore, I call it *Watashi-chan*, meaning another oneself.

I want to understand how many different sounds and noises exist around me and what kind of tunes there are in the sound at this very moment. I think listening to the sounds means hearing the voice of our heart.

I would like to have people realize that listening to sounds is proof of living life. Listening to sounds should be like eating and sleeping, a part of everyday life. A lot of experiences are nourishing, which is very important for people.

Watashi-chan divides the human audio range into six frequency bands and calculates the quantity and frequency of signals entering each frequency band. An electromagnetic valve is assigned to each frequency band. As a quantity of frequency signals is calculated, a signal is sent to an electromagnetic valve to open it. As a result, air is sent to a balloon corresponding to the quantity of frequency signals. After a balloon has been inflated for one second, air is released again, so that the balloon does not inflate too much.

Watashi-chan ist Kleidung, die unsichtbare Klänge im Raum sichtbar werden lässt. Sobald ein Klang im Umkreis einer Person entdeckt wird, blasen sich Ballons auf, die in die Kleidung integriert sind. Dies geschieht auch bei Klängen, die der Träger des Kleidungsstücks selbst überhaupt nicht wahrnimmt. Mit anderen Worten: Ich kann visuell erkennen, welche Klänge rund um dich existieren, wenn du „Watashi-chan" trägst.

„I" bedeutet im Japanischen „selbst", „chan" wird einem Nomen beigefügt, um eine Person zu bezeichnen; es ist ein Wort, das freundliche Gefühle einer Person gegenüber zum Ausdruck bringt und zur Anrede einer Person dient. *Watashi-chan* erkennt die Klänge, die der Träger selbst nicht bewusst erkennt, und vermittelt sie dem Träger – deshalb habe ich es auch „Watashi-chan" genannt, das etwa so viel wie „ein weiteres Selbst" bedeutet.

Ich möchte verstehen, wie viele Klänge und Geräusche im unmittelbaren Umfeld einer Person existieren und welche Stimmungen diese Klänge vermitteln. Ich glaube, den Klängen zu lauschen, bedeutet, die Stimme unseres Herzens zu hören.

Ich möchte, dass die Menschen erkennen, dass das Hören ein Beweis für unser Leben ist, dass das Hören von Klängen und Geräuschen ähnlich selbstverständlich und dennoch bewusst geschieht wie die Nahrungsaufnahme, die ja auch für den Menschen wichtig ist.

Watashi-chan zerlegt den vom Menschen wahrnehmbaren Bereich in sechs Frequenzbänder und berechnet den jeweiligen Nahfeldpegel für jedes dieser Frequenzbänder. Je nach Signalpegel werden dann sechs elektromagnetische Ventile angesteuert, über die Ballons aufgeblasen werden, und zwar in Abhängigkeit vom gemessenen Pegel. Wenn ein Ballon eine Sekunde lang aufgeblasen war, wird die Luft wieder ausgelassen, damit er nicht zu stark gefüllt wird und womöglich platzt.

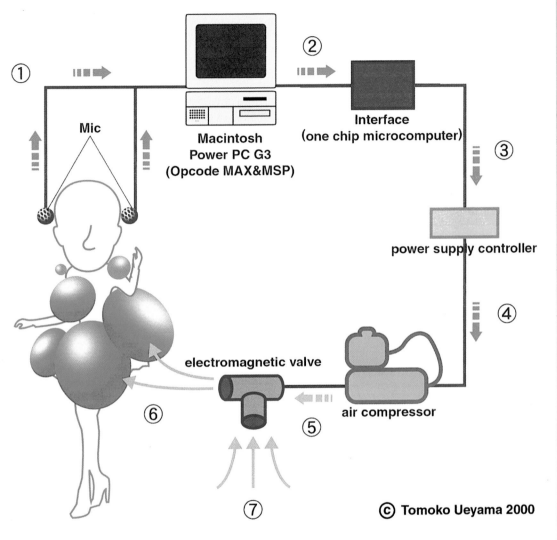

① ②

Mic

Macintosh
Power PC G3
(Opcode MAX&MSP)

Interface
(one chip microcomputer)

③

power supply controller

④

electromagnetic valve

air compressor

⑤

⑥

⑦

© Tomoko Ueyama 2000

Tomoko Ueyama (J), born 1972, B.A. Osaka University of Arts (Performing Arts, 1995). Television Sound Effects
Operator (1995—1998 NHK Osaka). Diploma, International Academy of Media Arts and Sciences (Arts and Media
Lab, 2000). ▬▬ Tomoko Ueyama (J), geb. 1972, 1995 B.A. in Darstellender Kunst an der Osaka University of Arts;
Tätigkeit als Sound Effects Operator bei NHK Osaka (1995–1998); 2000 Diplom der International Academy of Media Arts

A BODY OF WATER

Andrea Zapp / Paul Sermon

A *Body of Water* is an installation occupying three locations: firstly a chroma-key room at the Wilhelm Lehmbruck Museum in Duisburg, secondly the coal miners' changing room, or Waschkaue, at the disused Ewald/Schlägel und Eisen coalary in Herten, and thirdly the shower room at the Ewald/Schlägel und Eisen Waschkaue in Herten.

The audiences in Herten and Duisburg are connected in the following way: a video camera in Duisburg captures images of the audience standing in front of a chroma-key blue backdrop—this image is sent to Herten via an ISDN video conferencing link. The image is received in Herten and chroma-keyed together with a camera image of the audience standing in the Waschkaue changing room. The chroma-key mixed image is then video projected onto a fine wall of water, sprayed from high pressure shower heads in the Waschkaue shower room. A camera situated next to the projector captures an image of the projected image and feeds it to three monitors in the changing room space in the Waschkaue, and back via the ISDN video conferencing link to three video monitors surrounding the chroma-key space in Duisburg.

The water wall, or screen, is located in the centre of the shower room and has two different images projected onto it simultaneously from either side. The audience in the shower room is able to walk around the water screen and experience the images changing from a telematic link with Duisburg to black & white documentary footage of miners showering in the original Waschkaue. Floating independently on each side of the water wall, the two images are not mixed and appear as completely different scenarios from either side of the water screen.

This installation simply wouldn't exist without the water interface. It transports the public interaction and at the same time it reflects the urban area of the Ruhrgebiet as a network of rivers and waterways. The shower room forms the heart of the installation, all the visual and conceptual layers meet here, referring to the present changes of industrial culture in the region.

A *Body of Water* is an installation that draws an analogy between the disappearance of heavy industry and the disappearance of the human body and its telepresent reappearance in the digital network.

A *Body of Water* ist eine Installation, die drei Orte einnimmt: einerseits einen Chroma-Key-Raum im Wilhelm Lehmbruck Museum in Duisburg, zweitens den Umkleideraum – die Waschkaue – der Kohlenkumpel in der aufgelassenen Kohlengrube Ewald/Schlägel und Eisen in Herten und drittens den Duschraum der Waschkaue der Ewald/Schlägel und Eisen-Grube in Herten.

Das Publikum in Herten ist auf folgende Weise mit dem in Duisburg verbunden: Eine Videokamera in Duisburg nimmt Bilder des Publikums auf, das vor einem blauen Chroma-Key-Hintergrund steht. Dieses Bild wird über eine ISDN-Videokonferenzschaltung nach Herten übermittelt und mit einem Kamerabild des Hertener Publikums im Umkleideraum der Waschkaue über Chroma-Key verschmolzen. Das gemeinsame Videobild wird daraufhin auf einen feinen Wasservorhang projiziert, der aus Hochdruck-Brauseköpfen im Duschraum der Waschkaue heruntersprüht. Eine Kamera neben dem Videoprojektor nimmt das projizierte Bild auf dem Wasservorhang auf und überträgt es seinerseits auf drei Monitore im Umkleideraum der Waschkaue sowie über ISDN zurück nach Duisburg, wo drei Videomonitore den Chroma-Key-Raum umgeben.

Auf die Wasserwand – unsere Projektionsfläche – in der Mitte des Duschraums werden gleichzeitig von zwei Seiten Bilder projiziert. Das Publikum im Duschraum kann um die Wasserwand herumgehen und erleben, wie die Bilder der telematischen Verbindung mit Duisburg auf der einen zu Bildern mit duschenden Bergleuten aus einem alten Schwarzweißfilm auf der anderen Seite des Wasserbildschirms wechseln. Die auf die jeweilige Oberfläche des Wasservorhangs projizierten Bilder vermischen sich nicht, sondern erscheinen als völlig unterschiedliche Szenarien auf den entgegengesetzten Seiten der Wasserwand. Diese Installation könnte ohne das Wasser-Interface nicht existieren. Es transportiert die Interaktion des Publikums und reflektiert gleichzeitig die urbane Struktur des Ruhrgebiets als Netzwerk aus Flüssen und Wasserstraßen. Der Duschraum bildet das Herz der Installation; alle visuellen und konzeptuellen Ebenen treffen sich hier und beziehen sich auf die gegenwärtigen Veränderungen der Industriekultur dieser Region.

A *Body of Water* ist eine Installation, die eine Analogie zwischen dem Verschwinden der Schwerindustrie und dem Verschwinden des menschlichen Körpers und seinem Wiedererscheinen in telepräsenter Form im digitalen Netzwerk herstellt.

Commissioned by the Wilhelm Lehmbruck Museum Duisburg for Connected Cities, June 20 to August 1 1999

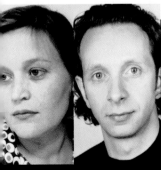

Andrea Zapp (D) studied film and media, Russian language and literature theory in Marburg, Germany, Moscow, Russia and Simferopol, Ukraine. 1991 M.A. on contemporary Russian film. Currently lecturer on new media at the Academy for Film and Television (HFF), Potsdam-Babelsberg: Theory of new media, film and new media, practical workshops on interactive multimedia applications and internet design and dramaturgy. Paul Sermon (GB), born 1966, studied at the University of Wales. Fine Art degree. Post-graduate MFA degree at The University of Reading, England. Artist in Residence at the ZKM in Karlsruhe (1993). Currently living in Berlin, working as Associate Professor for telepresence and telematic media in the department of media arts at the Academy of Graphic and Book Arts in Leipzig. ▬ Andrea Zapp (D) studierte Film und Medien, Russische Sprache und Literaturtheorie in Marburg, Moskau und Simferopol (Ukraine). 1991 graduierte sie zum M.A. mit einer Arbeit zum zeitgenössischen russischen Film. Derzeit ist sie Lektorin für neue Medien an der Hochschule für Film und Fernsehen (HFF) in Potsdam-Babelsberg und liest über Theorie der Neuen Medien, Film und Neue Medien, praktische Workshops zu interaktiven Multimedia-Anwendungen sowie Internet-Design und Dramaturgie. Paul Sermon (GB), geb. 1966, studierte an der Universtität von Wales (Abschluss in bildender Kunst). Post-graduate-Studium an der Universität Reading, England. Artist in Residence am ZKM in Karlsruhe (1993). Lebt zur Zeit in Berlin, außerordenticher Professor für Telepräsenz und Telematik an der Akademie für Grafik und

ROBOCUP
Hiroaki Kitano (President, The RoboCup Federation)

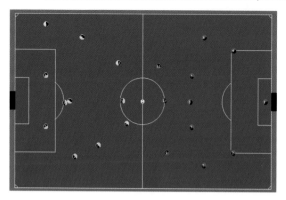

The Robot World Cup Initiative (RoboCup) is an international research and education initiative. It is an attempt to foster AI and intelligent robotics research by providing a standard problem where wide range of technologies can be integrated and examined, as well as being used for integrated project-oriented education.

For this purpose, *RoboCup* chose to use soccer game as a primary domain, and organizes RoboCup: The Robot World Cup Soccer Games and Conferences. In order for a robot team to actually perform a soccer game, various technologies must be incorporated, including: design principles of autonomous agents, multi-agent collaboration, strategy acquisition, real-time reasoning, robotics, and sensor-fusion. *RoboCup* is a task for a team of multiple fast-moving robots under a dynamic environment. *RoboCup* also offers a software platform for research on the software of *RoboCup*.

While soccer game is used as a standard problem where broad-range of efforts will be concentrated and integrated, competition is only a part of *RoboCup* acitivity.

Current activity of the *RoboCup* Initiative consists of technical conferences, Robot World Cup, RoboCup Challenge programs, education programs and infrastructure development.

Nevertheless, Robot World Cup competition is the central pillar of our activity, where researchers can get together and evaluate the research progress. Currently, *RoboCup* has: Simulator League, Small Roboter League, Full Set Small Robot League, which is 11 robots per team, Middle Size Robot League, Sony Legged Robot League (sponsored by Sony), Humanoid League (from 2002), TeleOperation Track, RoboCup Commentator Exhibition.

Die Robot World Cup Initiative *RoboCup* ist eine internationale Forschungs- und Bildungsinitiative. Sie ist ein Versuch, Künstliche Intelligenz und Roboterforschung weiterzubringen, indem sie ein Standardproblem stellt, zu dessen Lösung eine große Bandbreite von Technologien integriert und getestet werden kann, darüber hinaus bietet sie ein Betätigungsfeld für Projekte integrierter Ausbildung.

Zu diesem Zweck hat *RoboCup** das Fußballspiel als primäre Domäne gewählt und den RoboCup ausgerichtet – den Weltcup für Roboterfußballspiel und -konferenzen. Damit ein Roboterteam tatsächlich ein Fußballmatch durchführen kann, bedarf es der Integration verschiedener Technologien, z. B. der Designprinzipien autonomer Agenten, einer Multi-Agent-Zusammenarbeit, Strategieentwicklung, Echtzeit-Überlegungen, Robotertechniken und Sensor-Fusion. *RoboCup* ist eine Aufgabe für Teams aus mehreren schnellbeweglichen Robotern in einer dynamischen Umgebung. *RoboCup* bietet auch eine Software-Plattform für Forschungsarbeiten an *RoboCup*-Software.

Wenn auch das Fußballspiel als ein Standardproblemfeld eingesetzt wird, in dem eine große Bandbreite an Zielansprüchen konzentriert und integriert werden kann, so ist der Wettbewerb noch nur ein Teil der *RoboCup*-Aktivitäten. Die Initiative umfasst technische Konferenzen, den Roboter-Weltcup, diverse RoboCup Challenge Programme, (Aus-)Bildungsprogramme und die Entwicklung von Infrastruktur. Dennoch - der Roboter-Weltcupbewerb bleibt der Eckpfeiler unserer Aktivität, bei dem sich Forscher treffen und den Fortschritt in der Entwicklung bewerten können.

Derzeit umfasst *RoboCup* die Simulator-Liga, die Kleinroboter-Liga, eine Full-Set-Kleinroboter-Liga mit jeweils elf Robotern pro Mannschaft, eine Mittelroboter-Liga, die von Sony gesponserte Sony-Liga für Roboter mit Beinen, ab 2002 auch eine Humanoiden-Liga, einen TeleOperation-Track und die RoboCup-Kommentatoren-Ausstellung.

Soccer Server

THE ROBOCUP FEDERATION IS AN INTERNATIONAL ORGANIZATION, REGISTERED IN SWITZERLAND, TO ORGANIZE INTERNATIONAL EFFORT TO PROMOTE SCIENCE AND TECHNOLOGY USING SOCCER GAMES BY ROBOTS AND SOFTWARE AGENTS. ▬▬ Die RoboCup Federation ist eine internationale Organisation mit Sitz in der Schweiz, die die internationalen Bemühungen organisiert, Wissenschaft und Technologie

.NET
INTERACTIVE ART
COMPUTER ANIMATION
VISUAL EFFECTS
DIGITAL MUSICS
U19/CYBERGENERATION

TUNNELS AND FLIES
TUNNELS UND FLIEGEN

If tools are tools,
if craft relates to how well we use the tools,
if art relates to how and why we apply and manifest
our craft
then
is animation,
is visual effects,
is the body of work that was submitted,
is the collective work that was selected for prizes and
mentions
a tool, a craft or an art?

The opportunity to view a large and significant body
of work is wonderful—often regardless of context.
The chance to see where people are in their work,
what they are preoccupied with, what they value as
important, how far along they are in it is a gift. Creative expression of self in whatever form it is manifest is (obviously from my point of view) what life is
all about. In the case of these words, the context is
the Prix Ars Electronica 2000 competition and the
body of work the submissions for the "Computer Animation" and "Visual Effects" categories.
Mark Dippé, James Duesing, Barbara Robertson and
myself, as jury members watched some 260 pieces of
work. After three rounds of viewing, decisions, discussions, arguments and eventually consensus, we
awarded a Golden Nica, two Awards of Distinction
and fifteen Honorable Mentions in both categories.
The experience was stimulating and satisfying. The
jury process worked. The resulting collective of prize
winners and honorable mentions is a definite "must
see" for anyone interested in either of these areas.
The various and varying lines between "animation"
and "effects" continue to waver and move. We (the
jury) ended up moving some submitted work
between the "animation" and "effects" categories
when we felt that a piece fell under a better light in
the "other", but concisely summing up or numerating
the criteria for either category is beyond, certainly,
me alone. And, oddly enough, equalizing work to the
point where an incredibly successful (in all terms)
commercial piece can be compared to a student work
is actually easier than defining the distinction
between "animation" and "effects". In both categories
tools, craft (the application of the tools) and art (the
how and why of the applied craftsmanship) remain

Wenn Werkzeuge Werkzeuge sind,
wenn das Handwerkliche sich darauf bezieht, wie gut
wir diese Werkzeuge einsetzen,
wenn Kunst Bezug nimmt darauf, wie und warum wir
unsere handwerklichen Fähigkeiten anwenden und
manifestieren –
ist Animation,
sind Visual Effects,
ist die Gesamtheit der eingereichten Werke,
ist die Gesamtheit der prämierten und ausgezeichneten Arbeiten
dann eigentlich ein Werkzeug, ein Handwerk oder eine
Kunst?

Die Gelegenheit, ein großes und signifikantes
Ensemble von Arbeiten zu sehen, ist etwas Wunder-
bares – oft auch ungeachtet des Zusammenhangs.
Die Chance zu sehen, wo Menschen mit ihrer Arbeit
stehen, womit sie sich beschäftigen, was sie als
wichtig ansehen, wie weit sie auf ihrem Weg sind –
das ist ein Geschenk. Der kreative Ausdruck des
Selbst, in welcher Form es sich auch manifestieren
mag, ist – zumindest aus meiner Sicht – Kern des
Lebens. Im vorliegenden Fall ist der Kontext der Prix
Ars Electronica 2000 und das Ensemble der Arbeiten
die Einreichungen in den Kategorien „Computeranimation" und „Visual Effects".
Mark Dippé, James Duesing, Barbara Robertson und
ich haben als Jurymitglieder über 260 Arbeiten
angesehen. Nach drei Runden der Betrachtung, Diskussion, Argumentation und zuletzt auch Übereinstimmung haben wir in beiden Kategorien jeweils
eine Goldene Nica, zwei Auszeichnungen und 15
Anerkennungen vergeben – eine anregende und
befriedigende Aufgabe, der Juryprozess funktionier-
te gut. Die sich ergebende Sammlung von Preisträ-
gern und ausgezeichneten Werken ist ein absolutes
Muss für jeden, der sich für eines oder beide Gebiete
interessiert.
Die unterschiedlichen und veränderlichen Berüh-
rungslinien zwischen „Animation" und „Effects"
bleiben weiterhin verschwommen und in Bewegung. Wir – die Juroren – haben letztlich mehrmals
Arbeiten von der einen Kategorie in die andere ver-
schoben, wenn wir zur Überzeugung gelangt sind,
ein Stück sei in der jeweils anderen Kategorie besser
aufgehoben, aber eine Auflistung oder Zusammenfassung der Kriterien für eine solche Zuordnung ist
mehr, als ich jetzt (allein) zu leisten vermag. Und – so
seltsam das erscheinen mag – es ist leichter, ein (in
jeder Hinsicht) äußerst erfolgreiches kommerzielles
Werk auf jene Ebene zu reduzieren, auf der es mit
studentischen Arbeiten verglichen werden kann, als
die Unterschiede zwischen „Animation" und „Visual
Effects" zu definieren. In beiden Kategorien bleiben
letztlich die Werkzeuge, das Handwerkliche (also der
Einsatz der Werkzeuge) und das Künstlerische (das
Wie und Warum der angewandten Handwerkskunst)

alien in its portrayal of life. In *Maly Milos* all such shapes and motion paths were done away with as much as possible and irregularities and asymmetry were emphasized. The textures were all custom made using traditional materials such as acrylics and charcoal. The computer was then used to tweak and orchestrate elements such as color and texture. The set and characters were modeled and animated using 3D-Studio Max V2.0 from Discreet Logic. Pistecky's casting choices proved to be perfect for the film. Jitka Svestkova performed the voice of Babka.

Geschichtenerzählen hat. Der Betrachter erhält ausgiebig Gelegenheit, die einzelnen Charaktere zu studieren und der Erzählung in Versform zu folgen. Von Beginn an zeigt sich die Qualität dieser Arbeit: Das feine Detail, die Schärfe der Farben lassen an eine traditionelle Trickfilmarbeit denken. Ein erfrischender sanfter Touch in Beleuchtung und in der Auszeichnung der Konturen verstärkt den Eindruck von Tiefe und aktiver Bewegung im gesamten Kurzfilm, und nicht zuletzt erfasst auch die Musik (für Violine, Akkordeon und Schlagzeug) die Essenz des Films. *Maly Milos* wurde im Juni 1999 nach rund achteinhalbmonatiger Produktionszeit fertig gestellt. Der Kurzfilm erinnert an jene tschechischen Geschich-

Svestkova is a quiet and timid lady, who had to be deprived of nicotine to encourage and reveal a more frustrated and angst-filled delivery. Her rich and textured voice underscored by her strong Czech accent was quite fitting and helped shape the personality of Babka. The voice of the narrator turned out to be a little more difficult to find. Pistecky finally found his narrator, talented local actor Alex Williams, in the dark hallways of Emily Carr after several unsuccessful casting sessions. Pistecky discovered his violinist at a local Market Place: David Rabinovich, a short, pronounced Russian man, who was playing his violin to an overwhelmed crowd. His love and knowledge of the violin was evident. Pistecky knew he would be perfect for his film.

The film's name *Maly Milos* uses the Czech spelling as opposed to the English for a few reasons: the use of alliteration would foreshadow the poetic narrative, while the letters were visually more interesting and playful, and the Czech title reflected Pistecky's inspiration for the film.

ten, die Pistecky in seiner Jugend so beeindruckt haben. Das Werk bemüht sich erfolgreich, an die Puppentrickfilme anzuknüpfen, für die sein Heimatland so berühmt ist. Die Erzählung dreht das übliche Rollenschema um und wirft einen Blick auf die Grausamkeit und die Angst, die innerhalb einer Beziehung bestehen können. Die in diesem Werk eingesetzten Techniken sind eine Kombination aus traditionellen und Computer-Techniken. Es ging dabei auch darum, jenes „kalte" Gefühl zu vermeiden, das computergenerierten Filmen zumeist anhaftet. Computeranimation bedient Anforderungen nach geraden Linien, perfekten Bögen und glatten Bewegungen recht gut, aber dies bringt ein immer etwas unnatürliches Aussehen mit sich, wenn damit die Wirklichkeit dargestellt werden soll. In *Maly Milos* wurden all diese Formen und Bewegungen soweit wie möglich vermieden, Unregelmäßigkeiten und Asymmetrie wurden möglichst verstärkt. Auch alle Texturen entstanden auf traditionelle Weise mit Materialien wie Acrylfarben und Zeichenkohle. Der Computer wurde dann eingesetzt, um Elemente wie Farbe und Textur zu verzerren und zu orchestrieren. Szenerie und Charaktere wurden mit Hilfe von 3D-Studio Max V2.0 von Discreet Logic modelliert und animiert.

Maly Milos has been screened at over a dozen international festivals and has received several awards. Some of these accolades include; Best Canadian Film Award at The Student Animation Festival (Ottawa), Best BC Short Film at The Victoria Independent Film and Video Festival (Victoria, BC), First Runner Up for the Best Computer Animation Award at The Vancouver Effects and Animation Festival (Vancouver, BC), to name just a few. Critically acclaimed website Atomfilms.com picked up exclusive rights for worldwide Internet distribution.

Pisteckys Auswahl der Darsteller erwies sich als perfekt. Jitka Svestkova sprach die Stimme der Babka. Svestkova ist eine ruhige und eher schüchterne Dame, die auf Nikotinentzug gesetzt werden musste, um ein frustrierter und angsterfüllter klingendes Ergebnis zu erzielen. Ihre volle modulationsreiche Stimme mit dem deutlichen tschechischen Akzent erwies sich als genau passend und trug viel dazu bei, die Persönlichkeit der Babka zu formen. Schwieriger war es, eine Sprechstimme für den Erzähler zu finden. Nach mehreren erfolglosen Castings entdeckte Pistecky seinen Erzähler, den talentierten lokalen Schauspieler Alex Williams, in den dunklen Gängen des Emily-Carr-Instituts. Seinen Violinspieler fand Pistecky auf einem lokalen Markt – der wortkarge Russe David Rabinovich spielte vor einer überwältigten Zuhörermenge. Seine Liebe und sein Verständnis der Violine waren offensichtlich – Pistecky wusste von Anfang an, dass er für diesen Film ideal sein würde.
Der Filmtitel *Maly Milos* verwendet die tschechische Schreibweise ganz bewusst an Stelle der englischen: Die Alliteration weist bereits auf den Stil der Erzählung hin, die Buchstabenkombination ist optisch ansprechender und letztlich reflektiert der tschechische Titel auch Pisteckys Inspirationsquelle.
Maly Milos wurde bei über einem Dutzend internationaler Festivals gezeigt und hat zahlreiche Auszeichnungen erhalten, darunter: Bester Kanadischer Film beim The Student Animation Festival (Ottawa); Bester Kurzfilm aus British Columbia beim Victoria Independent Film and Video Festival (Victoria, B.C.), Zweiter beim Best Computer Animation Award des Vancouver Effects and Animation Festival (Vancouver), um nur einige zu nennen. Die von den Kritikern akklamierte Website <Atomfilms.com> hat die Exklusivrechte für eine weltweite Verbreitung übers Internet erworben.

JAKUB PISTECKY (CDN), BORN IN ZLIN, CZECHOSLOVAKIA IN 1975. IN 1981 PISTECKY AND HIS FAMILY ESCAPED TO BADEN, AUSTRIA, WAITING TO BE PERMITTED TO IMMIGRATE TO CANADA. IN 1994, AFTER HIGH SCHOOL, HE WENT BACK TO STUDY FILM AT THE ATELIER ZLIN FILM SCHOOL. BACK IN CANADA HE WENT ON TO STUDY ANIMATION AT VANCOUVER'S EMILY CARR INSTITUTE OF ART + DESIGN (MARTIN ROSE AND MARILYN CHERENKO). 1998 HE WAS GIVEN THE MOST PROMISING CANADIAN STUDENT AWARD AT THE OTTAWA ANIMATION FESTIVAL. BEFORE GRADUATING FROM THE FOUR-YEAR PROGRAM HE CREATED *MALY MILOS*. ▬▬▬ Jakub Pistecky (CDN), geboren 1975 in Zlin, Tschechien. 1981 flüchteten Pistecky und seine Familie nach Baden bei Wien, wo sie auf die Einreisegenehmigung nach Kanada warteten. 1994, nach Abschluss der High School, ging Jakub zurück, um an der Atelier Zlin Film School zu studieren. Nach seiner Rückkehr nach Kanada setzte er sein Studium im Fach Animation am Emily Carr Institute of Art + Design bei Martin Rose und Marilyn Cherenko fort. 1988 erhielt der den Most Promising Canadian Student Award beim Ottawa Animation Festival. *Maly Milos* entstand als Abschluss seiner vier-

TOY STORY 2

John Lasseter / Lee Unkrich / Ash Brannon / Pixar

Building a Better Sequel:	Eine bessere Fortsetzung entsteht:
The Return of	Die Rückkehr von
Buzz Lightjear and Woody	Buzz Lightning und Woody

Creating a sequel to one of the most successful and beloved animated films of all times is a daunting undertaking, but for John Lasseter and the creative team, the challenge was well worth it. It gave them a chance to work with established characters that they knew and loved as well as to create a cast of fresh new characters that would complement and add to the story possibilities. Ironically, some of the key plot points for the sequel (the garage sale, the kidnapping, the obsessive toy collector, a squeak toy penguin, etc.) date back to the development of the first feature. Lasseter hatched the idea for *Toy Story 2* one day over lunch with his colleague Pete Docter (who received a story credit on the first film). Andrew Stanton, who helped create the story and screenplay for

Eine Fortsetzung zu einem der erfolgreichsten und beliebtesten animierten Filme aller Zeiten zu machen, ist ein mühseliges Unterfangen, aber für John Lasseter und sein Kreativteam war die Herausforderung alle Mühen wert. Einerseits gab sie ihnen die Möglichkeit, mit bereits bekannten Gestalten zu arbeiten, die sie kannten und liebten, zum anderen konnten neue Charaktere eingeführt werden, die die Entwicklungsmöglichkeit der Geschichte ergänzten und erweiterten. Kurioserweise gehen einige der Schlüsselszenen dieses zweiten Teils (der Flohmarkt, die Entführung, der besessene Spielsachensammler, ein Quitschepinguin usw.) noch auf die Entwicklungsphase des ersten Films zurück. Lasseter besprach die Idee einer *Toy Story 2* eines Tages beim Lunch mit seinem Kollegen Pete Docter (der am Drehbuch des ersten mitgewirkt hatte). Andrew Stanton, der einer der „Geburtshelfer" des Storyboards und des Drehbuchs von *Toy Story* war und

STINKY PETE
THE
PROSPECTOR

POSEABLE TALKING DOLL · WITH
From the Wo...

the original *Toy Story* and went on to write and co-direct *A Bug's Life* with Lasseter, helped to flesh out the story and characters with a draft of the screenplay. A trio of other screenwriters—Rita Hsiao (*Mulan*), Doug Chamberlin & Chris Webb—are also credited with adding structure and dimension to the final film. Story development for the sequel officially began in the spring of 1996.

"When we were done with the first film," he continues, "we felt that there were so many more ideas and stories with these toys being alive that we hadn't dealt with. One of those was the notion of a toy being outgrown by its child. If you're lost, you can be found and everything will be okay. If you're broken, you can be fixed. But for a toy, being outgrown is the worst thing that can happen. That's it."

Another idea for the sequel came from Lasseter's personal experiences as a toy collector. He explains, "I have five sons and my four little ones love to come to

danach gemeinsam mit Lasseter *A Bug's Life* geschrieben und geleitet hat, half durch eine erste Skizze des Drehbuchs mit, dem Handlungsfaden und den Charakteren etwas Fleisch zu verpassen. Ein Trio von weiteren Drehbuchautoren – Rita Hsiao (*Mulan*), Doug Chamberlin und Chris Webb – verdient ebenfalls Erwähnung, denn auch sie haben wesentlich zur Struktur und Dimension des endgültigen Films beigetragen. Die Drehbucharbeit an der Fortsetzung begann offiziell im Frühjahr 1996.

Lasseter merkt an: „Als wir den ersten Film fertig hatten, spürten wird, dass es rund um diese lebendigen Spielsachen noch so viele weitere Ideen und Geschichten gab, die wir alle noch nicht behandelt hatten. Eine davon war die Vorstellung eines Spielzeugs, dem das zugehörige Kind davongewachsen ist. Wenn man – als Spielzeug – verloren geht, dann kann man gefunden werden, und alles ist wieder in Ordnung. Geht man kaputt, so wird man eben repariert. Aber für ein Spielzeug ist es am schlimmsten, wenn das Kind aus ihm herauswächst. Und das war es auch schon."

Eine weitere Idee für die Fortsetzung kam aus Lasseters persönlicher Erfahrung als Spielzeugsammler. Er erklärt: „Ich habe fünf Söhne, und meine vier kleineren kommen gerne in Vatis Büro und spielen dort mit meinen Spielsachen. Viele von diesen sind antik und in ihrer Art einmalig. Ich liebe meine Buben und ich möchte auch gerne, dass sie mit den Sachen spielen. Aber ich ertappte mich dabei, dass ich sagte: ‚Nein, mit dem da kannst du nicht spielen, nimm stattdessen dieses hier!' Und als mir das auffiel, musste ich lachen, weil schließlich Spielzeug genau deswegen hergestellt und auf die Welt gebracht wird – nämlich, damit Kinder damit spielen. Und das ist die Essenz von *Toy Story 2* und ist auch der Kern des Lebens unserer Charaktere. Alles, was sie davon abhält, von einem Kind als Spielzeug benutzt zu werden, ruft Ängste in ihrem Leben hervor."

Gestützt auf eine Gruppe talentierter Geschichtenerzähler nahm die Handlung von *Toy Story 2* bald Gestalt an. Der Drehbuchautor Andrew Stanton erklärt: „Unsere Verantwortung als Autoren ist es, die Geschichte zu analysieren, die Wahrheit zu finden und zu benutzen. Der schwierigste Teil in der Gestaltung eines Drehbuchs für einen abendfüllenden Film ist die Entwicklung der Charaktere. Sie müssen dreidimensional sein und genug Substanz für einen ganzen Film haben. Hier, bei der Fortsetzung, wusste ich schon, wer die Charaktere sind, und das war ein schönes Arbeiten, weil ich mich dransetzte und schon ging's los: Also, Buzz pflegt dies zu sagen, worauf Woody das sagen würde ... Man jongliert beim Schreiben sozusagen mit drei Bällen: Handlung, Figuren und etwas Drittes, das die ich „Drive" nennen möchte, jene Spannung, die die Zuseher bei der Stange hält. Da die Hauptgestalten bereits eingeführt waren, hatten wir die Freiheit, uns auf die beiden anderen Elemente zu konzentrieren."

Stanton fügt hinzu: „Ich bin der Überzeugung, dass

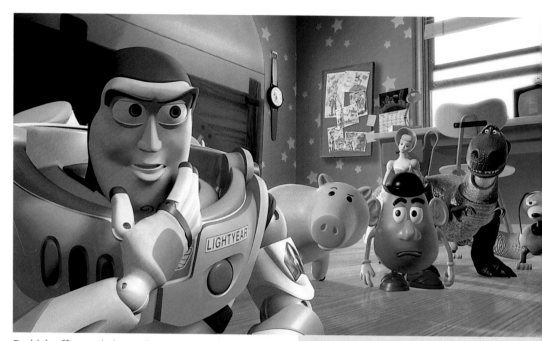

Daddy's office and play with my toys. A lot of them are antiques and one-of-a-kind items. I love my boys and I wanted them to play with these toys, but I found myself saying, 'No, no, you can't play with that one. Oh, here play with this one instead.' As I looked at myself I began laughing because toys are manufactured and put on this earth to be played with by a child. That is the essence of *Toy Story 2* and the core of the toys being alive. Everything that prevents them from being played with by a child causes them anxieties in their life."

Drawing on a talented group of storytellers, the plot for *Toy Story 2* began to take shape. Screenwriter Andrew Stanton observes, "our responsibility as writers is to analyze the story, discover the truth and utilize it. The hardest part of writing a feature is to come up with characters that are 3-dimensional and worth spending time with for the entire film. You pretty much spend every waking minute until the thing is in the can trying to make sure you've done it right. In the case of a sequel, I already knew who the characters were. It was great because I could say, 'Oh well, Buzz would say this' and 'Woody would say that.' There's three balls that you have to juggle when you're writing—plot, character and what I call drive, the thing that keeps an audience interested. With the main characters already established, we had the freedom to concentrate on the other two elements."

man eine Fortsetzung nicht erzwingen kann, nur weil allgemeine Nachfrage danach herrscht. Wenn der zweite Teil ebenso beliebt werden soll wie der Erste, dann muss die Begründung seiner Existenz als Film in ihm selbst liegen. Mit John und Joe Ranft und den anderen Co-Regisseuren (Lee und Ash) zusammenzukommen war wie ein Klassentreffen: Die Ideen begannen zu sprudeln, weil wir ja aus den selben Erinnerungsquellen schöpfen konnten. Wir haben uns gegenseitig mit Futter versorgt, so wie eine Gruppe wie Monty Python wohl funktioniert hat. Zusätzlich zu all seinem Humor und seiner Action hat *Toy Story 2* eine emotionale Tiefe, die uns besser gelungen ist als beim ersten Film. Ich bin stolz darauf und ich glaube, dass dies den Film einzigartig macht."

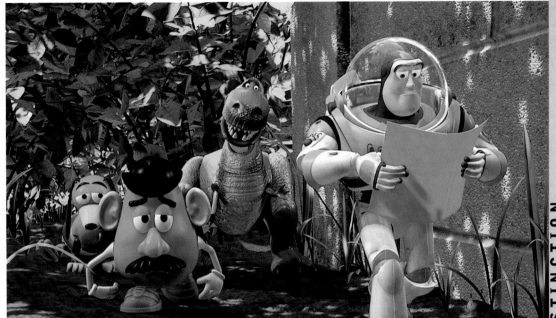

Stanton adds, "I'm a believer that you can't force a sequel just because of popular demand. If you want it to be liked as much as the first one, there's got to be a whole self-sufficient reason for the movie to exist. Getting together with John and Joe Ranft and the co-directors (Lee and Ash) was like a class-reunion. The ideas just started to flow and we began drawing from the same memory banks. We fed each other in the way a group like Monty Python must have worked. In addition to all the great humor and action, *Toy Story 2* has a depth of emotion that we were able to do better than in the original film. I'm proud of that and I think it makes the film unique."

JOHN LASSETER (USA) MADE MOTION PICTURE HISTORY IN 1995 AS THE DIRECTOR OF *TOY STORY*, THE FIRST-EVER COMPUTER-ANIMATED FEATURE FILM. HE HAS WRITTEN AND DIRECTED A NUMBER OF SHORT FILMS AND TELEVISION COMMERCIALS AT PIXAR, INCLUDING *LUXO JR.* (A 1986 OSCAR NOMINEE), *RED'S DREAM* (1987), *TIN TOY*, WHICH WON THE 1989 ACADEMY AWARD, FOR BEST ANIMATED SHORT FILM. LEE UNKRICH (USA). AS A TEENAGER HE SPENT A NUMBER OF YEARS ACTING WITH THE CLEVELAND PLAYHOUSE YOUTHEATER. HE GRADUATED FROM USC'S PRESTIGIOUS FILM SCHOOL IN 1991. HIS REPUTATION FOR BEING AN EXPERT ON THE AVID DIGITAL EDITING SYSTEM LED HIM TO HIS FIRST ASSIGNMENT AT PIXAR IN 1994. ASH BRANNON (USA) SEGUES INTO DIRECTING FOLLOWING A DISTINGUISHED CAREER AS A STORYMAN AND ANIMATOR. BRANNON DEVELOPED AN INTEREST IN ART WHILE ATTENDING A MAGNET SCHOOL FOR THE PERFORMING ARTS IN JACKSONVILLE, FLORIDA. THIS LED HIM TO PURSUE ANIMATION STUDIES AT CALARTS. HE JOINED PIXAR IN 1993. ■ John Lasseter (USA) schrieb 1995 als Regisseur des ersten abendfüllenden computeranimierten Films *Toy Story* Filmgeschichte. Er hat zahlreiche Kurzfilme und Werbespots bei Pixar produziert, darunter *Luxo jr.* (1986 für den Oscar nominiert), *Red's Dream* (1987), *Tin Toy* (Oscargewinner 1989 als bester Kurztrickfilm). Lee Unkrich (USA) war als Teenager einige Jahre beim Cleveland Playhouse Jugendtheater als Schauspieler aktiv. Er graduierte 1991 von der prestigeträchtigen Filmschule der USC. Sein Ruf als Experte am digitalen Avid-Editiersystem brachte ihm sein erstes Engagement bei Pixar 1994. Ash Brannon (USA) stieg nach einer erfolgreichen Karriere als Storyschreiber und Animator in die Regierolle. Brannon entwickelte sein Kunstinteresse bei einem Einführungskurs in darstellende Kunst in Jacksonville, Florida. Dies brachte ihn dazu, Animation am CalArts zu studieren. 1993 trat er bei Pixar ein.

ZEN
Yasuo Ohba

This work entitled *Zen* depicts what I think and feel, inspired by the environments and events surrounding me. The concept underlying this work is "the cherry blossoms season", "mutation", "in the fog", "serenity and dynamism", "moving vividly and lively", "fragrance of the spring season", "inconsistency between thought and action", "chain reactions", "an organized world", "unforeseeable world", "the inert self", "tension", and "explosion", etc. Those concepts were chosen, as they were best fit to what I felt in creating this work. Repeated observation of characters in the work in the course of creating it brought calmness in my mind, and the world that I tried to create was gradually shifted to the world of meditation.

In the first part of the work, many characters appear in a group and many hairy creatures appear and fade out. However, toward the end of the work, the scenes are getting rather simple on the screen. This may reflect my own mind.

In the last scenes, the accumulated tension is suddenly released, and the screen starts to show the essence of the works to the viewers as if they recall all the scenes as they appear and go in their mind's eye. This is when the viewers will experience the meditation that I have sought.

It was possible to complete the work thanks to an original tool developed especially for the production of the work, which was used to manipulate characters and a group of hairy creatures that appeared in the works at my fingertips. The tool is able to manipulate the unique movement of characters and the group of hairy creatures, create the concept of world peculiar to the work, control color arrangement and render the feeling of the hair volume of creatures.

In addition, the movement of characters and the group of hairy creatures can be controlled with a simple manipulation by changing various parameters in the tool that controls their movement. The tool is equipped with a GUI (Graphical User Interface), whereby each parameter is shown in the slide bar, so that any unexpected motion and unique world view

Diese Arbeit mit dem Titel *Zen* gibt wieder, was ich, inspiriert von der Umgebung und den Ereignissen rund um mich, denke und fühle. Das dieser Arbeit zu Grunde liegende Konzept lässt sich in Begriffe fassen wie „Kirschblütenzeit", „Mutation", „im Nebel", „Heiterkeit und Dynamik", „lebhafte Bewegung", „der Duft des Frühlings", „Inkonsistenz zwischen Gedanken und Handlung", „Kettenrekation", „eine organisierte Welt", „eine unvorhersehbare Welt", „das träge Selbst", „Spannung" oder „Explosion". Diese konzeptuellen Begriffe habe ich gewählt, weil sie am besten dem entsprachen, was ich bei der Schaffung dieses Werks gefühlt habe. Die wiederholte Beobachtung der Charaktere des Werks im Verlauf der Arbeit brachte mir Ruhe und die Welt, die ich zu schaffen versuchte, verschob sich langsam in den Bereich der Meditation hinein.

Im ersten Teil des Werks erscheinen viele Charaktere in Gruppen, und zahlreiche haarige Kreaturen tauchen auf und verschwinden wieder. Gegen Ende des Stücks hingegen sind die Bildschirmszenen relativ

may be achieved. The GUI, like a synthesizer to control the tone and quality of sound, enabled the tool to create the unique motion of various characters and the group of hairy creatures, emphasize their existence and compensate for the characters' movement quite easily and freely.

In developing the tool, I always had real time in mind. Thanks to the tool, the characters were imbued with life, and I was able to easily recognize their movement. It was these techniques that enabled me to view their world in the computer and shoot the camera. There were a number of steps in completing the work using this tool.

Firstly, I developed a basic program and added several algorithms to control the movement of 16 kinds of characters and the group of hairy creatures. Then motion tests were conducted for each one of the

klar strukturiert – dies mag durchaus auch meinen Zustand reflektieren.

In den Schlussszenen löst sich die Spannung, die sich aufgebaut hat, plötzlich wieder und der Bildschirm zeigt die Essenz der Arbeit, als würden die Betrachter das bisher Gesehene vor ihrem inneren Auge nochmals Revue passieren lassen. Und genau da können die Betrachter jene Form der Meditation erleben, nach der ich gesucht habe.

Das Werk konnte dank eines eigens dafür entwickelten Software-Werkzeugs realisiert werden, das die einfache Manipulation von Charakteren erlaubte – die Gruppe haariger Wesen erschien sozusagen auf Fingerdruck. Dieses Werkzeug erlaubt es, die Bewegung der Gestalten ebenso wie jene der haarigen Kreaturen individuell zu steuern, das Konzept dieser speziellen Welt zu schaffen, Farbarrangements zu steuern und dem Haar der Kreaturen Volumen zu geben.

Darüber hinaus kann die Bewegung der Figuren und haarigen Wesen mit einer einfachen Veränderung

algorithms, and I modified and tested it again as required. This series of steps resulted in producing an animation that gives the sense of "beautiful", "fun", "mysterious world" that has "never existed."
Through the validation of approximately 16 algorithms, two were selected as main algorithms for the work. The producer repeated additional motion tests and added improvements to the algorithms several times. Through an enormous number of motion tests, the final tools were completed for producing the work.
From the middle to the last stage of tool development, I sorted and categorized the motions of each character and the group of hairy creatures and organized the structure of the work. I sometimes had to reorganize it. And finally the work was completed. It would give me utmost pleasure if you could watch the work repeatedly and share what I experienced during the production period.

von Parametern im Steuerungswerkzeugs kontrolliert werden. Das Werkzeug enthält eine grafische Benutzeroberfläche, in der jeder Parameter über einen Schieberegler manipuliert werden kann, was die rasche Erstellung auch unerwarteter Bewegungen und einzigartiger Ansichten dieser Welt ermöglicht. Ähnlich wie ein Synthesizer zur Steuerung von Ton und Klangqualität verwendet wird, dient dieses grafische Interface zur einfachen Kontrolle von Bewegungen und Strukturen, wobei schon bei seiner Entwicklung der Echtzeitaspekt im Vordergrund stand. Die Charaktere wurden gleichsam mit Leben durchtränkt und ich konnte die erforderlichen Bewegungsabläufe schnell erkennen und realisieren. Dank dieser Techniken war es leicht, die Welt dieser Kreaturen im Computer zu erschaffen und wie mit einer Kamera festzuhalten.
Die Entwicklung dieses Tools erfolgte in mehreren Schritten: Zunächst habe ich ein Grundprogramm geschrieben und zahlreiche Algorithmen zur Steuerung von 16 verschiedenen Kreaturen und der haarigen Lebewesen geschrieben. Für jede davon wurden ausgiebige Bewegungstests durchgeführt, die je nach Bedarf Änderungen und weitere Tests

nach sich zogen. Diese Einzelschritte ließen eine Animation entstehen, die das Gefühl einer „schönen", „spaßigen" „mysteriösen Welt" vermittelt, die „nie existiert" hat.

Nach der Auswertung der 16 Algorithmen wurden zwei als Hauptalgorithmen für das Werk ausgewählt. Diese wurden in mehreren Etappen ausgebaut und weiterentwickelt, bis das Tool nach einer Vielzahl von Bewegungstests so weit gediehen war, dass das eigentliche Stück in Angriff genommen werden konnte. In den letzten Stadien der Werkzeugentwicklung wurden die diversen Bewegungen der einzelnen Charaktere sortiert und kategorisiert und die Struktur der Arbeit festgelegt (allerdings musste ich sie noch einige Male umorganisieren). Schließlich war die Arbeit fertig und es wäre meine größte Freude, wenn Sie das Stück mehrmals ansehen und an jenen Erfahrungen teilhaben könnten, die ich bei seiner Realisierung gemacht habe.

YASUO OHBA (J), BORN 1968 IN TOKYO, JAPAN, IS A RESEARCHER IN COMPUTER GRAPHICS WITH THE TECHNICAL RESEARCH & SUPPORT GROUP OF NAMCO LIMITED. HE GRADUATED FROM SOGO DENSHI COMPUTER COLLEGE IN 1989. HIS CG WORKS "FURBLE", "SUI", "WARASHI", "TANABATA" AND "KAZEMATSURI" WERE ACCEPTED BY SIGGRAPH 94, 95, 96, 97 AND 98 COMPUTER ANIMATION FESTIVAL, ELECTRONIC THEATER AND ART AND DESIGN SHOW, RESPECTIVELY. ▬▬ Yasuo Ohba (J) ist Computergrafik-Forscher bei Technical Research & Support Group bei NAMCO Limited. Er wurde 1968 in Tokio geboren und graduierte 1989 am Sogo Denshi Computer College. Seine computergrafischen Arbeiten *Furble*, *Sui*, *Warashi*, *Tanabata* und *Kazmatsuri* wurden bei der SIGGRAPH 94, 95, 96, 97 und 98 im Computer Animation Festival, im Electronic Theater und bei

MUSCA DOMESTICA

Denis Bivour

The film tells the old story of David against Goliath. Yet the end is neither black nor white, but somewhere in between. The main characters are an ordinary housefly (*Musca domestica*) and a person at a kitchen table. The person kills the fly and then eats the bread, in which the fly has just laid its eggs. It is left up to the audience to decide, who is the winner and who the loser.

3D animation was not used here for its own sake, but rather was intended to become integrated into a separate filmic aesthetic that was predetermined by the real video sequences. In order to avoid having the computer generated sequences become obtrusive, I refrained from an exaggerated characterization of the fly and concentrated on its true, realistic behavior.

The technical production of the film had to be accomplished with minimal means. Modelling, animation and rendering were done on a 604 200MHz MAC within a period of about three months.

Der Film erzählt die alte Geschichte von David gegen Goliath. Dabei ist das Ende weder schwarz noch weiß, sondern liegt irgendwo dazwischen. Die Hauptakteure sind die gemeine Stubenfliege (*Musca domestica*) und eine Person an einem Küchentisch. Die Person schlägt die Fliege tot und isst danach das Brot, auf das die Fliege vorher ihre Eier abgelegt hat. Wer dabei Sieger und Verlierer ist, bleibt der Entscheidung des Publikums überlassen.

3D-Animation wurde dabei nicht als Selbstzweck benutzt, sondern sollte sich in eine eigene filmische Ästhetik einfügen, die von den realen Videosequenzen vorgegeben wurde. Um eine Aufdringlichkeit der computergenerierten Sequenzen zu vermeiden, habe ich auf eine überzeichnete Charakterlichkeit der Fliege verzichtet und mich auf ihr wahres, realistisches Verhalten konzentriert.

Technisch musste der Film mit minimalen Mitteln realisiert werden. Modelling, Animation und Rendering fanden auf einem 604 200 MHz MAC in einem Zeitraum von ca. drei Monaten statt.

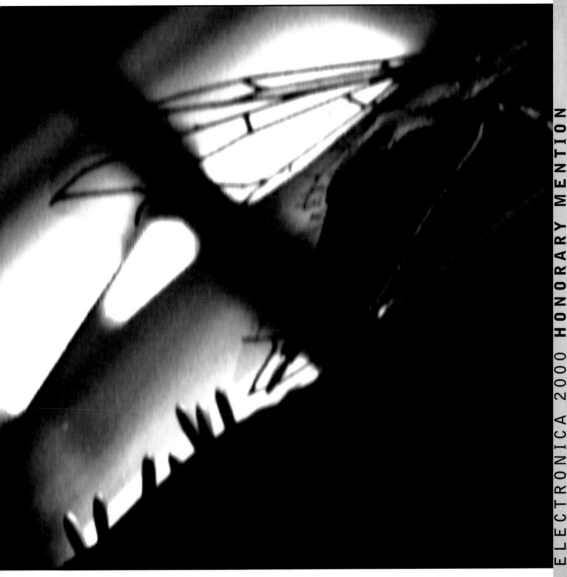

DENIS BIVOUR (D), BORN 1976, SPENT SIX MONTHS IN SOUTH AMERICA IN 1996. STUDIED FINE ARTS FROM 1997— 1999 AT THE ART COLLEGE OF BERLIN (PROF. W. STÖHRER). HAS BEEN STUDYING DIGITAL MEDIA SINCE 1998 AT THE ART COLLEGE OF BERLIN (PROF. J. SAUTER). 1999 PARTIAL GRADUATION REQUIREMENT WITH *MUSCA DOMESTICA*. FREELANCE WORK SINCE 1999 FOR VARIOUS COMPANIES (ART+COM, BERLIN; BVM BERLIN; EFFECTORY POTSDAM-BABELSBERG). COMPETITION SUBMISSIONS TO TRANSMEDIALE 2000, BERLIN, AND SEHSÜCHTE, POTSDAM. ■■■■ Denis Bivour (D), geb. 1976. 1996 sechsmonatiger Aufenthalt in Südamerika. 1997–1999 Studium der bildenden Künste an der HdK Berlin (Prof. W. Stöhrer). Seit 1998 Studium Digitale Medien an HdK Berlin (Prof. J. Sauter). 1999 Vordiplom mit *Musca domestica*. Seit 1999 freiberuflich tätig für verschiedene Firmen (art+com Berlin: bym Berlin: effectory Potsdam-Babelsberg) Wetthewerbsheiträge hei Transmediale

PAF LE MOUSTIQUE
Jean-François Bourrel / Jérôme Calvet

It is evening among the rooftops of Paris. A man chewing gum, lost completely in his music, captures the attention of a music-loving mosquito. An explosive encounter.

This short film was created in six months. It was produced entirely by Jean-François Bourrel and Jérôme Calvet, who also co-directed it. It was made with Maya 2.0 NT on two PCs, which were supported during the final two months by three other PCs, one of which was an SGI.

The main character in this little story is an exact copy of Jérôme, and the room where it takes place is none other than a copy of Jérôme's room. The apartment and the house are where the three of us live. (Kisses to Souziac! She is the third.)

So now you know everything, there are no more secrets ... or rather just one more: the logo for *Paf, le moustique* is by Christophe Delamare.

Paris, eines Abends, unter den Dächern. Ein Kaugummikauer, ganz in seine Musik vertieft, erregt die Aufmerksamkeit einer Musik liebenden Mücke. Eine explosive Begegnung.

Dieser Kurzfilm ist in sechs Monaten Arbeit entstanden. Er wurde zur Gänze von Jean-François Bourrel und Jérôme Calvet produziert, die beiden führten auch Regie. Er wurde mit Maya 2.0 NT auf zwei PCs realisiert, die während der letzten beiden Monate von drei anderen Computern unterstützt wurden. Einer davon war eine SGI.

Die Hauptperson dieser kleinen Geschichte ist eine exakte Kopie von Jérôme und der Raum, in dem sie spielt, ist nichts anderes als eine Kopie des Zimmers von Jérôme. Das Appartement und das Haus sind der Ort, an dem wir zu dritt leben (Bussi an Souziac! Sie ist die Dritte im Bunde.)

So, jetzt wissen Sie alles, es gibt keine Geheimnisse mehr ... Doch: Das Logo für *Paf, le moustique* stammt von Christophe Delamare.

jfl.oo@worldonline.fr
Copyright Calvet/Bourrel 1999
PAF Le Manufacture

JEAN-FRANÇOIS BOURREL (F), AGE 26, FREELANCE ON MAYA IN FRENCH COMPUTER GRAPHICS COMPANIES FOR ADVERTISING. AUTHOR OF *DINO IN THE BUSHES*, SELECTED AT THE ANNECY 97 FESTIVAL IN THE STUDENT CATEGORY. JÉRÔME CALVET (F), AGE 29, FREELANCE ON MAYA IN FRENCH COMPUTER GRAPHICS COMPANIES FOR ADVERTISING. AUTHOR OF *DERNIER DÉPART*, SELECTED AT IMAGINA 97 FOR THE VHS COMPILATION. ▬▬▬ Jean-François Bourrel (F), 26 Jahre alt, arbeitet freiberuflich auf Maya im Werbebereich für französische Computergrafikunternehmen. Autor von *Dino in the Bushes*, das beim Annecy Festival 97 in der Studentenkategorie ausgewählt wurde. Jérôme Calvet (F), 29 Jahre, Freiberufler im Werbebereich auf Maya für französische Computergrafikunternehmen. Autor von *Dernier Départ*, ausgewählt für die VHS-Sammlung der Imagina 97.

FIAT LUX

Paul Debevec

Fiat Lux features a variety of dynamic objects realistically rendered into real environments, including St. Peter's Basilica in Rome. The geometry, appearance, and illumination of the environments were acquired through digital photography and augmented with synthetic objects to create the animation.

Fiat Lux draws its imagery from the life of Galileo Galilei (1564—1642) and his conflict with the church. When he was twenty, Galileo discovered the principle of the pendulum by observing a swinging chandelier while attending mass. This useful timing device quickly set into motion a series of other important scientific discoveries. As the first to observe the sky with a telescope, Galileo made a number of discoveries supporting the Copernican theory of the solar system. As this conflicted with church doctrine, an elderly Galileo was summoned to Rome where he was tried, convicted, forced to recant, and sentenced to house arrest for life. Though honorably buried in Florence, Galileo was not formally exonerated by the church until 1992. Fiat Lux presents an abstract interpretation of this story using artifacts and environments from science and religion.

The objects in Fiat Lux are synthetic, but the environments and the lighting are real. The renderings are a

Fiat Lux enthält eine Vielzahl dynamischer Objekte, die in reale Umgebungen – auch in den Petersdom zu Rom – hinein gerendert wurden. Geometrie, Aussehen und Beleuchtung der Umgebung wurden mittels digitaler Fotografie erfasst und bei der Schaffung der Animation mit synthetischen Objekten angereichert.

Die Bilderwelt von Fiat Lux leitet sich aus dem Leben von Galileo Galilei (1564–1642) und seinem Konflikt mit der Kirche ab. Im Alter von 20 Jahren, als er während der Messe einen schwingenden Kerzenleuchter beobachtete, entdeckte Galileo das Prinzip des Pendels. Dieses nützliche Gerät zur Messung der Zeit zog schnell eine Menge anderer wichtiger wissenschaftlicher Entdeckungen nach sich. Er war nicht nur der Erste, der den Himmel mit einem Teleskop beobachtete, sondern machte auch zahlreiche Entdeckungen, die die kopernikanischen Theorien über das Sonnensystem stützten. Da dies mit den kirchlichen Theorien der damaligen Zeit nicht übereinstimmte, wurde der alternde Galileo nach Rom gerufen, wo man ihm den Prozess machte, ihn verurteilte, zum Widerruf zwang und für den Rest seines Lebens unter Hausarrest stellte. Obwohl er ein christliches Begräbnis in Florenz erhielt, dauerte es doch bis 1992, ehe er offiziell rehabilitiert wurde. Fiat Lux präsentiert unter Verwendung von Artefakten aus Wissenschaft und Religion eine abstrakte Interpretation dieser Geschichte.

Die Objekte in Fiat Lux sind synthetisch, aber Umgebung und Beleuchtung sind real. Die gerenderte Fassung stellt dar, wie die reale Umgebung aussähe, würden die simulierten Objekte tatsächlich in ihr

computed simulation of what the scenes would actually look like with the synthetic objects added to the real environments. The image-based lighting techniques we used represent an alternative to traditional compositing methods, in which the lighting on the objects is specified manually.

The environments were acquired in Florence and Rome. To record the full range of illumination, we used high dynamic range photography, in which a series of exposures with varying shutter speeds is combined into a single linear-response radiance image.

auftauchen. Die von uns verwendeten bildbasierten Beleuchtungstechniken stellen eine Alternative zu den traditionellen Kompositionsmethoden dar, in denen die Objektbeleuchtung von Hand eingestellt wird.

Die Umgebungsaufnahmen entstanden in Florenz und Rom. Um das ganze Spektrum der Beleuchtung aufzuzeichnen, wurde hochdynamische Fotografie eingesetzt, wobei eine Serie von Aufnahmen mit unterschiedlichen Belichtungszeiten zu einer einzigen Linear-Response-Abbildung der Leuchtkraft kombiniert wird.

PAUL DEBEVEC EARNED DEGREES IN MATH AND COMPUTER ENGINEERING AT THE UNIVERSITY OF MICHIGAN IN 1992 AND COMPLETED HIS PH.D. AT THE UNIVERSITY OF CALIFORNIA AT BERKELEY IN 1996, WHERE HE IS NOW A RESEARCH SCIENTIST. DEBEVEC HAS WORKED ON A NUMBER OF IMAGE-BASED MODELING AND RENDERING PROJECTS. HIS WORKS HAVE BEEN SCREENED AT SIGGRAPH 96, 97 AND 98. ▬▬▬ Paul Debevec (USA). 1992 Diplom in Mathematik und Computer Engineering an der University of Michigan. 1996 PhD an der University of California, Berkeley, wo er jetzt als wissenschaftlicher Assistent arbeitet. Er hat an einer Reihe von bildbasierten Modeling- und Rendering-Projekten gearbeitet. Projektmitarbeit bei Interval Research in Palo Alto. Seine Arbeiten wurden bei der SIGGRAPH 96, 97 und 98 gezeigt.

FISHING
David Gainey

Fishing is the first short film to be rendered entirely in a watercolor simulation process developed at PDI. Using techniques innovated by Cassidy Curtis, technical director on *Fishing*, shallow computer-generated watercolor begins with a 2D shadow matte from a 3D object, i.e. the Fisherman. Once composited, the model is made to simulate the optical effect of super-imposed glazes in complimentary colors. The final effect is varied over time to give the images a hand-painted, wet and vibrant look.

Fishing ist der erste Kurzfilm, der zur Gänze in einer von PDI entwickelten Aquarell-Simulation gerendert wurde. Die als „flaches computergeneriertes Aquarell" bezeichnete Technik stützt sich auf Prozesse, die von Cassidy Curtis – dem technischen Leiter bei *Fishing* – verbessert wurden und geht vom 2D-Schattenmodell eines 3D-Objekts – etwa des Fischers – aus. Nach Abschluss der Bildkomposition werden im Modell die optischen Effekte übergelegter Farbschattierungen in Komplementärfarben simuliert, deren endgültiger Effekt im Laufe der Zeit variiert wird, um den Bildern einen feuchten, handgemalten und

When the story in PDI's latest short film featured a "tsunami of fish," effects animators went to work using Nick Foster's Academy Award winning fluid dynamics simulation system. They were able to adjust the parameters of the simulation to feel heavier and denser than any prior water simulation. Combined with thousands of fish models and a watercolor post process, the "tsunami" or "fluid dynamics sequence" is transformed into a tidal wave of murderous wiggling fish.

The story of *Fishing* is simple and colorful and supported by a similar visual style. Gainey kept the animation, character design and final look as "loose and gestural" as possible. The motion animation is broad with extreme holds and poses that accentuate the character's features, i.e. big hands, big feet, heavy

vibrierenden Ausdruck zu verleihen.
Als klar war, dass die Handlung in PDIs neuestem Kurzfilm nach einem „Tsunami von Fischen" verlangte, griffen die Effektanimatoren nach Nick Fosters (mit einem Oscar ausgezeichneten) „Fluid Dynamics"-Simulationssystem. Dieses System erlaubt es, die Simulationsparameter so einzustellen, dass das Ergebnis schwerer und dichter wirkt als je eine Wassersimulation zuvor. In Kombination mit Tausenden von Fischmodellen und einer Aquarell-Nachbarbeitung verwandelte sich die „Tsunami-" oder „Flüssigkeitsdynamik-Sequenz" in eine Flutwelle aus mörderisch schwänzelnden Fischen.
Die Geschichte von *Fishing* ist simpel und farbenfroh und wird durch einen entsprechenden visuellen Stil getragen. Gainey bemühte sich, Animation, Charaktere und Gesamteindruck so „locker und gestisch" zu halten wie möglich. Die Bewegungsanimation ist breit angelegt, mit extremen Haltezeiten und Posen, um hervorstechende Eigenschaften der Figuren – große Hände, große Füße, schwerer Bauch – zu

stomach. The watercolor CG process is adjusted throughout the film to ensure that subtlety is not lost in the abstract paint style. Finally, the colors are animated to reflect the changing time in day as well as the mood of the Fisherman. Monet's "Cathedral" paintings were inspiration for the changes in the color palette as the day progresses from dawn to dusk.

unterstreichen. Der Aquarell-Effekt wird über die ganze Animation hinweg stets nachgeregelt, um sicherzustellen, dass die Feinheiten nicht im abstrakten Malstil untergehen. Abschließend wurden noch die Farben animiert, um sowohl dem Tagesablauf als auch der Stimmung des Fischers Ausdruck zu verleihen. Monets Kathedralenbilder dienten als Inspiration für die Veränderungen der Farbpalette im Verlauf des Tages zwischen Morgendämmerung und Abendrot.

DAVID GAINEY (USA) ATTENDED THE RICE UNIVERSITY SCHOOL OF ARCHITECTURE, INTERNED AT LUCASFILM THX, AND STUDIED AT THE YALE SCHOOL OF DRAMA BEFORE JOINING PDI. AS PART OF PDI'S CHARACTER GROUP, DAVID ANIMATED ON ANTZ AND DIRECTED AND ANIMATED THE SHORT FILM FISHING, WHICH PREMIERED AT THE ELECTRONIC THEATER 1999. DAVID IS CURRENTLY DEVELOPING HIS OWN PROJECTS AND WORKING ON A FREELANCE BASIS. ▬▬ David Gainey (USA) besuchte die School of Architecture der Rice University, war Praktikant bei Lucasfilm THX und studierte an der Yale School of Drama, bevor er sich PDI anschloss. Als Teil der Character Group von PDI animierte David bei Antz, danach leitete und animierte er den Kurzfilm Fishing, der beim Electronic Theater 1999 der Öffentlichkeit vorgestellt wurde. Derzeit entwickelt David seine

PLUME
Cécile Gonard

Plume is a short animated film that draws its inspiration from the collection of poems, Plume, by Henri Michaux. Without taking any narrative principles into consideration, I have taken excerpts from poems and interpreted them graphically in a very personal way. A poetic universe is created, based on elements related to writing and what is written—books, pages, pens, lines, words. Here the monitor may be seen as both a page and a theater stage, where persons come and go, taking a walk and meeting each other in the most absurd situations.

The sound element increases a sense of alienation and a feeling of displacement, which is already present in the image, without falling into redundancy and synchronism.

Technically the film attempts to coherently unite 2D and 3D animations.

Plume ist ein kurzer animierter Film, der seine Inspiration aus der Gedichtesammlung Plume von Henri Michaux bezieht. Ohne auf irgendwelche narrativen Prinzipien zu achten, stütze ich mich auf einige Auszüge von Gedichten und interpretiere sie grafisch auf sehr persönliche Weise.

Ausgehend von Elementen, die sich auf das Geschriebene und die Schrift beziehen – Bücher, Blätter, Federn, Striche, Worte – entsteht ein poetisches Universum, in dem der Monitor sowohl als Seite als auch Theaterbühne gesehen wird, wo die Personen kommen und gehen, spazieren gehen und einander in den absurdesten Situationen begegnen. Die klangliche Seite verstärkt die Fremdheit und das Gefühl des Verschobenseins, das schon im Bild präsent ist, ohne in Redundanz und Synchronismus zu verfallen.

Technisch versucht der Film auf kohärente Weise 2D und 3D-Animationen zu verbinden.

CÉCILE GONARD (F) RECEIVED A EUROPEAN MEDIA MASTER OF ART (EMMA) IN 2D/3D ANIMATION AT THE LABORATOIRE D'IM-AGERIE NUMÉRIQUE (LIN) AT THE CENTRE NATIONAL DE LA BANDE DESSINÉE ET DE L'IMAGE (CNBDI) IN ANGOULÊME. THE FILM *PLUME*, WHICH SHE CREATED AS HER FINAL DEGREE REQUIREMENT, WAS NOMINATED FOR SEVERAL FESTIVALS, INCLUDING THE FESTIVAL D'ANIMATION D'ANNECY. CÉCILE IS CURRENTLY ARTIST IN RESIDENCE AT LIN (CNBDI). ▬▬ Cécile Gonard (F) hat einen European Media Master of Art (EMMA) in 2D/3D-Animation am Laboratoire d'Imagerie Numérique (LIN) im Centre National de la Bande Dessinée et de l'Image (CNBDI) in Angoulême erhalten. Ihr Abschlussfilm, *Plume*, wurde für mehrere Festivals nominiert, u. a. für das Festival d'animation d'Annecy. Cécile ist zurzeit Artiste in Residence bei LIN (CNBDI).

AU LOUP
Jean Hemez / Sébastien Rey

This film deals with a pedophile crime. The story is based on Grimm's "Little Red Riding Hood". Using this fairy tale has made it easier to treat the purpose of the film. The tale includes the principal message of the film, which warns children about the dangers that may threaten them.

We chose to take a humorous approach first. But then, the atmosphere gradually becomes very worrying. The film ends very tragically. We decided to adopt a child-like aesthetic, similar to children's books or to puppet shows. This choice reinforces the difference between the first part of the film (animation) and the end (video).

Au loup! is a graduation film. It is a 3D computer animation and video film. The scenario and the storyboard were written during the first year of study, and the film was made during the second year. This animation film is made with 3DSmax and After Effects software. We worked for five months on the design first and the rendering, and then four months on production (animation, rendering, shooting, compositing and editing (sound and image)).

Dieser Film beschäftigt sich mit Sexualverbrechen an Kindern. Die Geschichte basiert auf dem Grimmschen „Rotkäppchen". Die Verwendung des Märchens erleichtert die Behandlung des Filmzwecks sehr, da das Märchen selbst die Hauptbotschaft schon enthält – Kinder vor den ihnen möglicherweise drohenden Gefahren zu warnen.

Wir haben einen zunächst leicht humoristischen Ansatz gewählt. Aber mit der Entwicklung der Geschichte wird auch die Atmosphäre beklemmender. Der Film endet tragisch. Wir haben bewusst eine kindliche Ästhetik gewählt, die an Bilderbücher oder Puppentheater erinnert. Dies unterstreicht den Kontrast zwischen dem ersten Teil des Films (Animation) und dem Ende (Video).

Au Loup! ist eine Diplomarbeit, die sowohl 3D-Computeranimation wie Video umfasst. Das Szenario und das Drehbuch wurden im ersten Studienjahr verfasst, die Realisierung erfolgte im zweiten. Die Animation entstand mit 3DSmax und After Effects-Software. Zunächst wurden fünf Monate für Design und Rendering, danach vier weitere auf die Produktion (Animation, Rendering, Compositing und Schnitt von Ton und Bild) verwandt.

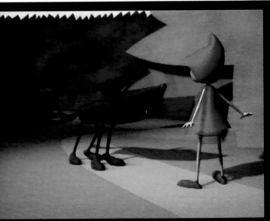

Sébastien Rey decided to do a 3 d training in Supinfocom (Valenciennes) after a drawing and fine art school. He now works for both Mac Guff line and Buf Compagnie in Paris and also directs his own productions. Jean Hemez joined Supinfocom after having studied arts and movies. He is now a 3D animator at Xilam Animation (Paris) and is still continuing his creative projects. ▬▬ Sébastien Rey hat sich nach Abschluß einer Zeichen- und Kunstschule zu einer Ausbildung in 3D am Supinfocom (Valenciennes) entschlossen. Jetzt ist er sowohl für MacGuff als auch für die Buf Compagnie in Paris tätig und führt bei seinen eigenen Produktionen Regie. Jean Hemez trat nach einem Film- und Kunststudium ins Supinfocom ein. Jetzt ist er 3D-Animator bei Xilam Animation (Paris) und arbeitet weiter an

UNTERWERK

Dariusz Krzeczek

superimposed horizontal and vertical patterns in dark, saturated colors cross the screen in pleasantly slow tempos. the starting material for these abstract, floating image impressions were—contrary to the contemplative effect of the video—recordings of speed. Krzeczek achieved the effect of unfocused movement by filming exclusively underground from moving subway trains. in this way, the eye of the digital video camera was no longer able to supply clear recordings.

in Unterwerk the "nihilism of speed" (paul virilio) is stylized into a formal, aesthetic experience. the abstraction is achieved by purposely crossing the boundaries of what can be filmed (or perceived). whereas the multilayered combinations of fades and superimpositions continuously result in new image patterns at the visual level, the soundtrack is formed by a monotone loop. this sound—a loop consisting of unprocessed noise from a ventilating system lasting only a few seconds—attains clearly musical qualities through its constant repetition. one is forced to listen more closely to this "acoustic waste product" and has an impression of beginning to notice minimal variations. yet these are not present on the tape, they are solely the result of the reception situation.

in this way, props from a real urban environment are used to create a clearly structured, abstract scenario, which, in addition to its convincing aesthetic quality, offers a penetrating reflection on everyday schemata of perception. (norbert pfaffenbichler)

übereinandergelagerte horizontal- und vertikalmuster in dunklen, satten farben überziehen in angenehm langsamen tempi den bildschirm. ausgangsmaterial dieser abstrakt schwebenden bildeindrücke waren – im gegensatz zur kontemplativen wirkung des videos – aufzeichnungen von geschwindigkeit. Krzeczek erzielt den effekt der bewegungsunschärfe dadurch, dass er ausschließlich unterirdisch aus fahrenden u-bahnzügen filmte, wodurch das digitale videokameraauge nicht mehr in der lage war, klare aufnahmen zu liefern.

der „nihilismus der geschwindigkeit" (paul virilio) wird in Unterwerk zum formal-ästhetischen erlebnis stilisiert. die abstraktion wird durch die willentliche überschreitung der grenzen des aufzeichenbaren (bzw. wahrnehmbaren) erzielt. während sich auf der visuellen ebene auf grund der vielschichtig kombinierten blenden und überlagerungen ständig neue bildmuster ergeben, wird die tonspur von einer monotonen schleife gebildet. dieses geräusch – ein unbearbeiteter, nur wenige sekunden dauernder loop des rauschens einer belüftungsanlage – bekommt durch die ständige repetition durchaus musikalische qualitäten. man ist gezwungen, genauer auf dieses „akustische abfallprodukt" zu hören, und vermeint, minimale variationen zu bemerken. diese kommen jedoch keineswegs vom band, sondern entstehen ausschließlich in der rezeptionssituation.

aus versatzstücken der real-urbanen umwelt wird so ein klar strukturiertes, abstraktes szenario kreiert, das neben seiner überzeugenden ästhetischen qualität eine eindringliche reflexion über alltägliche wahrnehmungsschemata bietet. (norbert pfaffenbichler)

Dariusz Krzeczek (A), born 1971 in Cracow/Poland, emigrated to Austria in 1991, has been studying at the University of Applied Arts Vienna since 1996, master-class Peter Weibel / Karel Dudesek. ▬ Darius Krzeczek (A), geb. 1971 in Krakau/Polen, 1991 Auswanderung nach Österreich. Seit 1996 Studium an der Universität für Angewandte Kunst Wien, Meisterklasse für Visuelle Mediengestaltung bei Prof. Peter Weibel / Karel Dudesek.

SENTINELLES

Guy Lampron

It is dawn in a dreamlike, surrealistic city evoking New York. Its design is inspired by the twenties' Art Deco style. The city is constantly smothered under a heavily clouded sky. An appearance of the sun is a very rare occurence in this world. Our attention is drawn to a magnificent skyscraper, a replica of the Chrysler Building. A cornice is then revealed, to which two metallic eagle heads are attached, staring invariably in the distance with all the solemnity of living sentinels … The statues then take life, as they awake suddenly and exchange tender loving gestures with one another: they form a couple.

In the horizon, light just above them. As his partner weakens more and more under the "weight" of the light, the other eagle pulls out a metallic wing from the building's wall, thus creating an explosion of mortar and bricks. The eagle then spreds this newly discovered limb above its beloved partner, casting a protective shadow over it. The unscathed eagle attempts to extract its whole body from the buildings's structure but to no avail: to do so would simply go against its nature.

The eagle experiences growing difficulties in keeping its wing spread above its suffering partner: this new limb is too heavy. Exhausted, the protecting eagle must let its wing fall back. The light beam again strikes its partner now lying unconscious. The end is near, happening right before the spared eagle's eyes, as it gazes helplessly.

Some spirited little shadows start to move nervously among the steel beams left unconvered by the extraction of the eagle's wing. A metallic eaglet suddenly comes out and starts ascending towards the sky. Another follows it, then a third and then up to a dozen of them start to fly above the Chrysler Building, totally unaffected by the ray of light. The spared eagle is stunned by the discovery of this progney, as it emerges from the Chrysler Building.

Dämmerung in einer an New York erinnernden, traumähnlichen surrealistischen Stadt. Ihr Design is inspiriert vom Art-Déco-Stil der 20er-Jahre. Die Stadt liegt ständig unter einer dichten Wolkendecke – ein Sonnenstrahl ist in dieser Welt eine seltene Erscheinung. Unsere Aufmerksamkeit wird auf einen großartigen Wolkenkratzer gelenkt, eine Nachbildung de Chrysler Building. Aus einem umlaufenden Gesims blicken zwei metallene Adlerköpfe unablässig in die Ferne, feierlich wie lebende Wachtposten … Die Statuen erwachen plötzlich zum Leben und tauschen zärtliche Gesten aus: Sie sind ein Paar. Vom Horizont ein Lichtstrahl, genau über ihnen. Und als der eine Partner unter dem ungewohnten „Gewicht" des Sonnenlichts immer schwächer und schwächer wird, zieht der andere Adler plötzlich eine Schwinge aus der Wand des Gebäudes und verursacht damit eine Explosion aus Mörtel und Ziegeln. Der Adler hält seinen neu entdeckten Körperteil schützend über seinen Partner, um ihm Schatten zu spenden. Und er versucht sogar, seinen ganzen Körper aus der Fesselung durch das Mauerwerk zu befreien, aber umsonst – das würde einfach gegen seine Natur verstoßen. Immer schwerer fällt es dem Adler, seinen Flügel über dem leidenden Partner zu halten: Der neue Körperteil wiegt einfach zu viel. Erschöpft mus er den Flügel hängen lassen und wieder brennt das Licht auf seinen jetzt bewusstlosen Partner. Das Ende ist nah und hilflos muss der Adler dem Tod des anderen zusehen. Da beginnen sich einige lebhaften Schatten zwischen den Stahlträgern, die durch das Herausziehen des Flügels freigelegt worden sind, zu bewegen. Und plötzlich kommt ein kleiner metallener Adler heraus und beginnt, sich in den Himmel zu erheben. Bald folgt ein Zweiter, dann ein Dritter und zuletzt steigt ein ganzes Dutzend in den Himmel über dem Chrysler Building – unbeeindruckt und ungestört vom Sonnenlicht. Erstaunt über diese seine Nachkommenschaft blickt der überlebende Adler den Kleinen nach, wie sie nach oben entschweben.

GUY LAMPRON (CDN) HOLDS A BA IN FILM STUDIES FROM CONCORDIA UNIVERSITY IN MONTREAL. HE WORKED AS A 3D ANI-MATOR AT GROUPE IMAGES BUZZ DEVELOPING SPECIAL EFFECTS FOR FEATURE FILMS AND ADVERTISEMENTS. HE THEN BECAME A SENIOR LEVEL ANIMATOR AT BEHAVIOUR ENTERTAINMENT. SINCE JOINING PYGMÉE PRODUCTIONS, THE ENTERTAINMENT DIVISION OF CDMED, HE HAS FOCUSED EXCLUSIVELY ON THE DEVELOPMENT OF ANIMATION PROJECTS FOR FILM AND TELE-VISION. ■■■■ Guy Lampron (CDN) graduierte an der Concordia University in Montreal zum BA aus Filmstudien. Er hat als 3D-Animator bei der Groupe Images BUZZ gearbeitet, wo er Special Effects für Spielfilme und Werbespots entwickelt hat. Später wurde er leitender Animator bei Behaviour Entertainment. Seit er zu Pygmée Productions – der Unterhaltungsdivision von CDMED – gekommen ist, konzentriert er sich ausschließlich auf die Entwicklung von Animationsprojekten für Film

LOWRIDER CRAB

Charlotte Manning

These mutated crustaceans, known locally as "Los Crafty Cangrejos", are highly intelligent and have shed their legs in favor of wheels and a hydraulic suspension. To maintain their modifications, they need a continual supply of autoparts. They are constantly tinkering.

They live near bridges, where they forage for autoparts from the submerged wrecks of occasional unfortunate cars which have met an unexpected end.

In times of scarcity, they become more aggressive and are known to climb bridges and go looking for cars (fancy custom cars particularly attract their attention ...).

Software used in the project was Softimage 3D v. 3.8 for animation and mental ray v2.1 for rendering.

This animation employs the most recently developed features of mental ray 2.1, such as the full simulation of global illumination and of caustics to create a realistic lighting situation both above water and beneath.

It also makes extensive use of the mental ray shader interface with a collection of shaders written especially for this project, including: reaction-diffusion patterns, cellular basis textures, realistic carpaint /varnish, irridescence, scales, atmosphere, and a special "sparkle" shader for the crabs' shell to model paint or plastic containing metal shavings.

Diese mutierten Krustentiere – in ihrer Heimat bekannt als „Los Crafty Cangrejos" – sind hoch intelligent und haben ihre Beine zu Gunsten von Rädern und einem hydraulischen Fahrgestell abmontiert. Sie sind ständig am Basteln. Um ihre Umbauten in Schuss zu halten, sind sie auf kontinuierlichen Nachschub an Autoteilen angewiesen.

Sie leben in der Nähe von Brücken, stets auf der Jagd nach Ersatzteilen aus den untergegangenen Wracks jener Fahrzeuge, die – wie das ja manchmal vorkommt – einem unerwarteten Unglück zum Opfer gefallen sind.

In Zeiten der Not, wenn die Versorgungslage schlecht ist, werden sie aggressiver – es wird berichtet, dass sie dann auch Brücken erklettern und selbst auf die Suche nach Autos gehen (und aufgemotzte Muscle-Cars ziehen ihr Interesse in erhöhtem Maße auf sich ...)

Als Software wurde für die Animation Softimage3D v3.8 verwendet, für das Rendering mental ray v2.1. Für diese Animation wurden die neuesten Features von mental ray v2.1 eingesetzt, z. B. die umfassende Simulation globaler Beleuchtungseffekte und Kaustiken, um eine realistische Lichtsituation sowohl über als auch unter dem Wasser zu schaffen.

Ausgiebiger Gebrauch wird auch vom mental ray Shader Interface gemacht. Wir haben extra für dieses Projekt eine ganze Reihe von Shadern geschrieben, darunter Reaktions-Diffusions-Muster, zelluläre Basistexturen, realistische Lacke und halbtransparente Beschichtungen, Irisieren, Schuppen, Atmosphäre und einen speziellen „Funkel"-Shader für die Krabbenpanzer, mit dem wir Farb- oder Kunststoffschichten mit eingeschlossenen Metallpartikeln simulieren können.

Charlotte Manning (USA), born on 15 November 1968 in Cleveland/Ohio. BA in Neurobiology/Physiology, Northwestern University of Chicago, 1991. Dec 1994—Dec. 1999: Technical Director, mental images, Berlin. Since Jan. 2000: Techn. Director, Digital Domain., Venice/CA. ■■■ Charlotte Manning (USA), geb. 1968 in Cleveland, Ohio. B.A. aus Neurobiologie und Physiologie der Northwestern University of Chicago 1991; Dezember 1994 bis Dezember 1999: Technische Leiterin bei Mental Images, Berlin; seit Januar 2000 technische Leiterin bei Digital Domain, Venice.

LES DEPOSSEDES

Juliette Marchand

The film *Les dépossédés* combines 3D and 2D techniques. The cold, smooth character of 3D is mitigated by borrowing from paper cutting techniques. The plot involves a couple and quite comically describes how the man takes possession of the woman and the woman of the man in very different ways.

Der Film *Les dépossédés* mischt 3D- und 2D-Techniken. Dank Anleihen an der Scherenschnitttechnik wird der kalte, glatte Charakter des 3D gemildert. Die Geschichte des Films handelt von einem Paar und beschreibt auf recht komische Weise, wie der Mann die Frau und die Frau den Mann auf die unterschiedlichsten Weisen in Besitz nimmt.

JULIETTE MARCHAND (F), BORN 1972. 1993 ATELIER PRÉPARATOIRE PENNIGHEN. 1994 BEGAN STUDIES AT ENSAD. 1997 DEGREE IN VIDEO FROM ENSAD. 1999 POSTGRADUATE DEGREE FROM ENSAD (COMPUTER GRAPHICS AND SPECIAL EFFECTS). ▬
Juliette Marchand (F), geb. 1972. 1993 Atelier Préparatoire Pennighen. 1994 Beginn des Studiums an der ENSAD. 1997 Diplom in Video an der ENSAD. 1999 Postgraduate Diplom an der ENSAD (Computergrafik und Special Effects).

USE A DOODLE ON THE NOODLE

Timm Osterhold / Max Zimmermann

the story

a new toon-day in doodleland. the sun comes up, not really actually.

nevertheless, our voyeur wakes up and discovers the flower FLOUA. a charming, delightful creature, equipped with a very feminine aura. as soon as she catches a glimpse of our P*man in the distance, she hatches a plan.

the innocent P*man comes wandering along and just as he wants to play with the butterflies, FLOUA grabs the little P*man and goes heedlessly right between his legs. a miracle seems to happen: there, where there was not even a bump before, suddenly a giant P(enis) shoots out. we see: the world's first award-winning toon3D porno sequence! sperm flies through space, races past the sun! nearly collides with a satellite, heading for our voyeur. the voyeur: it is a live "M"! runs to the "CONDOMAT", a robot for putting on a condom fast, sperm splatters onto "M"—and it is protected.

background

we wanted to produce a pro-condom shot that is not preachy about when and how to use a condom. kids today already know that!

we wanted to motivate people to use a condom with a new kind of wacky and chaotic story. and to do it with humor. because if you think about it: putting a rubber thing like that on your prick is actually very weird.

zum ablauf

einer neuer toontag im doodleland. die sonne geht auf, eigentlich nicht wirklich.

trotzdem wacht auch unser voyeur auf und entdeckt die blume FLOUA. ein reizendes, wunderschönes geschöpf, ausgestattet mit einer sehr weiblichen ausstrahlung. als sie unseren P*mann schon aus der ferne erspäht, hat sie bereits einen plan.

der unschuldige P*mann läuft seines weges und als er mit schmetterlingen spielen will, schnappt sich FLOUA den kleinen P*mann und geht ihm rücksichts-los zwischen die beine. ein wunder scheint zu pas-sieren: da wo noch nicht mal eine beule war, schießt ein riesiger P(immel) raus. wir sehen: die erste awardwinning-toon3D-pornosequenz der welt! sperma fliegt durchs all, rast an der sonne vorbei! trifft beinahe einen satelliten, auf unseren voyeur zu. der voyeur: es ist ein lebendes „M"! rennt zum „CON-DOMAT", ein robot zum kondom-schnell-überzug, das sperma klatscht auf das „M" – und es ist geschützt.

hintergrund

wir wollten einen pro-kondom-shot produzieren, der nicht mit dem zeigefinger darstellt, wann und wie man kondome benutzt. die jungs von heute wissen das!

wir wollten auf eine neue art mit einer total über-drehten und chaotischen story leute zur kondom-nutzung bringen. und das mit humor. denn wenn man sich überlegt: so ein gummiding über seinen löres zu ziehen, ist eigentlich eine verdammt komi-sche angelegenheit.

TIMM OSTERHOLD (D), BORN 1972 ON PLANET EARTH. AFTER SECONDARY SCHOOL TRAINED FOR TWO YEARS AS A MULTIMEDIA DESIGNER. SUBSEQUENTLY WORKED FOR THREE YEARS IN MUNICH AS ANIMATOR AND SUPERVISOR FOR THE 3D AND SPECIAL EFFECTS COMPANY SILVERHAZE FRANKFURT. SEPTEMBER 1998 FOUNDED THE COMPANY 58, FOR WHICH HE HAS SINCE THEN BEEN MANAGING PARTNER AND CHARACTER ANIMATOR. MAX ZIMMERMANN (D). BORN 1974 ON PLANET MARS. FOLLOWING GRADUATION FROM SECONDARY SCHOOL, SLEPT FOR HALF A YEAR AND THEN WORKED THREE YEARS FOR THE COMPANY MEIRÉ & MEIRÉ AS ILLUSTRATOR AND GRAPHICS ARTIST. WORKED ON THE SIDE AS FREELANCE ILLUSTRATOR AND DESIGNER FOR AGENCIES THROUGHOUT GERMANY. EMPLOYED BY SILVERHAZE FOR ONE YEAR, MOSTLY IN THE AREA OF STORYBOARD, CONCEPTION, DESIGN AND 3D MODELLING; THEN FOUNDED THE COMPANY FIFTYEIGHT3D WITH TIMM OSTERHOLD, MARC ECKART. ■■■ timm osterhold (D), geb. 1972 planet erde. nach dem abi zweijährige ausbildung als multi-media-designer. in münchen danach dreijährige berufserfahrung als animator und supervisor bei 3D und specialeffect-firma silverhaze frankfurt. september 1998 gründung der firma 58, bei der er seitdem als geschäftsführender gesellschafter und charakter animator tätig ist. max zimmermann (D). geb. 1974 planet mars. nach dem abi ein halbes jahr schlafen, daraufhin drei jahre bei der firma meiré und meiré gearbeitet als illustrator und grafiker. nebenher als freier illustrator und designer für agenturen in ganz deutschland tätig. ein jahr festanstellung bei silverhaze, vorwiegend im bereich storyboard, konzeption, design und 3D modelling. daraufhin gründung der firma FIFTYEIGHT3D mit timm osterhold, marc eckart.

LOOP
Makoto Sugawara

A man is peacefully working in a small room on the creation of his new robot. Suddenly the room starts to warp and tries to squeeze him. The man immediately takes the robot out the door and commands the robot to re-shape the room back to normal. However, the robot does not follow his command and instead starts creating his own new robot. A very mysterious and ominous story which illustrates a very distinct character.

Ein Mann arbeitet friedlich in einem kleinen Raum an der Konstruktion seines neuen Roboters. Plötzlich beginnt der Raum sich zu verschieben und versucht, ihn zu zerdrücken. Der Mann nimmt den Roboter, stellt ihn vor die Türe und trägt ihm auf, das Zimmer wieder in den Normalzustand zurückzuführen. Der Roboter aber kümmert sich nicht darum, sondern beginnt, seinen eigenen neuen Roboter zu bauen ... Eine sehr mysteriöse und eigenwillige Geschichte, die einen ausgeprägten Charakter illustriert.

MAKOTO SUGAWARA (J) STUDIED PHYSICS AND STARTED WORKING FULL-TIME AT LINKS, CO., ONE OF THE LARGEST COMPUTER GRAPHICS STUDIOS IN JAPAN, SOON AFTER GRADUATION. HE BECAME INDEPENDENT AFTER WORKING THERE FOR FIVE YEARS AS AN ARTIST ON MANY TV COMMERCIALS, FEATURING HIS OWN UNIQUE, ORIGINAL CHARACTERS. CURRENTLY, HE WORKS AT THE DIGITAL ENGINE LABORATORY (AFFILIATED WITH BANDAI) CLOSELY UNDER THE SUPERVISION OF MAMORU OSHII (DIRECTOR OF *GHOST IN THE SHELL*). ▬▬▬ Makoto Sugawara (J) studierte Physik und begann kurz nach Studienabschluss bei Links Co., einem der größten Computergrafikstudios Japans, zu arbeiten. Nach fünfjähriger Tätigkeit machte er sich als Künstler selbstständig und produzierte zahlreiche TV-Werbespots, alle mit seinen einzigartigen originellen Figuren. Derzeit arbeitet er unter der Leitung von Mamoru Oshii (dem Regisseur von *Ghost in the Shell*) bei Digital Engine Laboratory (einer

MAAZ
Christian Volckman

Christian Volckman has entered the "painterly phase" of his film. Let us call to mind that this involves a unique fiction project, the ninety scenes of which were created entirely with *Flame*, whereby the individual pictures were subsequently painted over by the artist and a friend. In his words, this procedure has become quite simple today: "We export the individual pictures to the hard disk as TIFFs and use JAZ to transport them home, where Lionel and I retouch them. This enables us to maintain the same resolution. After processing, I put them all back on the *Flame* computer, it is all transferred to a Digital Betacam and finally—after montage—brought to Eclair, in order to transfer it back to film."

Before they reached this point, though, Christian Volckman and Lionel Richerand faced a full month of retouching work. Lionel, who will soon have completed an animation himself, cleverly chose the same path as his partner. According to Aton Soumache, the producer of the film, "it was not really easy to find someone, who could paint together with Christian—it is ultimately a very personal work. You have to be able to feel the movement, and Lionel understood that very well." Lionel has been working on the project for several months. Many tests were made involving both artistic and technical aspects. "We went on step by step," explains Lionel. "First we wanted to see whether the pictures were even suitable for a high resolution. A projection showed us that the transition could be made easily—so well that the retouching was not even visible in the scenes we had painted over. So we decided to take an even more radical course, deforming the persons even more. Since then,

Christian Volckman ist in die „malerische Phase" seines Films eingetreten. Erinnern wir uns daran, dass es sich um ein einzigartiges Fiktionsprojekt handelt, dessen 90 Szenen zur Gänze auf Flame geschaffen wurden, wobei die Einzelbilder anschließend vom Künstler und einem Freund übermalt wurden. Seinen Worten nach ist diese Prozedur heute ganz einfach: „Aus Flame exportieren wir die Einzelbilder als TIFFs auf die Festplatte, transportieren sie über JAZ nach Hause, wo Lionel und ich sie retouchieren. Das erlaubt uns, die gleiche Auflösung zu wahren. Nach der Bearbeitung bringe ich das alles zurück auf den Flame-Rechner; das Ganze wird auf eine Digital Betacam übertragen und schließlich – nach der Montage – zu Eclair gebracht, wo alles wieder auf Film übertragen wird."

Bevor es aber so weit war, wartete ein guter Monat voller Retouchierarbeiten auf Christian Volckman

Ebene erdacht, aber wenn dann einmal der Film als Ganzes abgespult wird, dann „muss der Zuschauer wissen, was er zu sehen hat – wenn es überall irgendwas zu sehen gäbe, würde das vom eigentlichen Film ablenken. Man muss also an den richtigen Stellen malen können und Christian hat diese Entscheidungen nach dem Schnitt treffen könnten", meint Aton.

Die Malerei wird also eingesetzt, weil und soweit sie die Emotionen unterstützt. *Maaz* ist gerade in den dramatischsten Momenten auch am stärksten nachbearbeitet. Christian arbeitet momentan an einem Bild, in dem er versucht, zwei Maaz zu morphen. Die Person verdoppelt sich. Um dies zu erreichen, arbeitet der Künstler auf einem beschränkten Hintergrund, um die Illusion der Teilung der Person zu verstärken – eine Szene, die die Panik der Gestalt transportieren soll.

Die Übermalung verstärkt den irrealen Aspekt des Films, aber – wie Aton erklärt – „würde man alles übermalen, so fände man sich ganz schnell erst wieder in einer Computeranimation wieder. Und das hatten wir nicht vor. Es ging darum, die Balance zu finden zwischen dem Realen und dem Imaginären, wir mussten also eine Hierarchie der Retouchen festlegen."
(*Emanuelle Achard*)

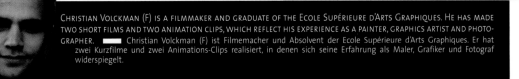

CHRISTIAN VOLCKMAN (F) IS A FILMMAKER AND GRADUATE OF THE ECOLE SUPÉRIEURE D'ARTS GRAPHIQUES. HE HAS MADE TWO SHORT FILMS AND TWO ANIMATION CLIPS, WHICH REFLECT HIS EXPERIENCE AS A PAINTER, GRAPHICS ARTIST AND PHOTOGRAPHER. ▬▬ Christian Volckman (F) ist Filmemacher und Absolvent der Ecole Supérieure d'Arts Graphiques. Er hat zwei Kurzfilme und zwei Animations-Clips realisiert, in denen sich seine Erfahrung als Maler, Grafiker und Fotograf widerspiegelt.

FIGHT CLUB
Pierre Buffin / BUF

Sex Sequence (5 shots)

This sequence shows us the mental picture that Jack (Edward Norton) imagines of the couple intertwined on the bed. The director, David Fincher, required that the action be in fluid movement and the actors just barely recognizable. The sequence was technically challenging because it involved recreating camera movements around the couple, Helena Bonham Carter and Brad Pitt. The innovative technique that BUF developed to create this sequence makes it possible to infinitely control the movements of the camera, framing, focus and of the degree of streak/blur effect vs. focus with freedom from physical limitations.

BUF created two tests to convince Fincher of the feasibility of this technique. Because the room where this scene took place was relatively small, it was not possible to film directly in the environment. The bed and the couple were filmed in studio, which made it possible to comfortably install the lights and cameras. The room was then entirely recreated in 3D, using precise plans and measurements. Still photographs of the bedroom allowed BUF to reapply all the textures on the 3D geometry. By reconstituting the room in 3D, it allowed the director to use a camera motion on the room that was slightly different from the camera on the couple, thus accentuating the vertigo effect.

During filming, the synchronization of the cameras was carefully monitored. This involved the synchronization of two camera groups. Working from an animatic directed by Fincher and executed by Pixel Liberation Front, BUF determined the ideal camera positions.

Sex-Sequenz (fünf Einstellungen)

Diese Sequenz zeigt uns das geistige Bild, das Jack (Edward Norton) sich von dem sich im Bett umschlungen haltenden Paar macht. Der Regisseur David Fincher wollte, dass die Handlung in einer flüssigen Bewegung ablaufen und die Schauspieler nur vage erkennbar sein sollten. Dies war technisch anspruchsvoll, weil wir Kamerabewegungen rund um das Paar – Helena Bonham Carter und Brad Pitt – nachempfinden mussten. Die von BUF für diese Sequenz entwickelte innovative Technik erlaubt es, die Bewegung der Kamera unendlich fein zu steuern und Bildausschnitte, Tiefenschärfe und Verwischungs-/Bewegungseffekte völlig frei von physikalischen Einschränkungen festzulegen.

Um Fincher von der Machbarkeit zu überzeugen, wurden zwei Testeinstellungen konstruiert. Wegen der Enge des Aufnahmeorts dieser Szene war es unmöglich, direkt vor Ort zu filmen. Das Bett und das Paar wurden deshalb im Studio aufgenommen, wo eine bequeme Installation aller Lichter und Kameras möglich war. Dann wurde der gesamte umgebende Raum auf der Basis einer genauen Vermaßung in 3D nachgebildet. Mit Hilfe von Standfotos konnten alle Texturen des Raums auf die 3D-Geometrie abgebildet werden. Der virtuelle 3D-Raum erlaubte dem Regisseur, Kamerabewegungen im Raum einzusetzen, die sich leicht von den Einstellungen auf das Paar unterscheiden, was den Schwindeleffekt verstärkt.

Während der Filmaufnahmen wurde die Synchronizität der Kameras genau überwacht, was die Synchronisation von zwei Kameragruppen erforderte. Ausgehend von einem *Animatic*-Modell (Regie: Fincher, Ausführung Pixel Liberation Front) legte BUF dann die idealen Kamerapositionen fest.

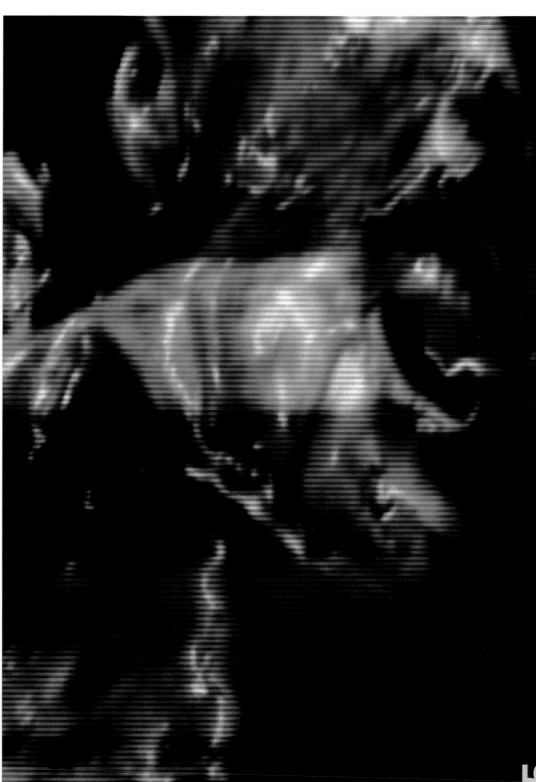

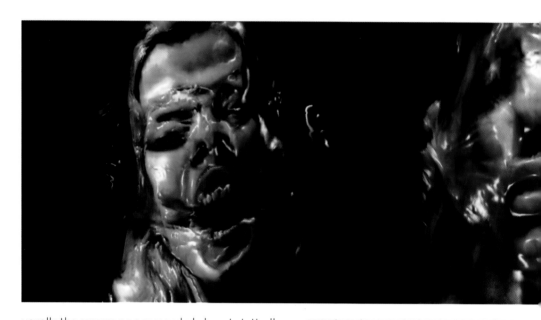

usually the camera was suspended almost statically over the bodies (the camera movements were created later in the 3D program). Using 3D software, 3D objects were created from the video material and distorted with specially developed shaders that use a combination of displacement, bump mapping and light warping by pressing in shadow surfaces and pulling out points of light. Because of the virtual objects' dependency on the light of the real situation, the body form mutates constantly in 3D space and creates a bizarre scenario. One problem with this procedure was finding the right light technique, because the rendered material did not become visible until weeks after the test shots (many, many months ...), and it is hardly possible to tell from the original material how each light situation will affect the emerging body form. Another problem was the complexity of the shaders that interwove the video material through 8 to 14 layers and required constantly changing the parameters. To achieve a realistic effect, it was also necessary for the main light source in the 3D software to follow the same motion curve as the real light source. All the other visual effects were also created with 3D software to generate a unique 3D body world with a homogeneous effect.

meist beinahe statisch über den Körpern (die Kamerabewegungen wurden erst im 3D-Programm erzeugt). Aus dem Videomaterial wurden dann in der 3D-Software (Alias Power Animator) 3D-Objekte durch eigens entwickelte Shader, die eine Kombination aus Displacement, Bumpmapping und Lightwarping verwenden, erzeugt und verzerrt, indem Schattenflächen hineingedrückt und Lichtpunkte herausgezogen wurden. Durch die Abhängigkeit der virtuellen Objekte vom Licht der Realsituation mutiert die Körperform stetig im 3D-Raum und erzeugt ein bizarres Szenario. Die Schwierigkeit bei dieser Vorgangsweise bestand zum einen in der Findung der richtigen Lichttechnik, da das gerenderte Material meist erst Wochen nach den Testshots sichtbar wurde (viele, viele Monate ...) und es am Ausgangsmaterial kaum erkenntlich ist, wie sich die jeweilige Lichtsituation auf die entstehende Körperform auswirkt. Zum anderen lag sie in der Komplexität der Shader, die das Videomaterial über 8 bis 14 Layer ineinander verflechten und ständige Parameteränderungen verlangen. Um eine realistische Wirkung erzielen zu können, musste außerdem die Hauptlichtquelle in der 3D-Software die gleichen Bewegungskurven wie die reale Lichtquelle vollführen. Alle weiteren visuellen Effekte wurden ebenfalls in der 3D-Software kreiert, um eine eigene, homogen wirkende 3D-Körperwelt zu erzeugen.

Markus Degen (A), born 1973, studied visual media design at the University for Applied Art. 1996 International Video Art Prize - Talent Promotion Award with *lightformed*. 1996—98 exhibitions in Austria, Germany, Spain, Australia and France. 1996—98 projects in advertising, web design and music video. 1999 diploma for Visual Media Design with honors. ▬ Markus Degen (A), geb. 1973. Studium Visuelle Mediengestaltung an der Universität für angewandte Kunst. 1996 Internationaler Videokunstpreis-Förderpreis für *lightformed*. 1996–98 Ausstellungen in Österreich, Deutschland, Spanien, Australien und Frankreich. 1996–98 Projekte in Werbung, Webdesign und Musikvideo.

CHICKEN SUPPER

Zach Bell / Chris Gallagher / Steven Schweickart / Scott Smith

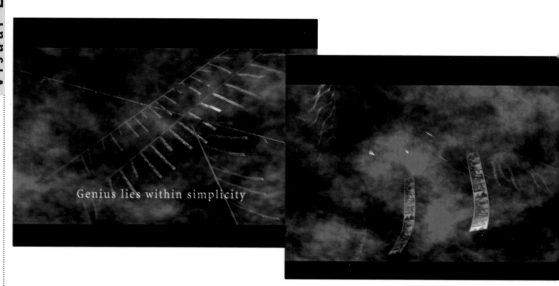

Genius lies within simplicity

The first step in the creating, *Chicken Supper*, was to travel to Tampa, Florida and capture footage of Sean Schweickart. Sean is a thirty seven year old man with Down syndrome. We chose Sean as the leading character for the piece, because he has proven to be a great inspiration in both his nephew, Steven, who is a member of this project, and those he encounters by the way he lives his life. Amidst all his disavantages he still remains positive and finds joy in things the common person would find to be trivial. This makes him the perfect candidate. While visiting with Sean we set up a series of interviews.

This piece, as it stands now, is a work in progress. In the final version we will have slower moving, more realistic clouds and animated texture maps along each line of film. We will also add the final scenes that we had to leave out due to running out of time. The idea of this piece is that you, the viewer, are floating to these vast levels of hanging film and learning about the life of this particular man. Each filmstrip represents a period of Sean's life. As you get closer, the once still image of the frame comes to life and plays out a video sequence of that particular event. In some instances the viewer will pass directly through the frame of film like a ghost passing through a wall. Ascending higher and higher we gaze

Der erste Schritt bei der Produktion von *Chicken Supper* war eine Reise nach Tampa, Florida, um dort Aufnahmen von Sean Schweikart zu machen. Sean ist ein 37-jähriger Mann, der am Down-Syndrom leidet. Wir haben Sean als die Hauptfigur des Werks ausgesucht, weil er sich als Inspirationsquelle sowohl für seinen Neffen Steven erwiesen hat, der Mitarbeiter dieses Projekts ist, als auch für alle anderen, die er auf seinem Lebensweg getroffen hat. Trotz seiner Behinderungen bleibt er immer positiv eingestellt und findet Freude an Dingen, die anderen vielleicht trivial erscheinen. Dies machte ihn zum perfekten Kandidaten. Während unseres Besuchs bei Sean haben wir eine Reihe von Interviews mit ihm gemacht.

Das Stück ist im Moment Work-in-progress. In der endgültigen Fassung wird es in allen Sequenzen des Films sich langsamer bewegende und realistischere Wolken und Texturen geben. Wir werden auch noch jene Szenen einfügen, die wir bisher aus Zeitmangel nicht einbauen konnten. Die Grundvorstellung dieses Films ist, dass der Besucher sozusagen durch viele Ebenen von hängenden Filmstreifen schweben kann und dabei etwas über das Leben dieses speziellen Mannes lernt. Jeder Filmstreifen steht für einen Abschnitt in Seans Leben. Wenn man näher kommt, erwachen die Einzelstandbilder zum Leben und spielen eine Videosequenz über die jeweilige Lebensphase ab. In einigen Fällen wird der Betrachter direkt durch den Filmstreifen hindurchtauchen, wie ein Geist, der durch Wände geht. Indem wir immer höher hinaufsteigen, blicken wir auf Seans Leben und reflektieren zuletzt über unser eigenes. Eine

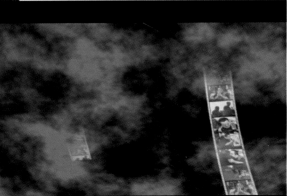

I like all the chicken suppers

I always watch my movies every day and every morning

upon Sean's life and in the end reflect on our own. Another way to view the piece is to take it from the understanding that Sean has passed away and is rising to heaven. In his ascension he reflects on his life. This is seen from the hanging film and the audio clips that have been mixed with the background music. Higher and higher he rises and he sees himself as one not with regrets, but full of inner peace and joy. Finally in the end he rises above the lines of film and looks down upon them one last time and then moves on. Please understand that Sean is still alive, but considering that the life expectancy of someone with Down syndrome is forty, he may not be with us much longer.

andere Interpretation des Films könnte von der Annahme ausgehen, dass Sean gestorben sei und zum Himmel auffährt. Während seiner Himmelfahrt reflektiert er sein Leben. Höher und höher steigt er und blickt auf sich selbst zurück, nicht mit Bedauern oder Reue, sondern voll innerem Frieden und Freude. Am Ende steigt er über die einzelnen Filmstreifen empor und blickt ein letztes Mal auf sie zurück, bevor er weiterschwebt. Man darf dabei allerdings nicht aus den Augen verlieren, dass Sean in Wirklichkeit noch am Leben ist, aber da die durchschnittliche Lebenserwartung von Patienten mit Down-Syndrom 40 Jahre beträgt, wird er möglicherweise nicht mehr lange bei uns sein.

ZACH BELL, CHRIS GALLAGHER, STEVEN SCHWEICKART, AND SCOTT SMITH ARE STUDENTS, CURRENTLY ENROLLED AT THE SAVANNAH COLLEGE OF ART AND DESIGN. CHRIS AND STEVEN ARE ATTENDING SCAD TO GAIN THEIR MFA. CHRIS GAINED HIS BFA IN GRAPHIC DESIGN FROM CENTRAL CONNECTICUT UNIVERSITY, WHILE STEVEN GRADUATED FROM THE UNIVERSITY OF CINCINNATI WITH A DEGREE IN MUSICAL THEATRE. ZACH AND SCOTT ARE UNDERGRADUATES AT SCAD. BOTH ZACH AND CHRIS ARE ON COURSE TO RECEIVE A DEGREE IN 3D MODELING AND ANIMATION. ■ Zach Bell, Chris Gallagher, Steven Schweickart und Scott Smith sind derzeit Studenten am Savannah College of Art and Design. Chris und Steven bereiten sich auf ihren Master of Fine Arts vor; Chris hatte zuvor den Bachelor in Graphics Design an der Central Connecticut University erworben, während Steven von der University of Cincinnaty mit einem Abschluss in Musiktheater abgegangen war. Zach und Scott bereiten sich auf ihren ersten akademischen Grad im Bereich 3D-Modellierung und Animation vor.

CHEMICAL BROTHERS

<u>Pierre Buffin / BUF</u>

This music video was created in honor of the 1970's (caleidoscope and mosaic effects) using in-camera effects rather than in-post. This video was shot using two different media, video for the real world and 35 mm for the effects that needed to be created in post-production. Our work was to create a transition and in-between to marry both worlds. Starting with one talent, our challenge was to multiply the talent, or starting with multiple talents and ending with just one. The last image of the dream sequence becomes the first image of the real world and vice versa. Frame after frame, talents and objects have to transform to become the resulting image. The challenging part of the work for the artists was in creating in-between frames. The artists created 23 transitions. Each transition was from 15 to 50 frames long with an average of 30. Technically the challenge was in matching film and video and creating in-between frames on images that were completely different.

Dieses Musikvideo entstand in Anlehnung an und zu Ehren der 70er-Jahre-Techniken (Kaleidoskop- und Mosaikeffekte), die eher auf Kameraeffekten denn auf Postproduktion basierten. Das Video wurde unter Verwendung zweier unterschiedlicher Medien gedreht: Video für die realen Bilder und 35-mm-Film für die Effekte, die im Schnitt nachbearbeitet werden mussten. Unsere Aufgabe war es, den Übergang und die Verschmelzung der beiden Welten zu schaffen. Dafür musste entweder eine Person vervielfacht oder mehrere Personen zu einer zusammengefasst werden. Das letzte Bild der Traumsequenz wird zum ersten der Wirklichkeit und umgekehrt. Kader um Kader müssen sich Personen und Objekte verwandeln, um zum fertigen Bild zu werden. Die Herausforderung an die Animatoren bestand eben darin, die dazwischen liegenden Kader zu erzeugen. Insgesamt wurden 23 Übergänge geschaffen, jeweils zwischen 15 und 50 Kader lang, durchschnittlich also 30 Kader pro Übergang. Die technische Herausforderung lag darin, Film und Video aneinander anzupassen und die Übergangskader zwischen den völlig verschiedenen Anfangs- und Endbildern zu kreieren.

BUF
COMPAGNIE

AFTER HAVING SEEN THE FILM "TRON", TWO FRIENDS, PIERRE BUFFIN AND HENRY SEYDOUX BOUGHT AN APPLE. THE VERY FIRST PC USED BY BSCA WAS A 8088 (WHICH BELONGED TO GEORGES LAGARDERE). WITH 640 KBYTES, A 8087 CO-PROCESSOR AND TWO 5-INCH FLOPPY DISKS, THEY SHIFTED RAPIDLY ONTO AN INTEL 286 MACHINE RUNNING ON INTEL UNIX AND DID THEIR FIRST ANIMATIONS. OVER ITS 14 YEARS OF EXISTENCE SINCE THAT TIME, "BUF COMPAGNIE" HAS EARNED A SOLID REPUTATION IN THE CREATION OF 3D COMPUTER-GENERATED SPECIAL EFFECTS. THEY HAVE DONE DIGITAL EFFECTS FOR FEATURE FILMS, ADVERTISING AND TV. ▬ Nachdem sie den Film *Tron* gesehen hatten, beschlossen die beiden Freunde Pierre Buffin und Henry Seydoux, einen Apple zu kaufen. Der erste PC von BSCA war ein 8088 (der Georges Lagardère gehörte) mit 640 kBytes, einem 8087-Coprozessor und zwei 5¼-Zoll-Floppylaufwerken. Bald schon stiegen sie auf eine Intel 286er-Maschine um, die unter einem Intel Unix lief, und machten ihre ersten Animationen. In den seither vergangenen 14 Jahren hat die „Buf Compagnie" sich einen soliden Ruf im Bereich computergenerierte 3D-Effekte erworben. BUF hat Digital Effects für Spielfilme, Werbung und Fernsehen produziert.

CAPTIVES
N+N Corsino

Three women characterized by their capacity to say no. Their movements and their gestures are applied to their respective clones. Their dances and their actions reject any attachments, compromises, or approximations. They do not pretend. Their dreams, their imaginary world are not simulations of the real world: they extend their bodies while trying to thwart the traps of a formal modernity. This 12 minute film is made entirely in 3D. The movement of the bodies of the dancers was recorded in motion captures and applied to their clones. The choreography falls under a scenography divided into imaginary worlds and traversed by the virtual movements of cameras.

Drei Frauen, die sich dadurch auszeichnen, dass sie „nein" sagen können. Ihre Bewegungen und Gesten werden auf die ihnen zugehörigen Clones übertragen. Ihr Tanz, ihre Handlungen lehnen jede Anlehnung, jeden Kompromiss, jede Annäherung ab. Sie täuschen nichts vor. Ihre Träume, ihre imaginäre Welt - das ist keine Simulation der Wirklichkeit: Sie erweitern ihre Körper, während sie versuchen, sich den Fallen einer formalen Modernität entgegenzustellen. Dieser Zwölf-Minuten-Film wurde zur Gänze in 3D erstellt. Die Bewegung der Tänzerinnen wurde durch Motion Capture eingefangen und auf ihre Clones übertragen. Die Choreografie fällt unter eine Szenografie, die in imaginäre Welten aufgeteilt ist und von den virtuellen Bewegungen der Kameras durchschnitten wird.

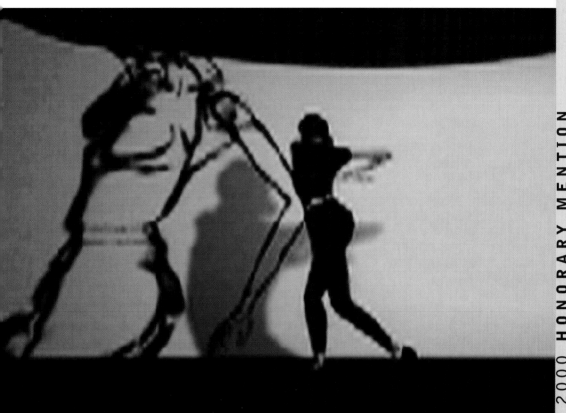

N+N CORSINO (F) WORK AS CHOREOGRAPHERS AND DIRECTORS. THEY EXPLORE THE TERRITORY, IN WHICH DANCE IS CREATED AND ABLE TO EXPRESS ITSELF, IN ORDER TO SHOW HOW MOVING THE BODY TRANSFORMS IT. THEY CHANGE THE SPACE IN WHICH DANCE IS DISPLAYED, BY SHOWING THEIR CHOREOGRAPHIC FICTIONS AS FILM AND MULTIMEDIA INSTALLATIONS. WINNERS OF THE PRIX VILLA MÉDICIS 1993 FOR AN INTERACTIVE CHOREOGRAPHY THAT THEY DEVELOPED WITH THE PROGRAM "LIFE FORMS"; IN 1996 THEY CREATED THE COMMISSIONED WORK *TRAVERSÉES*. ▬▬ N+N Corsino (F) arbeiten als Choreografen und Regisseure. Sie erkunden das Territorium, in dem Tanz entstehen und sich ausdrücken kann, um zu zeigen, wie die Bewegung der Körper ihn verwandeln. Sie verändern den Raum der Darstellung des Tanzes, indem sie ihre choreografischen Fiktionen als Film und Multimedia-Installationen zeigen. Preisträger beim Prix Villa Médicis 1993 für eine interaktive Choreografie, die sie mit dem Programm „Life Forms" erarbeitet haben; 1996 entstand das Auftragswerk *Traversées*.

AUTOTRADER WHOOSH

Ray Giarratana / Digital Domain

Autotrader Whoosh was directed by Buddy Cone for the Donner agency outside Detroit. Digital Domain's Ray Giarratana handled the visual effects supervisor role as well as serving as animation director.

Over the last four years, Digital Domain has created a distinctive array of CG ads for various clients. *Autotrader Whoosh* brings together Digital Domain's CG work with its move into Internet related advertising. The CG cars were produced digitally in Lightwave. Foreground plates were shot on stage in Los Angeles, and the spot was composited in both NUKE, Digital Domain's proprietary program, and Flame.

Autotrader Whoosh wurde unter der Regie von Buddy Cone für die Agentur Donner außerhalb Detroits produziert. Ray Giarratana von Digital Domain agierte als Visual Effects Supervisor und führte Regie bei der Animation.

In den letzten vier Jahren hat Digital Domain eine ganze Serie von individuellen Werbespots für verschiedenste Kunden gestaltet. *Autotrader Whoosh* kombiniert die digitale Computergrafik von Digital Domain mit einem Schritt in die Internet-Werbung. Die computergenerierten Autos entstanden in Lightwave; die Vordergrundaufnahmen wurden in Los Angeles gedreht. Die Komposition des Spots erfolgte sowohl in Nuke, Digital Domains hauseigenem Programm, wie in Flame.

PRIX ARS ELECTRONICA 2000 HONORARY MENTION

Ray Giarratana (USA) is director and visual effects supervisor at Digital Domain. His responsibilities as director and/or visual effects supervisor have included advertising campaigns for Coca Cola (*Skydiver*, Panda *Picnic*), Budweiser (*Recycled Ants*, *Tongue Lashing*), for Mercedes Benz (*Rhinos*) and the visual effects for Michael Jackson's video *Ghost*. ▬▬▬ Ray Giarratana (USA) ist Regisseur und Visual Effects Supervisor bei Digital Domain. In seinen Verantwortungsbereich als Regisseur und/oder Visual Effects Supervisor fallen Werbekampagnen für Coca Cola (*Skydiver, Panda Picnic*), Budweiser (*Recycled Ants, Tongue Lashing*), für Mercedes Benz (*Rhinos*) sowie die Visual Effects für

PROTEST
Steve Katz / Josselin Mahot

Protest is a dreamlike meditation on the plight of the elephant, whose natural habitat shrinks each year. It is set to the recognizable aria from the opera *La Wally*.

I did not set out to create a film that had an environmental theme. *Protest* came about because of a simple image I had of an Elephant falling. In working out the narrative, it became clear that the conclusion to the story had to be tragic, because that is the actual situation for these magnificent animals.

In that sense you could say that the design and research process for *Protest* forced me to extract a more universal message than would have been the case with a project that concentrated on design alone.

Protest was entirely done using 3D Studio Max, Photoshop and After Effects, and only 6 seconds were shot in 35 mm. This project was done on a HD format and transferred onto film by Sony High Definition. It is an independent project.

Protest ist eine traumähnliche Meditation über die Lebensumstände des Elefanten, dessen natürlicher Lebensraum von Jahr zu Jahr mehr schwindet. Untermalt ist der Film mit der Arie aus der Oper *La Wally*.

Ich hatte nicht vor, einen Film zu einem Umweltthema zu machen – *Protest* entstand, weil ich ein einfaches Bild von einem fallenden Elefanten hatte. Als ich die zu erzählende Geschichte ausarbeitete, wurde mir klar, dass das Ende tragisch sein musste, weil dies einfach der gegenwärtigen Lage dieser großartigen Tiere entspricht. In diesem Sinne könnte man behaupten, dass die Gestaltung und die Vorarbeiten zu *Protest* mich gezwungen haben, eine viel allgemeinere Botschaft zu transportieren als bei einem Projekt, das sich ausschließlich auf Design-Anliegen konzentriert.

Protest wurde durchgängig mit 3D Studio Max, Photoshop und After Effects realisiert, einzig sechs Sekunden wurden auf 35-mm-Film gedreht. Das Projekt wurde im HD-Format angelegt und mit Sony High Definition auf Film übertragen. *Protest* ist eine unabhängige Produktion.

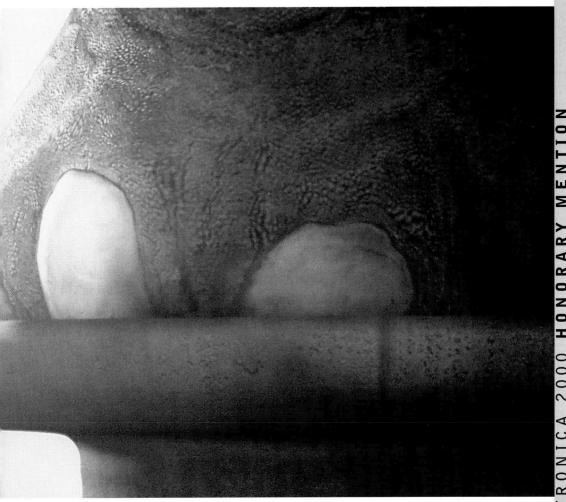

STEVE KATZ (US) IS A FILMMAKER/WRITER AND A PARTNER IN PITCH STUDIO, A NYC BASED PRODUCTION AND CONTENT COMPANY. HE IS THE AUTHOR OF THE BEST-SELLING BOOKS *SHOT BY SHOT* AND *CINEMATIC MOTION* AND A CONTRIBUTING EDITOR TO *MILLIMETER MAGAZINE*. JOSSELIN MAHOT (US) IS SCENIC DESIGNER AND DIGITAL MODEL MAKER AND WAS ALSO HEAD OF THE TEXTURE TEAM FOR THIS PRODUCTION. ▬▬ Steve Katz (USA) ist Filmemacher/Autor und Partner bei Pitch Studio, einer in New York beheimateten Produktionsfirma. Er ist Autor der Bestseller *Shot by Shot* und *Cinematic Motion* und ständiger Mitarbeiter des *Millimeter Magazine*. Josselin Mahot (USA) ist szenischer Gestalter und digitaler Modellbauer

AUDI FEUERFILM

Manfred Laumer

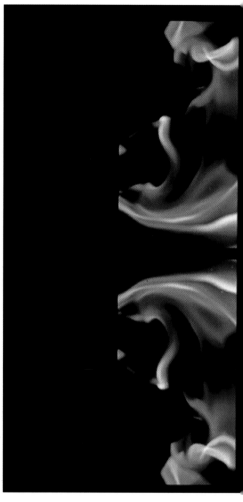

The piston of an 8-cylinder, the primeval power of combustion, and the mystic force of fire are symbolized with eight plasma screens arranged above one another in two rows.

Ten years of the turbo diesel engine, Audi's achievement, are presented in a video and sound installation in the form of a moving poster pillar. The drum is arranged with 3 x 8 plasma screens in a circle, resulting in 24 plasma screens set up like a poster pillar. The scenes were composed in such a way that the fire, filmed at a certain speed (300 frames per second and above), results with each respective symmetrical mirroring in a broad range of fire figures. These vary from demonic fire-faces to harmonious shapes.

The power of the fire and the harmonious course of the piston are underscored by the surround-cinema experience of the sound design.

Certain terms were interwoven into the contents of the film: primeval power; acceleration; range; performance, 10 years of TDI.

Der Kolbenlauf eines 8-Zylinders, die Urkraft der Verbrennung, die mystische Macht des Feuers werden symbolisiert durch acht Plasmascreens, die in zwei Reihen übereinander platziert wurden.

Zehn Jahre Turbodieselmotor, die Errungenschaft von Audi, wird über eine Video- und Soundinstallation in Form einer bewegten Litfasssäule inszeniert. Das Rondel ist mit 3 x 8 Plasmascreens im Kreis angeordnet. Somit erhalten wir 24 Plasmascreens, die wie eine Litfasssäule aufgebaut sind.

Die Szenen wurden so komponiert, dass sich bei einer gewissen Drehgeschwindigkeit des Feuers (ab 300 frames per second) mit der jeweiligen symmetrischen Spiegelung eine Vielfalt von Feuergestalten ergeben. Diese variieren von dämonenhaften Feuergesichtern bis zu harmonischen Formfindungen.

Die Kraft im Feuer und der harmonische Lauf des Kolbens werden im Surround-Kinoerlebnis vom Sounddesign unterstützt.

Folgende Begriffe wurden inhaltlich in den Film eingewebt: Urkraft. Beschleunigung. Reichweite. Leistung. Zehn Jahre TDI.

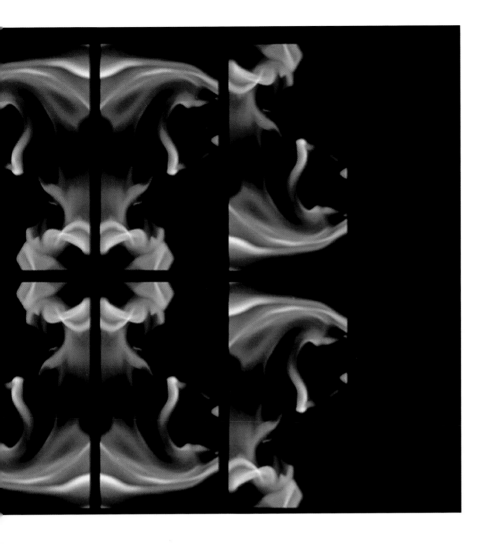

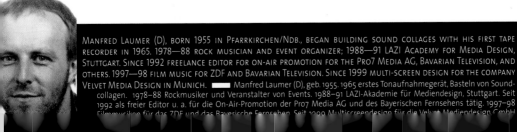

MANFRED LAUMER (D), BORN 1955 IN PFARRKIRCHEN/NDB., BEGAN BUILDING SOUND COLLAGES WITH HIS FIRST TAPE RECORDER IN 1965. 1978—88 ROCK MUSICIAN AND EVENT ORGANIZER; 1988—91 LAZI ACADEMY FOR MEDIA DESIGN, STUTTGART. SINCE 1992 FREELANCE EDITOR FOR ON-AIR PROMOTION FOR THE PRO7 MEDIA AG, BAVARIAN TELEVISION, AND OTHERS. 1997—98 FILM MUSIC FOR ZDF AND BAVARIAN TELEVISION. SINCE 1999 MULTI-SCREEN DESIGN FOR THE COMPANY VELVET MEDIA DESIGN IN MUNICH. ▬▬ Manfred Laumer (D), geb. 1955. 1965 erstes Tonaufnahmegerät, Basteln von Sound-collagen. 1978–88 Rockmusiker und Veranstalter von Events. 1988–91 LAZI-Akademie für Mediendesign, Stuttgart. Seit 1992 als freier Editor u. a. für die On-Air-Promotion der Pro7 Media AG und des Bayerischen Fernsehens tätig. 1997–98 Filmmusiken für das ZDF und das Bayerische Fernsehen. Seit 1999 Multiscreendesign für die Velvet Mediendesign GmbH

LEVIS INVISIBLE MAN

Fred Raimondi / Digital Domain

For director Michael Bay, Propaganda Films and agency client TBWA Chiat Day, Digital Domain was tasked with creating a provocative and photo-real couple in various stages of undress for Levi's latest advertising campaign. Digital Domain is known for its "invisible" effects work. "Socks" effects not only required rendering the couple invisible but more importantly required creating photo-real 3D CG clothing to be tracked over the existing plate photography in such a way as to make the invisible gag seamless. Visual effects supervisor Fred Raimondi and his team clearly accomplished the task at hand.

Im Auftrag von Regisseur Michael Bay, Propaganda Films, und des Agenturkunden TBWA Chiat Day wurde Digital Domain die Herstellung eines provokanten und dennoch fotorealistischen Paares in verschiedenen Stadien der Entkleidung für die neueste Werbekampagne von Levi's übertragen. Digital Domain ist für seine „unsichtbaren" Effects-Arbeiten bekannt. Die Visual Effects für *Socks* verlangten nicht nur, dass das Paar unsichtbar gemacht wurde, das noch viel wesentlichere Element waren die fotorealistischen 3D-Kleidungsstücke, die derart über die existierende Fotografie gezogen wurden, dass der sich ergebende Gag in jeder Hinsicht nahtlos war. Der Visual-Effects-Supervisor Fred Raimondi und sein Team haben diese Aufgabe überzeugend gelöst.

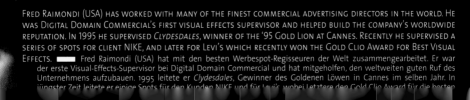

FRED RAIMONDI (USA) HAS WORKED WITH MANY OF THE FINEST COMMERCIAL ADVERTISING DIRECTORS IN THE WORLD. HE WAS DIGITAL DOMAIN COMMERCIAL'S FIRST VISUAL EFFECTS SUPERVISOR AND HELPED BUILD THE COMPANY'S WORLDWIDE REPUTATION. IN 1995 HE SUPERVISED *CLYDESDALES*, WINNER OF THE '95 GOLD LION AT CANNES. RECENTLY HE SUPERVISED A SERIES OF SPOTS FOR CLIENT NIKE, AND LATER FOR LEVI'S WHICH RECENTLY WON THE GOLD CLIO AWARD FOR BEST VISUAL EFFECTS. ■■■ Fred Raimondi (USA) hat mit den besten Werbespot-Regisseuren der Welt zusammengearbeitet. Er war der erste Visual-Effects-Supervisor bei Digital Domain Commercial und hat mitgeholfen, den weltweiten guten Ruf des Unternehmens aufzubauen. 1995 leitete er *Clydesdales*, Gewinner des Goldenen Löwen in Cannes im selben Jahr. In jüngster Zeit leitete er einige Spots für den Kunden NIKE und für Levi's, wobei Letztere den Gold Clio Award für die besten

WEGE ZUR QUALITÄT

Lisa Slates / Chitra Shriram

Wege zur Qualität, a 70mm ride film for Volkswagen's upcoming theme park "Autostadt", departs from the usual themes found in the genre with its non-narrative approach. Instead of the typical "thrilling chase/ escape" theme, the client wanted an exploratory, experimental journey.

The "Quality Department" at Volkswagen views the search for quality as a never-ending, ever-improving spiral. One never reaches an ultimate point in the search for quality, and creative energies propel the journey. Our ride reflects this philosophy. It plunges the audience into chaotic disorientations from which emerge passages of order and harmony.

Our pathway of innovative improvement is driven by oppositional energies. Within this conceptual context, we explore various interpretations of a spiraling journey in a fun ride through a variety of computer-generated environments. The initial environment is the most rational of the spaces, set in a somewhat recognizable warehouse space. From here we enter the first of three refrains. The refrains reveal infinite worlds of geometric perfection, and we return to see several variations of it.

The infinite arrays of cubes, hexagons and spheres provide a soothing counterpoint to the overwhelming complexity of the following scenes. The first of these is a mechanical "forest." Here we experience a repetition of form and movement in a cacophony of motion, texture and sounds—an obstacle course of various seemingly familiar but strange objects. The simulator platform weaves, banks and dives along with our camera, giving viewers a visceral understanding of this chaotic environment.

In the sky/water environment the simulator experience is integral to conveying the turbulence of the scene. A fantastical creature/mechanism greets us in this churning, elemental environment, processing the energy of its environment into a digitized world of swirling data. Wind currents ripple through strips of "high tech" fabric and we experience the chaotic nature of this energy intimately. XAOS' custom cloth simulation software use physically-based forces and surface tensions to recreate the inherent order and beauty found in chaotic turbulence. The turbulence increases as we are dropped into the swells of an ocean and plunge through a collapsing tunnel of water into the calm of the refrain once more.

Wege zur Qualität – ein 70mm-Ride-Film für den jüngst eröffneten Themenpark „Autostadt" von Volkswagen – unterscheidet sich durch seinen nicht-narrativen Ansatz von den üblichen Arbeiten des Genres. Statt der typischen Sequenz von spannender Verfolgungsjagd und knappem Entkommen wollte der Kunde eine experimentelle Forschungsreise.

Die Abteilung Qualitätssicherung bei VW sieht das Bemühen um Qualität als eine unendliche Spirale ständiger Verbesserungen an. Ein Schlusspunkt wird nie erreicht, getrieben von kreativer Energie geht die Reise stets weiter. Unser Film reflektiert diese Philosophie. Er stürzt das Publikum in eine chaotische Orientierungslosigkeit, aus der Passagen voller Harmonie und Ordnung auftauchen.

Unser Weg zur innovativen Verbesserung wird von gegensätzlichen Energien angetrieben. Innerhalb dieses Kontexts erforschen wir zahlreiche Interpretationen in einer spiralförmig kreisenden Reise durch eine Vielzahl computergenerierter Umgebungen. Die Anfangsszene ist ein höchst rationaler Raum, der in einem als solchem erkennbaren Lagerraum angesiedelt ist. Von dort aus begeben wir uns in den ersten von drei „Refrains". Diese wiederkehrenden Themen zeigen unendliche Welten geometrischer Perfektion und bei jeder Wiederkehr sehen wir neue Variationen.

Die unendlichen Arrangements von Würfeln, Sechsecken und Kugeln bieten einen beruhigenden Kontrapunkt zur umwerfenden Komplexität der folgenden Szenen, deren erste ein mechanischer „Wald" ist. Hier erleben wir die Wiederholung von Formen in einer Kakophonie aus Bewegung, Texturen und Klängen – ein Hindernislauf durch scheinbar vertraute, aber dennoch seltsame Objekte. Die Simulator-Plattform schwingt, schwankt und kippt parallel zur Kamerafahrt und vermittelt den Besuchern dieses chaotische Environment auch in physischer Unmittelbarkeit.

Bei der Himmel/Wasser-Umgebung ist die Simulatorerfahrung tragendes Element in der Vermittlung der Turbulenz der Szene. Eine fantastische mechanische Kreatur grüßt uns in dieser aufgewühlten Elementarumgebung und verwandelt die Energie ihrer Umwelt in eine digitalisierte Welt aus wirbelnden Daten. Windböen ziehen durch Streifen von High-Tech-Geweben – wir erleben die chaotische Natur dieser Energie hautnah mit. Die Custom-Cloth-Simulationssoftware von XAOS verwendet physikalische Modelle von Kräften und Oberflächenspannungen, um jene Ordnung und Schönheit zu schaffen, die der chaotischen Turbulenz inhärent ist. Das Wirbeln ver-

The final environment is the one furthest from any known reality. The internal workings and external curvilinear surfaces of a new VW Beetle provided the raw material for this colorful "roller coaster". The ride on the simulator platform has become more controlled, as if we have mastered the motion of our vehicle. The end of the ride contrives to give the impression that there is even more to the ride beyond the portion we have just witnessed/experienced as there is no real "end" to this journey. The journey is the destination.

Software: 3D Studio MAX, mit Afterburn und Digital Nature Tools als Plugins; eigens entwickeltes Partikelsystem und Textilsimulator von XAOS, Nothing Real Shake, Eyeon Digital Fusion, Adobe After Effects, Hardware: Intergraph Workstations, Render-Server und HD Animationsrecorder, Simex Bewegungssimulator Steuerungs- und Programmiersoftware.

stärkt sich, bis wir in die Wogen eines Ozeans eintauchen und durch einen zusammenbrechenden Tunnel aus Wasser wiederum in die Ruhe des Refrains zurückkehren.

Das letzte Environment ist am weitesten von jeder bekannten Wirklichkeit entfernt. Die internen Strukturen und die kurvigen Außenflächen des VW New Beetle lieferten das Rohmaterial für diese farbenfrohe Achterbahn. Die Fahrt auf der Simulationsplattform wird strukturierter – als hätten wir die Bewegung unseres Vehikels in den Griff bekommen. Das Ende der Fahrt verstärkt zusätzlich den Eindruck, dass noch eine weit längere Fahrt vor uns liegt als jener kurze Ausschnitt, den wir gerade erlebt/erfahren haben – diese Reise hat kein wirkliches „Ende". Der Weg ist das Ziel.

LISA SLATES (USA) JOINED XAOS, INC. AS AN ANIMATOR IN 1995 AND BEGAN HER CURRENT ROLE AS CO-CREATIVE DIRECTOR IN 1998. LISA WAS ALSO AN ANIMATION ASSISTANT FOR JAMES DUESING ON HIS "LAW OF AVERAGES", SHOWN AT THE SUNDANCE FILM FESTIVAL. SHE RECEIVED AN EMMY FOR HER CONTRIBUTION TO MSNBC'S 1996 ID CAMPAIGN. CHITRA SHRIRAM (INDIA) TRAINED AND WORKED AS A TRADITIONAL ANIMATION FILMMAKER IN INDIA. IN THE UNITED STATES SHE RECEIVED A MASTERS DEGREE FROM THE OHIO STATE UNIVERSITY, SPECIALIZING IN COMPUTER GRAPHICS AND ANIMATION. SHE JOINED XAOS IN 1993. WORKED ON THE EMMY AWARDED MSNBC ID CAMPAIGN, SCI-FI CHANNEL, PAGEMASTER, JONNY QUEST, ISLAND OF THE SHARKS AND DOLPHINS. ▬

Lisa Slates (USA) trat 1995 als Animatorin bei Xaos, Inc., ein und begann ihre derzeitige Tätigkeit als Co-Creative Director im Jahre 1998. Sie war Animationsassistentin bei James Duesings *Law of Averages*, das beim Sundance Film Festival präsentiert wurde. Sie hat einen Emmy Award für ihren Beitrag zur 1996er-ID-Kampagne von MSNBC erhalten. Chitra Shriram (Indien) hat als traditionelle Filmemacherin in Indien begonnen und gearbeitet. In den USA graduierte sie an der Ohio State University zum Master und spezialisierte sich auf Computergrafik und Animation.1993 trat sie bei Xaos ein. Mitarbeit an der preisgekrönten MSNBC-Kampagne, an *Sci-Fi Channel, Pagemaster, Jonny Quest, Island of the Sharks and Dolphins.*

SUPERNOVA
Mark Stetson / Digital Domain

Originally, the movie was slated for completion by February 1999, less than a year from the start of principal photography. The scheduling left just six months for the production of 250 large-scale visual effects shots. Those shots broke down to approximately seventy-five shots of the Nightingale medical spaceship, a handful of digital deep-space environments—including the blue giant sun, the rogue moon and the surrounding galaxy—the exterior of the mining facility, a series of "dimension jump" shots, and the alien object effects.

Responsible for those original 250 shots was Digital Domain. Visual effects supervisor Mark Stetson, executive producer Nancy Bernstein and visual effects producer Julian Levi immediately launched into effects preproduction, creating a breakdown, initiating storyboarding and establishing a visual effects presence at Raleigh Studios. "We very quickly put together a production staff and got some key creative heads in place," Stetson recalled, "including art director Ron Gress, director of photography Bill Neil and digital effects supervisor Jonathan Egstad—basically the same core team I'd had on *Fifth Element*. I also engaged George Trimmer as an art director for miniature effects, and my long-time associate Scott Schneider as miniature effects supervisor. We were in a real hurry, so we did all of this in a matter of days." Also instrumental was Digital Domain coordinator Rochelle Gross, who would eventually move into the role of visual effects producer, acting as liaison between production and all of the effects units.
(*Jody Duncan*)

Ursprünglich sollte der Film bis zum Februar 1999 fertiggestellt sein – weniger als ein Jahr nach Beginn der Dreharbeiten. Diese Terminplanung ließ gerade sechs Monate für die Produktion von insgesamt 250 großen Visual-Effects-Einstellungen, und zwar gut 75 Einstellungen des Nightingale Medizin-Raumschiffs, eine Handvoll digitaler Ansichten des fernen Weltraums – darunter die gigantische blaue Sonne, der Einzelmond und die umgebende Galaxie –, das Äußere des Bergwerkkomplexes, eine Serie von „Dimensions-Sprung"-Einstellungen und die Effekte mit außerirdischen Objekten.

Verantwortlich für diese 250 Effektaufnahmen war Digital Domain. Der Visual Effects Supervisor Mark Stetson, die ausführende Produzentin Nancy Bernstein und der Visual-Effects-Produzent Julian Levi stürzten sich sofort in die Vorproduktionsphase, schufen einen Detailplan, brachten die Drehbücher auf den Weg und installierten eine Kontaktstelle in den Raleigh Studios. „Wir haben ganz schnell eine Produktionsmannschaft zusammengestellt und dabei einige der wichtigsten kreativen Köpfe gewinnen können", erinnert sich Stetson, „darunter den Art Director Ron Gress, den Chefkameramann Bill Neil und den Digital Supervisor Jonathan Egstad – im Wesentlichen also jenes Kernteam, das ich bei *The Fifth Element* hatte. Ich habe auch George Trimmer als Art Director für Miniatureffekte engagiert und meinen langjährigen Partner Scott Schneider als Supervisor für Miniatureffekte. Wir haben es wirklich eilig gehabt, deshalb musste das alles in wenigen Tagen abgehandelt werden." Eine Schlüsselfunktion nahm auch die Koordinatorin Rochelle Gross bei Digital Domain ein, die später in die Rolle der Produzentin schlüpfen sollte, zunächst aber die Verbindung zwischen der Produktion und den ganzen Effects-Einheiten zu halten hatte.
(*Jody Duncan*)

Mark Stetson (USA), born 1952, studied industrial design at the University of Bridgeport in Connecticut, and trained for a year as an industrial model-maker at General Electric Co. in Bridgeport. Mark joined Digital Domain in July 1995 to supervise the visual effects for Luc Besson's *The Fifth Element*. By that time, he had worked on over 100 films. He had created miniature scenes of New York City for six projects, including *The Hudsucker Proxy*. He had created spaceships for humans and seven other alien species in eight films. He had worked as a special effects designer, visual effects art director, ride film production designer, even as a property master and prop designer. For his efforts on *The Fifth Element*, Stetson received the British Academy Award (BAFTA) for Best Visual Effects. ▬ Mark Stetson (USA) geb. 1952, studierte Industriedesign an der University of Bridgeport in Connecticut und lernte ein Jahr Industriemodellbau bei General Electric Co. Er kam im Juli 1995 zu Digital Domain, um die Visual Effects für Luc Bessons *The Fifth Element* zu leiten. Zu diesem Zeitpunkt hatte Mark Stetson bereits an über 100 Filmen mitgewirkt. Er hat die Miniaturszenerie von New York für insgesamt sechs Projekte gebaut, darunter für *The Hudsucker Proxy*, und Raumschiffe für Menschen und sieben verschiedene Arten von Außerirdischen in acht Filmen konstruiert. Er war als Special Effects Designer ebenso tätig wie als Art Director für Visual Effects, hat Ride-Filme konzipiert und selbst als Requisiteur und Bühnendesigner gearbeitet. Für seine Leistungen bei *The Fifth Element* hat Stetson den

VIENNA 1402

Alexander Szadeczky / Marcus Salzmann

As an integral part of the permanent exhibition of the Museum on Judenplatz in Vienna, Nofrontiere is designing four interactive mediatectural installations and a 3-D visualisation and animation of the medieval Jewish Quarters (ghetto) and its synagogue. Through illuminating selected aspects of medieval Jewish life, the mediatectural installations will create a cultural-historical context for the excavated synagogue and the exhibited archaeological finds. Each installation will be centred upon a specific conceptual world *(Ideenwelt)*.

The installations will allow visitors to make their own historical (re)discovery of the medieval ghetto through advanced interface technologies. The project's intention is to approach the prohibition of images in Jewish culture through abstraction. It is an attempt to render things through reduction, not to narrow down the visitors' perception, but rather to bring light to their imagination.

The outcome of the intense collaboration with Jewish scholars and public officials, a 12-minute visualization allows visitors to experience the urban fabric of the medieval ghetto. On a screen measuring 6 by 10 feet, viewers will make their way through the Jewish Quarter, visiting a hospital, a money-changers' place of work, a school, and will finally arrive at the medieval synagogue. From an initially rather vague

Nofrontiere entwickelte vier interaktive „mediatekturale" Installationen sowie eine 3D-Visualisierung und Animation des mittelalterlichen Judenviertels (Gettos) und dessen Synagoge, die einen Teil der Dauerausstellung des Museums am Judenplatz in Wien bilden. Indem ausgewählte Aspekte des jüdischen Lebens im Mittelalter beleuchtet werden, schaffen die Installationen einen kulturhistorischen Kontext, in den sich die ausgegrabene Synagoge und die ausgestellten archäologischen Funde nahtlos integrieren. Jede Installation bezieht sich auf eine ganz spezifische Ideenwelt.

Die Installationen ermöglichen es den Besuchern mittels hoch entwickelter Interface-Techniken, sich auf eine individuelle historische (Wieder-)Entdeckungsreise durchs mittelalterliche Getto zu begeben. Das Projekt bemüht sich, auch das Bilderverbot der jüdischen Kultur durch Abstraktion erfahrbar zu machen – es ist ein Versuch, Gegenstände durch Reduktion darzustellen, dabei aber nicht die Wahrnehmung des Betrachters einzuschränken, sondern vielmehr Licht in seine Imagination zu bringen.

Das Ergebnis der intensiven Zusammenarbeit mit jüdischen Gelehrten und öffentlichen Stellen ist eine zwölfminütige Visualisierung, die den Besuchern die Gelegenheit gibt, die urbane Struktur des mittelalterlichen Gettos zu erfahren. Auf einer Projektionsfläche von 2 mal 3 Metern können Besucher das Judenviertel durchwandern, ein Spital, die Stube eines Geldwechslers oder eine Schule besuchen, bis sie schließlich bei der mittelalterlichen Synagoge ankommen. Aus der anfänglich eher vagen Idee, den Besuchern mehr zu bieten als nur die abstrakten

idea of providing the visitors with more than just the abstract fundamentals of central themes of the exhibit, there emerged the concrete concept of utilizing state-of-the-art visualization techniques to reawaken jaded consumers of mass-media's ephemeral meanings and repetitions. The objective was not to didactically pulverize the material into easy-to -digest pablum but rather to provide the tools for the viewer to personally rediscover the realm of Viennese Jewry.

From the source reference material of simple hand sketches, roughly drawn city maps and detailed plans from the excavation site, a detailed 3-D model was meticulously generated for 320 objects, including not only the 38 buildings but every single object the visitor could experience in their passage.

Grundlagen der zentralen Themen der Ausstellung, entstand das konkrete Konzept, die eingelullten Konsumenten von ephemeren, repetitiven Inhalten der Massenmedien mittels modernster Visualisierungstechniken aufzurütteln. Das Ziel war nicht, das vorhandene Material didaktisch in leicht verdauliche Wissenshappen aufzuteilen, sondern den Besuchern das Werkzeug an die Hand zu geben, das untergegangene Reich des Wiener Judentums selbst neu zu entdecken.

Aus dem Grundmaterial einfacher Handzeichnungen, grob skizzierter Stadtpläne sowie detaillierter Pläne des Ausgrabungsortes wurde ein exaktes 3D-Modell für insgesamt 320 Objekte erstellt, nicht nur für die 38 Gebäude, sondern auch für alle Gegenstände, denen der Besucher auf seinem Weg begegnen kann.

ALEXANDER SZADECZKY-KARDOSS (A) STUDIED MECHANICAL ENGINEERING AND INDUSTRIAL DESIGN, AND HAS WORKED FOR MATHEO THUN (AMONG OTHERS). IN 1990 HE FOUNDED THE "NOFRONTIERE" AGENCY, PRODUCING AWARD-WINNING WORK (EMMA AWARD, AUSTRIAN NATIONAL MULTIMEDIA D'ART PRIZE, GRAND PRIX MILIA 2000, CCA-AWARD.) MARCUS SALZMANN (A) IS THE 3D SUPERVISOR AND HAS BEEN INVOLVED IN SEVERAL AUSTRIAN FILMS AND DOCUMENTARIES.

Alexander Szadeczky-Kardoss (A) studierte Maschinenbau und Industriedesign und arbeitete u. a. für Matheo Thun. 1990 gründete er die Agentur „Nofrontiere", deren Schaffen mit vielen Preisen ausgezeichnet wurde (EMMA-Award, öst. Staatspreis Multimedia d'Art, Grand Prix Milia 2000, CCA-Award.) Marcus Salzmann (A) ist 3D-Supervisor und hat an mehreren österreichischen Spiel- und Dokumentarfilmen mitgewirkt.

AUDI TELEMATIC

Cornelia Unger

The task that was posed for the Telematic film was to design a vision for the future based on Audi's existing navigational system. It was to illustrate the further developments of this system and how it could be used.

The starting point for the concept was the question of which functions a navigation and service system of the future could have. Situations that a driver might experience were described with three fictive examples.

In this case, the three monitors of the trade fair installation stand for the three elementary levels of the telematic system:

The monitor on the right symbolizes the connection to the outside, for example Internet access.

The monitor on the left visualizes data processing and the evaluation of information.

The center monitor represents communication between the driver and the telematic system.

These functions are represented abstractly by the silhouette of a woman, played by a mime.

This silhouette is linked with the rubber-like surface via the hands and is thus able to soak up information, so to speak. This results in associations with a data glove that collects, processes and passes on virtual data. The work is intended to be reminiscent of a dance.

In conjunction with the telematic task, the film addresses a vision of future communication and networking. The tangible and intuitively comprehensible representation conveys the feeling that this system can do everything for the user.

Yet elements that inspire critical reflection are also purposely integrated. They are intended to contrast a vision of the future that is all too euphoric with doubts, for instance those involving data protection and individual freedom of movement and decision.

Die Aufgabenstellung für den *Telematic*-Film bestand darin, auf Basis des bereits existierenden Navigationssystems von Audi eine Zukunftsvision zu gestalten, die Weiterentwicklung und künftige Einsatzmöglichkeiten dieses Systems veranschaulicht. Ausgangspunkt des Konzeptes war die Überlegung, welche Funktionen ein Navigations- und Servicesystem der Zukunft haben könnte. Anhand von drei fiktiven Beispielen wurden Situationen beschrieben, die ein Autofahrer erleben könnte.

Die drei Monitore der Messeinstallation stehen dabei für die drei elementaren Ebenen des Telematic-Systems: Der rechte Monitor symbolisiert die Verbindung nach außen, beispielsweise den Zugang ins Internet. Der linke Monitor visualisiert Datenverarbeitung und Auswertung von Informationen. Der mittlere Monitor stellt die Kommunikation zwischen Fahrer und Telematic-System dar.

Die abstrakte Darstellung dieser Funktionsweise erfolgt in Gestalt einer Frauensilhouette, gespielt von einer Pantomimin. Diese Silhouette ist über die Hände direkt mit der gummiartig wirkenden Oberfläche verbunden und kann so Informationen gleichsam in sich aufsaugen. So entsteht die Assoziation eines Datenhandschuhes, mit dem virtuelle Daten gesammelt, verarbeitet und weitergegeben werden. Die Arbeit soll an einen Tanz erinnern.

Im Rahmen der Aufgabenstellung Telematic thematisiert der Film eine Vision zukünftiger Kommunikation und Vernetzung. Die greifbare und intuitiv erfassbare Darstellung vermittelt das Gefühl, dass dieses System alles für den Nutzer leisten kann. Bewusst werden aber auch nachdenklich stimmende Elemente integriert, die einer allzu euphorischen Zukunftsvision auch Zweifel entgegenstellen, beispielsweise Themen wie Datenschutz und individuelle Entscheidungs- und Bewegungsfreiheit.

CORNELIA UNGER (D), BORN 1969, TRAINED AS AN EDITOR FROM 1990—1991. WORKED 1991—1992 AS DESIGNER FOR TELE 5, 1992 - 1994 AS DESIGNER AND ART DIRECTOR AT THE BROADCASTING CENTER MUNICH (PROSIEBEN, DSF, KABEL), 1994—1995 AS A FREELANCE ART DIRECTOR. 1995—1996 ART DIRECTOR FOR DMC DESIGN FOR MEDIA AND COMMUNICATION, VIENNA, SINCE 1997 CREATIVE DIRECTOR AT VELVET MEDIA DESIGN, MUNICH. ▬▬ Cornelia Unger (D), geb. 1969. 1990–1991 Ausbildung als Editor. 1991–1992 Designer bei Tele 5. 1992–1994 Designer und Art Director im Sendezentrum München (ProSieben, DSF, Kabel). 1994–1995 Freier Art Director. 1995–1996 Art Director bei DMC Design for Media and Communicaton Wien. 1997–jetzt Creative Director bei velvet Mediendesign München.

FAUX PLAFOND

François Vogel

Faux Plafond: On a night with a full moon, a couple who cannot get to sleep start working on their flat. A domestic distraction and science fiction journey under the starry sky ...

In einer Vollmondnacht kann ein Paar nicht schlafen und so beginnt es, an seiner Wohnung zu arbeiten. Eine häusliche Ablenkung und Science-Fiction-Reise unter dem Sternenhimmel beginnt ...

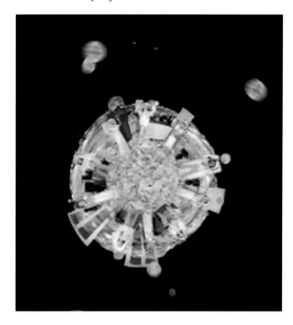

FRANÇOIS VOGEL (F), BORN 1971. UNTIL 1993 HE STUDIED IN THE VIDEO DEPARTMENT OF THE FINE ARTS FACULTY AT THE ECOLE NATIONALE SUPÉRIEURE DES ARTS DÉCORATIFS (ENSAD) WHILE TAKING COURSES IN APPLIED COMPUTER PROGRAMMING AT THE CONSERVATOIRE NATIONAL DES ARTS ET MÉTIERS. 1994—95 EDITOR AND CARTOON ANIMATOR FOR THE FEARLESS STUDIO, PARIS. 1995—96 STUDIED AT THE ATELIER D'IMAGES ET D'INFORMATIQUE DE L'ENSAD, 2D AND 3D ANIMATION SPECIALIST AT MIKROS IMAGE (PARIS) IN 1996. HIS ARTISTIC WORK WITH THE MEDIUM VIDEO, BEGUN IN 1991, HAS GRADUALLY EVOLVED INTO AN EXPERIMENTATION CROSSING BORDERS BETWEEN VIDEO, ANIMATION AND PHOTOGRAPHY. ▬ François Vogel (F), geb. 1971. Bis 1993 studierte er Video an der Fakultät für Bildende Kunst der Ecole Nationale Supérieure des Arts Décoratifs (ENSAD), während er gleichzeitig Kurse in angewandtem Programmieren am Conservatoire National des Arts et Métiers belegte. 1994–95 war er Cartoon-Animator beim Fearless Studio, Paris. 1995–96 studierte er am Atelier d'Images et d'Informatique de l' ENSAD und war 1996 2D- und 3D-Spezialist bei Mikros image (Paris). Seine 1991 begonnene künstlerische Auseinandersetzung mit dem Medium Video hat sich langsam hin zu einem Experimentieren an der Grenze zwischen Video, Animation und Fotografie entwickelt.

.NET
INTERACTIVE ART
COMPUTER ANIMATION
VISUAL EFFECTS
DIGITAL MUSICS
U19/CYBERGENERATION

FORWARD TO THE WORLD
WEITERLEITEN AN DIE WELT

It is amusing and bizarre to read the many emails berating both the 1999 Digital Musics Winners and the Digital Musics Jury for their choices. Before elaborating on this years decisions, it is necessary, not to defend our decisions, which we have already done in the 1999 Jury Statement, but rather to analyse the assumptions that guide those criticisms.

Hostility especially centred on last year's Golden Nica winners—the music video director Chris Cunningham and the producer Aphex Twin. It's audible that our critics in the academic electroacoustic community have a visceral dislike of popular culture. For this sector, the entire value of the now defunct Computer Music Prize stemmed from it's historical role as a refuge from, a direct opponent of, and a zone of aesthetic superiority over the inescapable vulgarity of popular music.

Last year, the Jury definitively broke with this position, one which had led the Computer Music Category down a dead end of irrelevance and terminal boredom. To our critics, no music video has a place in a serious digital music competition, because by definition the form epitomises entertainment culture—unserious, frivolous and unworthy.

But dismissing Cunningham and Aphex, let alone the Mego label and Ikue Mori for their frivolous commercialism displays a staggering ignorance of their influence and importance in contemporary digital music. Crude terms such as commercialism, entertainment or popular music are irrelevant because they are unable to analyse the winning entries to any meaningful degree. And this is their real purpose; not to understand this digital moment but to prevent analysis, to block thought, to replace it with sneers and ridicule.

Scrolling down the emails posted, emails seething with wounded sensitivities, you'd think we had

Es ist ebenso amüsant wie bizarr, die vielen E-Mails zu lesen, die sowohl die Sieger der Kategorie Digital Musics 1999 wie die Jury wegen ihre Entscheidungen kritisieren. Bevor wir uns über die diesjährigen Entscheidungen auslassen, erscheint es uns deshalb angebracht, nicht etwa unsere vorjährigen Entscheidungen zu verteidigen – diese haben wir im Jury-Statement 1999 ausführlich begründet –, sondern jene Annahmen zu analysieren, die den Kritiken zu Grunde liegen.

Die Feindseligkeit konzentrierte sich vor allem auf die Gewinner der Goldenen Nica 1999, den Musik-video-Regisseur Chris Cunningham und den Produzenten Aphex Twin. Es wird deutlich, dass unsere Kritiker in der akademischen Elektroakustik-Gemeinschaft eine tiefgehende Abneigung gegen populäre Kultur hegen. Aus Sicht dieses Bereichs lag der gesamte Wert des inzwischen begrabenen „Computermusik"-Preises des Prix Ars Electronica in seiner Existenz als Zuflucht vor, als direkter Gegenpol zu und als Bereich garantierter ästhetischer Überlegenheit über die allgegenwärtige Vulgarität der Pop-Musik.

Im vergangenen Jahr wurde diese Haltung von der Jury ganz grundlegend durchbrochen, weil sie die Computermusik-Kategorie in eine Sackgasse aus Irrelevanz und tödlicher Langeweile geführt hat. Für unsere Kritiker ist in einem ernsthaften Wettbewerb für digitale Musik kein Platz für Musikvideos, weil diese Form *ex definitione* die Unterhaltungskultur verkörpert – unseriös, frivol, wertlos.

Aber Cunningham und Aphex – und ganz zu schweigen vom Label Mego und von Ikue Mori – wegen ihres frivolen Kommerzialismus anzuprangern, beweist nur eine erstaunliche Unkenntis ihres Einflusses und ihrer Bedeutung in der zeitgenössischen digitalen Musik. Krude Begriffe wie „Kommerzialismus", „Unterhaltung" und „populäre Musik" sind deswegen irrelevant, weil ihnen die Fähigkeit fehlt, die preisgekrönten Einreichungen in irgendeiner sinnstiftenden Weise zu analysieren. Und das ist auch ihr Hauptzweck: nicht dieses digitale Moment zu verstehen, sondern eine Analyse zu verhindern, ein Nachdenken über sie zu vermeiden und stattdessen Verhöhnung und Verachtung an ihre Stelle zu setzen.

Liest man sich so durch die eingelangten E-Mails –

awarded the Golden Nica to Christine Aguilera. If only. Ars Electronica is as far from the worlds of teenpop, boygroups, alt.country, pop artificielle, 2 Step garage, broken beats, militant R&B, turntablism, nu electronic dub, playa hiphop, postrock, avant-rap, nu Afrobeat and Eurotrance as it ever was.

In fact, the Prix Ars Winners epitomise a digital innovation and seriousness that's as far from Britney Spears as it's possible to get. And to believe otherwise, as our critics do, is to proudly flaunt their ignorance of today's musical landscape. Sonically speaking, our critics actively work hard at dismissing the world the rest of us live in. Academic composers, bless them, insist on maintaining a distance from the extreme complexities of modern digital music.

The Jury was invited specifically because of their familiarity with this complexity. You might think our enemies' squeamish revulsion at today's digital music disqualifies them from judging our decisions but, paradoxically, the more demonstratively ignorant our critics reveal themselves to be, the more entitled they felt to launch weirdly personal attacks at the Judges.

This can only be because their extreme minority isolation is experienced not as a fatal disability but as a valued attribute. Which raises the point—why are the worlds of popular music, entertainment culture and commercial music so automatically detested? What is so inherently terrible about these fields? Why are they wielded as self evident insults? Are these fields really inimical to originality? Of course they are not. To assume otherwise is to adopt a position of reactionary avant-gardism, long discredited in a post Greenbergian art world.

This defensive, fearful and sterile attitude has incubated and bloomed in the narrow world of academic music composition. So we can hear the heirs of elec-

Mails, die von verwundeter Eitelkeit triefen –, könnte man fast meinen, wir hätten die Goldene Nica an Christine Aquilera verliehen. Hätten wir das nur! Ars Electronica ist genauso weit von der Welt des Teenypop, von Boygroups, alt.country, pop artificielle, 2Step Garage, Broken Beats, von militantem Rhythm & Blues, vom Turntablism, Nu Electronic Dub, Playa Hiphop, Post-Rock, Avant-Rap, Nu Afrobeat und von Eurotrance entfernt wie eh und je.

Im Gegenteil: Die Gewinner des Prix Ars Electronica stehen für eine digitale Innovation und Ernsthaftigkeit, die so weit von Britney Spears entfernt ist, wie es nur möglich ist. Und wenn jemand anderes glaubt oder behauptet – wie das unsere Kritiker tun –, dann heißt das, die eigene Unkenntnis der heutigen Musikszene stolz hinausposaunen. Klanglich betrachtet geben sich unsere Kritiker Mühe, die Welt auszuklammern, in der der Rest von uns lebt. Akademische Komponisten bestehen darauf, Distanz zur extremen Komplexität der modernen digitalen Musik zu halten.

Die Jury wurde eingeladen, weil sie mit dieser Komplexität vertraut ist. Man könnte glauben, allein schon der zimperliche Ekel unserer Gegner vor der heutigen digitalen Musik disqualifiziere sie davon, unsere Entscheidungen zu beurteilen, aber paradoxerweise fühlten sie sich umso mehr zu recht seltsamen persönlichen Attacken gegen die Juroren berufen, je demonstrativ ignoranter sie sich erwiesen.

Dies kann nur daraus herrühren, dass ihre Isolation als extreme Minderheit nicht als fatale Behinderung empfunden wird, sondern als wertvolle Eigenschaft. Was wiederum die Frage aufwirft, warum die Welt der populären Musik, der Unterhaltungskultur und der kommerziellen Musik so automatisch verachtet wird? Was ist an diesen Bereichen denn so Schreckliches? Warum werden sie mit absoluter Selbstverständlichkeit als Beleidigung betrachtet? Stehen diese Gebiete tatsächlich jeder Originalität feindlich gegenüber? Aber natürlich nicht. Nähme man das Gegenteil an, so würde man die Position eines reaktionären Avantgardismus beziehen, der in einer Nach-Greenberg'schen Kunstwelt schon längst diskreditiert ist.

Diese defensive, ängstliche und sterile Haltung wurde in der engen Welt der akademischen Musikkomposition herangezüchtet und zur Blüte gebracht. So

troacoustic, musique concrete and acousmatic tradition behaving as if their (illusory) distance from the market automatically bestows an aesthetic superiority on their music. Conversely, music delivered through the market is automatically inferior. But commercialism doesn't corrupt music as academics believe. In fact it multiplies and mutates all media into networks of audiosocial desire.

Our critics turn networks into hierarchies of value. Prix Ars Electronica is not naively anti-hierarchical at all; on the contrary, we are keen to rank, grade and evaluate. It's a question of cultural power—whose evaluation, whose definition counts here? All these complaints about commercialism really point to an academic community that is losing the cultural authority and historical privilege it has taken for granted. The virulent hostility we have received only indicates what we all know: time is up for the ancient regime of electroacousticians. Astonish us or fade away.

Because the leading edges of 21st Century digital music are elsewhere. As Naut Humon, who has done more than anyone to engineer this change, says, "Ars Electronica is catching up to the changes that are already going on. It's crucial if Ars wants to be cutting edge." For our enemies even these baby steps are one too many. But we haven't heard anything yet—All the adventures are still ahead.

For the Digital Musics Prize to live up to its founding principles it must honour the most influential living producers, the composers who have actively shaped today's approaches and thought processes. Only active recognition can dissolve the terrible burden of academicism it still labours under. Zeena Parkins the composer, musician and now Judge of this years Prize remarked, "I always thought of Ars as conservative and academic. My impression didn't change until Ikue won—that for me was a profound change."

The pioneering figures are not at all mysterious. Anyone who has paid close attention to the music of the last decade will recognise them. At this stage, those pioneers have spawned generations of helpless acolytes thus making it all the more necessary to recognise those originators.

This year 288 entries were submitted. Moving the closing entry date forward a month decreased numbers, while other composers doubtless decided not to

erleben wir, dass die Erben der Elektroakustik, der Musique Concrète und der akusmatischen Tradition sich so verhalten, als würde die (illusorische) Distanz, die ihre Musik zum Markt einnimmt, dieser automatisch eine ästhetische Überlegenheit verleihen. Und umgekehrt wäre dann eine über den Markt vertriebene Musik automatisch minderwertig. Aber der Kommerzialismus korrumpiert die Musik nicht, wie die Akademiker glauben, im Gegenteil: Er multipliziert und verwandelt alle Medien in Netzwerke audiosozialen Verlangens.

Unsere Kritiker verwandeln Netzwerke in Werthierarchien. Der Prix Ars Electronica ist ganz und gar nicht naiv anti-hierarchisch; im Gegenteil, wir freuen uns darauf, zu reihen, einzustufen, zu bewerten. Es ist eine Frage der kulturellen Macht – wessen Bewertung, wessen Definition zählt hier? All diese Beschwerden über Kommerzialismus weisen auf eine akademische Gemeinschaft hin, die dabei ist, jene kulturelle Autorität und jenes historische Privileg zu verlieren, die sie für garantiert und immerwährend gehalten hat. Die uns entgegenschlagende virulente Feindseligkeit ist nur ein Indikator für etwas, was wir alle wissen: Die Zeit des *Ancien Régime* der Elektroakustiker ist abgelaufen. Erstaunt uns oder verschwindet!

Die führenden Kräfte der digitalen Musik des 21. Jahrhunderts sind nämlich woanders. Wie Naut Humon – der mehr als irgendjemand anderer dazu beigetragen hat, diese Änderung herbeizuführen - sagt: „Der Prix Ars Electronica holt schön langsam auf gegenüber jenen Veränderungen, die überall im Gange sind. Und das ist entscheidend, wenn Ars Electronica selbst am Puls der Zeit bleiben will." Aber für unsere Gegner ist selbst dieser winzige Schritt einer zu viel. Dabei haben wir noch nicht einmal wirklich etwas gehört. Die ganzen Abenteuer liegen noch vor uns.

Damit der Preis für Digitale Musik seinen Gründungsprinzipien gerecht werden kann, muss er die einflussreichsten lebenden Produzenten ehren, jene Komponisten, die aktiv die heutigen Ansätze und Denkprozesse geformt haben. Nur eine aktive Anerkennung kann die schreckliche Bürde des Akademismus auflösen, unter der sie noch immer leidet. Die Komponistin und Musikerin Zeena Parkins, selbst Jurorin beim diesjährigen Wettbewerbs, bemerkte: „Ich habe den Prix Ars Electronica immer als konservativ und akademisch angesehen. Mein Eindruck hat sich erst gebessert, als Ikue gewonnen hat – das war für mich eine tiefgreifende Veränderung."

Die Pioniergestalten sind keineswegs mysteriös – jeder, der der Musik der letzten Dekade aufmerksam gelauscht hat, wird sie erkennen. Und aus heutiger Sicht haben diese Pioniere Generationen von hilflosen Akoluthen in die Welt gesetzt, was es umso nötiger macht, diese Innovatoren entsprechend hervorzuheben.

In diesem Jahr wurden 288 Werke eingereicht. Die

submit because of the rightwing turn in Austrian politics. On listening carefully to the electroacoustic entries, the producer Peter Rehberg offered a damning judgement. "That kind of music has become a formula, it's become formalised. It's a cover version of people like Pierre Schaeffer and Francois Bayle. It's been done before. It's documented. It's in museums." These composers often work by minor alterations and miniscule improvements to their forebears. It often feels as if the more subtle or indetectible the change, the more applauded they expect to be. It's a short step to concluding that the blame lies, not with their inability to reinvent, but with us, the Judges, for being too coarse and stupid to appreciate them. The music we honoured in fact updates the extreme traditions of 20th Century electroacoustic and the future shock of musique concrete—reinstigating a new awareness of "the perception of processes and processes of perception." It sheds the institutional structures and substitutes independent labels, networks of informal support systems. In the passage from techniques of cut & mix back in the 80s to early 00s click & cut, "the digital routing of ideas based on sound" leads us towards today's digital minimalism. Analysed as glitch, microsound or lowercase and "characterised by colossal shifts in dynamics, tone and frequency" digital audio now feeds off "the technical errors and unplanned outcomes of electrical society." Think of it as "an urban environmental music —the cybernetics of everyday life, that reflects the depletion of natural rhythms in the city experience and in the striated plateaux of the virtual domain." The Honourable Mentions for the Digital Musics Prize were awarded to veteran composers such as Tone, Amacher and Troyer and encouragingly to new producers such as snd, Dat Politics, Errorsmith and Radian.

We were unanimous in our regard for the endless immersive drones of Maryanne Amacher's *Sound Characters* (*making the third ear*) (US) to Kaffe Mathews for *Cécile*'s mesmeric sustain (UK), the amplitude tautenings of Yasunao Tone's *Wounded Man*(Japan) to snd for the warm minimal house of *makesnd cassette* (UK) and Radian for the fuzz scoured small jazz trio of TG11 (Austria).

The harmonic swells of Markus Schmickler's *Sator Rotas* (Germany) impressed us as did the extreme

Vorverlegung des Einreichungstermins um einen Monat hat die Zahl reduziert, und zweifellos haben auch einige Komponisten angesichts der Rechtswendung in der österreichischen Politik bewusst auf Einreichungen verzichtet. Bei intensivem Anhören der elektroakustischen Einreichungen kam der Produzent Peter Rehberg zu einem vernichtenden Urteil: „Diese Art von Musik ist zu einer Formel geworden, sie ist formalisiert worden. Das ist eine Cover-Version von Leuten wie Pierre Schaeffer und François Bayle. Das ist alles schon da gewesen, das ist alles dokumentiert, das ist alles im Museum." Diese Komponisten verändern und verbessern das Erbe, auf das sie zurückgreifen können, oft nur minimalst und häufig hat man den Eindruck, je subtiler und unmerklicher die Änderungen, desto größeren Applaus erwarten sie. Es ist nur ein kleiner Schritt zur Schlussfolgerung, dass der Fehler nicht bei ihrer Unfähigkeit zur Neu-Erfindung liegt, sondern bei uns, den Juroren, die wir zu grob und zu dumm sind, sie zu würdigen. Jene Musik aber, die wir tatsächlich gewürdigt haben, modernisiert die extremen Traditionen der Elektroakustik und den Zukunftsschock der Musique Concrète des 20. Jahrhunderts, indem sie „die Wahrnehmung der Prozesse und die Prozesse der Wahrnehmung" steigert. Sie verwirft die institutionellen Strukturen und ersetzt sie durch unabhängige Labels, durch Netzwerke informeller Unterstützungssysteme. Im Übergang von den Cut-and-Mix-Techniken der 80er-Jahre zu den Click-and-Cut-Techniken zu Anfang des neuen Millenniums führt uns „die digitale Umleitung von auf Klang basierenden Gedanken" zum heutigen digitalen Minimalismus. Als Glitch, als Microsound oder schlichtweg als „von kolossalen Verschiebungen in der Dynamik, Tonalität und Frequenz" bestimmte digitale Musik verstanden, ernährt sich das heutige Audio von „den technischen Fehlern und ungeplanten Ergebnissen einer elektrischen Gesellschaft". Man stelle sich das vor als „Musik einer urbanen Umwelt – die Kybernetik des Alltagslebens, die den Verfall der natürlichen Rhythmen in der Stadterfahrung ebenso reflektiert wie in den gefurchten Hochebenen des Virtuellen". Die Anerkennungen der Kategorie „Digital Musics" gehen an Veteranen unter den Komponisten wie Tone, Amacher und Troyer, aber als Zeichen der Unterstützung auch an junge Produzenten wie snd, Dat Politics, Errorsmith und Radian. Es herrschte Einstimmigkeit bei unserer Wertschätzung für die endlose immersive Monotonie von Maryanne Amachers *Sound Character* (*making the third ear*; USA); für das mesmerisierende Sustain in Kaffe Mathews *Cécile* (GB); für das Straffziehen der Amplituden in Yasunao Tones *Wounded Man* (J); für das warme minimalistische Umfeld der *makesnd cassette* von snd (GB) ebenso wie Radian für das fusselfrei polierte Small-Jazz-Trio von TG11 (A). Das An- und Abschwellen der Obertöne in Markus

fluctuations of Dat Politics' *Villiger* (France), Kevin Drumm's disruptive switchbacks on *Three* (US), the granular splinters of Uli Troyer's *NOK* (Italy) and error-smith (Germany) for the jarring jumpsplices and sudden frequency shifts of *EP1*—the first ever 12" vinyl to be honoured in the Digital Musics Award.

Our first Distinction for the Digital Musics Prize went to GESCOM—the English trio of producers Rob Brown, Sean Booth and the designer/artist Russell Haswell for *MiniDisc*. Firstly for its innovation—*MiniDisc* is the world's first independently produced MiniDisc only release. It integrates the digital compression of the format into it's aesthetics by presenting 45 micro-tracks spread over 88 cue points which you play in shuffle mode. In the abbreviated perforations and pock marks of *Is We* and *Vermin* and the brooding meditations of *Shoegazer* and *Dan Dan Dan*, *MiniDisc* demonstrated a bewitching range of digital signal treatments, transformations and effects.

The second deciding factor is that GESCOM is the alter ego of Booth and Brown who record together as the immensely influential Autechre, the duo whose non-linear programming and extreme digital signal processing pioneered the post techno world of glitch and microsound we now inhabit. Like Aphex Twin, they emerged in the early 90s and record for Warp Records, one of the world's most revered electronic music labels. From experience Rehberg asserted that "There are so many Autechre copyists these days. If we had to choose our favourite Autechre tracks we'd be fighting for 3 days. Someone would choose LP5, someone else Chiastic Slide, but this is okay, here's a serious concept and we can all agree on its importance."

Our other Distinction went to the sound recordist Chris Watson for his CD *Outside the Circle of Fire*. Once part of the groups Cabaret Voltaire and Hafler Trio, Watson's winning CD consists of 22 field recordings of mammals, birds and insects from Mozambique nightjars circling the Zambesi rivers to the scritch-scratch of deathwatch beetles in ageing floorboards. By precision miking these sounds Watson records in close up, recording events which impact the listener with their ferocious, ambient unfamiliarity.

The mystery, Parkins observed, is "how music emerges from field recordings." Rimbaud pointed out

Schmicklers *Sator Rotas* (D) beeindruckte uns ebenso wie die extremen Fluktuationen in *Villiger* von Dat Politics (F), wie Kevin Drumms abrupte Rücksprünge in *Three* (USA), wie die körnigen, brutzelnden Klänge von Uli Troyers *NOK* (I) und wie errorsmith (D) mit seinen plötzlichen Jumpsplices und Frequenzverschiebungen in *EP1* – übrigens die erste 12er-Vinylscheibe, die in der Kategorie Digital Musics Anerkennung gefunden hat.

Unsere erste Auszeichnung geht an GESCOM, das englische Trio der Produzenten Rob Brown und Sean Booth und dem Designer/Künstler Russell Haswell, für *MiniDisc*. Zum ersten, weil es innovativ ist – *MiniDisc* ist die erste unabhängige Produktion, die ausschließlich auf Minidisc erscheint. Es integriert die digitale Kompression des Formats in seine Ästhetik, indem es 45 Micro-Tracks mit 88 Cue-Punkten bringt, die alle im Shuffle-Modus abgespielt werden können. In den abgekürzten Perforationen und Pockennarben von *Is We* und *Vermin* und in den brütenden Meditationen von *Shoegazer* und *Dan Dan Dan* zeigte *MiniDisc* eine bezaubernde Bandbreite von digitaler Signalverarbeitung, Transformationen und Effekten.

Der zweite entscheidende Faktor ist, dass GESCOM das Alter Ego von Booth und Brown ist, die zusammen als das immens einflussreiche Duo Autechre aufnehmen, das mit seiner nicht-linearen Programmierung und extremen digitalen Signalverarbeitung ganz wesentliche Impulse für jene Post-Techno-Welt des Glitch und Microsound gegeben hat, die wir jetzt bewohnen. Wie Aphex Twin entstand das Duo in den frühen 90-ern und nahm bei Warp Records auf, einem der ehrwürdigsten Elektronik-Musik-Labels der Welt. Aus Erfahrung merkte Rehberg an: „Es gibt so viele Autechre-Kopierer heutzutage. Müssten wir unseren Lieblingsstrack von Autechre festlegen, würden wir drei Tage lang streiten. Die einen würden *LP 5* wählen, andere vielleicht *Chiastic Slide* – aber das ist in Ordnung, denn hier finden wir ein ernsthaftes Konzept und wir sind uns alle einig, dass es wichtig ist."

Unsere zweite Auszeichnung geht an den Aufnahmekünstler Chris Watson für seine CD *Outside the Circle of Fire*. Watson, der früher Mitglied von Cabaret Voltaire und dem Hafler Trio war, präsentiert auf seiner preisgekrönten CD 22 Vor-Ort-Aufnahmen von Säugetieren, Vögeln und Insekten, die von über dem Sambesi kreisenden Ziegenmelkern bis zum Klopfen der Holzwürmer in alternden Bodenbrettern reichen. Mit Hochleistungsmikrofonen hat Watson diese Klänge von ganz nahe aufgenommen und verblüfft den Hörer mit dem wild-aggressiven Klang des Ungewöhnlichen.

Das Geheimnis liegt – wie Parkins angemerkt hat – darin, „wie aus Außenaufnahmen Musik entsteht". Rimbaud wies darauf hin, dass *Outside the Circle of Fire* „wie digitale Klangbearbeitung klingt, es aber nicht ist". Die Jury anerkannte, wie Watsons über

that *Outside the Circle of Fire* "sounds like digital signal processing but isn't." The Jury appreciated how Watson's radio-linked audio-events had arrived at electroacoustic composition from the other direction, by an opposite method of exact miking rather than signal processing. Watson, as Humon suggested, had in fact "stripped electroacoustics away" only to turn a microphone and a vulture into inadvertently brilliant composers. Parkins also drew attention to Watson's obsessive notes - a case where information multiplied mystique instead of draining fascination.

Finally, we, the Digital Musics Jury awarded the 2000 Golden Nica for Digital Musics to *20' to 2000*, the 12 CD series of 20 minute compositions conceived by the Berlin producer and sound artist Carsten Nicolai for Noton, the label he runs with the producer Olaf Bender. On one hand the decision was not difficult— the ambition of the concept eclipsed the other entries.

On the other, we were uncertain as to whether we should honour a project which included the judge, Robin Rimbaud in his Scanner persona. None of us, least of all Scanner, wanted to be flamed for nepotism. Scanner abstained from the final vote. The rest of us pondered and argued, comparing the situation to previous years then referring back to precedents in other categories. Finally we decided the scope, the scale and the brilliant execution of *20' to 2000* was such that it absolutely deserved the Golden Nica.

20' to 2000 started in January 1999 with Frank Bretschneider/Komet's pinpoint nanosyncopation. In February came the whine, scree and juddering frequencies of Pan Sonic's Ilpo Vaisanen. Ryoji Ikeda presented his *Variations for modulated 440hz sinewaves* in March—a painful treble pulsating like submarine sonar. The powerstation hum of Ivan Pavlov/CoH's *memories of s-tone for Gavin Bryars* followed in April. Next came Olaf Bender/Byetone's project of swooping frequencies while Jens Massel/Senking's issued baleful drones in June. Thomas Brinkmann's *Ester Brinkmann* released his steady state minimal house music in July, while Scanner's Cystic-electric blue looped an anxious, prim Jenny "Walkabout" Agutter voice in August. Noto aka Nicolai's noto.time.dot-sea blue arrived in September-all rips, pocks and shreds. In October Mika Vainio of Pan Sonic released flash flood roar and melting brass band tones. The immer-

Funk übertragene Audio-Events sozusagen von hinten zu einer elektroakustischen Komposition wurden – indem sie höchst präzise und frei von Fremdeinflüssen per Mikrofon aufgenommen wurden und nicht durch elektronische Bearbeitung entstanden sind. Laut Naut Humon hat Watson „die Elektroakustik abgestreift", um ein Mikrofon und einen Geier in unvermutet brillante Komponisten zu verwandeln. Parkins wies auch auf Watsons fast schon obsessive Anmerkungen hin – ein Fall, wo die Information zur Vertiefung der Mystik beiträgt, anstatt die Faszination wegzunehmen.

Und schließlich hat die Jury die Goldene Nica 2000 für Digital Musics an *20' to 2000* des Berliner Produzenten und Klangkünstlers Carsten Nicolai für noton, dem Label, das er mit dem Produzenten Olaf Bender betreibt, verliehen. Einerseits fiel die Entscheidung nicht schwer – das ambitionierte Projekt stellte alle anderen in den Schatten.

Andererseits waren wir unsicher, ob wir ein Projekt auszeichnen sollten, an dem einer der Juroren – Robin Rimbaud als Scanner – mitgewirkt hat. Keiner von uns – und am allerwenigsten Scanner selbst – wollte wegen Nepotismus angeprangert werden. Scanner enthielt sich deshalb auch bei der Schlussabstimmung der Stimme. Wir anderen überlegten, diskutierten, argumentierten und verglichen die Situation mit der früherer Jahre und ähnlich gelagerten Fällen in anderen Kategorien. Letztlich entschieden wir, dass Ausrichtung, Umfang und die brillante Durchführung von *20' to 2000* so großartig waren, dass sie die Goldene Nica absolut verdiente.

20' to 2000 begann im Jänner 1999 mit Frank Bretschneiders bzw. Komets nadelspitzer Nanosyncopation. Im Februar kamen dann die jaulenden, gleitenden, vibrierenden Frequenzen von Pan Sonics Ilpo Vaisanen. Ryoji Ikeda präsentierte seine *Variations for modulated 440Hz sinewaves* im März – ein fast schmerzendes Treble, das wie ein U-Boot-Sonar pulsiert. Das Kraftwerksgewumme von Ivan Pavlov / CoH's *memories of s-tone for Gavin Bryars* folgte im April.

Als nächstes kam Olaf Benders bzw. Byetones Projekt der abstürzenden Frequenzen, während im Juni Jens Massel / Senking bösartiges Dröhnen losließen. Thomas Brinkmanns *Ester Brinkmann* brachte seine stetige Minimal-House-Musik im Juli heraus, während Scanners zystischer *electric blue* die ängstlich-präzise Stimme von Jenny „Walkabout" Agutter in Schleifen laufen ließ (August). Noto aka Nicolais *noto.-time.dot-sea blue* kam im September heraus – ganz rips, pocks and shreds.

Im Oktober publizierte Mika Vainio von Pan Sonic *flash flood roar* und schmelzende Blasmusikklänge. Die immersiven Surges und Spitzen von Wolfgang Voights *20 minuten gas* im November waren sofort erkennbar, während John Balance, Drew MacDowell und Peter Christopherson von Coil unter dem Namen Elph mit dem brillanten *Zwölf* die Serie

sive surges and peaks of Wolfgang Voight's *20 minuten gas* in November was immediately recognisable while John Balance, Drew Macdowell and Peter Christopherson of Coil under the name Elph concluded with the brilliant *Zwölf*, a ravine of audio turbulence.

By synthesising product and artwork, concept and design, serial form and individual signature into a single event-structure, *20' to 2000* became the ultimate curated series. A different production for each month of the last year of the 20th Century, it encapsulates and probably concludes the 90s fascination for serial projects familiar from producers such as Richie Hawtin and Thomas Brinkman and labels such as Kompakt. It's difficult to see how anyone can exceed the light touch of this sober limited edition series.

"For me" , Parkins commented, "its like seeing a series on TV and when you reach the conclusion it's bigger than the sum of its parts." "Curators," Rehberg suggested "are very important at the moment because there's so much information out there." The serial nature of the project makes it a sound art collaboration between some of the key producers in 21st Century post techno minimalism.

Rehberg also drew the Jury's attention to the striking design conceived by Desiree Heiss and Ines Kaag of Bless, the brilliant Berlin based conceptual fashion duo. Each transparent CD is identified by information arranged around the rim and by different coloured dots arranged around the central hole, then encased in a transparent CD shaped case with a thin base held together by a magnet. "The magnet did it for me," Rehberg said. "The 12 magnets that fit in the hole and snap together. It's something you can put on your desk. It's become an object."

As an object that radiates a lowercase mystique, *20' to 2000* epitomises several tendencies in mid 90s to early 00s microsonic audio: the tendency towards anonymity, pseudonymity and equivalence. Towards a frequent flyer internationalism that embraces producers from Tokyo, London, Cologne, Berlin and Barcelona. Towards a signature, branded sound. Towards audio austerity. Towards visual reduction in its lower case typeface. Towards sound art: Nicolai, Ikeda, Pan Sonic and Scanner all manufacture instal-

beschlossen – einem Abgrund auditiver Turbulenz. Indem es Produkt und Kunstwerk, Konzept und Design, die Reihenform und individuelle Signatur in eine gemeinsame Event-Struktur umgießt, ist *20' to 2000* die ultimative Form einer kuratierten Serie. Mit einer unterschiedlichen Produktion für jeden Monat des letzten Jahres des 20. Jahrhunderts schließt es die Faszination der 90-er Jahre für Serienprojekte ein (und wahrscheinlich ab), wie wir sie von Produzenten wie Richie Hawtin und Thomas Brinkman und von Labels wie Kompakt kennen. Es ist schwer vorstellbar, wie jemand den lockeren Touch dieser ernsten Serie in beschränkter Auflage noch übertreffen könnte.

„Für mich", kommentierte Parkins, „ist das, wie wenn man eine Serie im Fernsehen anschaut und bei der letzten Folge draufkommt, dass sie mehr ist als nur die Summe der Einzelfolgen." Rehberg meinte, dass Kuratoren derzeit sehr wichtig seien, weil es da draußen so viel Information gebe. Die Reihen-Natur dieses Projekts macht daraus eine Klangkunst-Kollaboration zwischen einigen der Schlüsselgestalten unter den Produzenten des Post-Techno-Minimalismus des 21. Jahrhunderts.

Rehberg lenkte die Aufmerksamkeit der Jury auch auf das hervorragende Design von Desirée Heiss und Ines Kaag von Bless, dem brillanten Konzept-Mode-Duo aus Berlin. Jede der transparenten CDs wird durch Information, die um die Nabe angeordnet ist, und durch farbige Punkte rund um das Mittelloch identifiziert; die CDs selbst sind in transparente CD-förmige Boxen mit schmaler Basis eingelegt, die von einem Magneten zusammengehalten werden. „Der Magnet war's letztlich bei mir", meinte Rehberg. „Die zwölf Magneten, die in das Loch passen und aneinander haften. Das ist etwas, das man auch auf den Schreibtisch stellen kann – das ist ein Objekt geworden."

Als Objekt, das eine gewisse Mystik ausstrahlt, fasst *20' to 2000* zahlreiche Tendenzen des mikroskopischen Audios der End-Neunziger und des beginnenden 21. Jahrhunderts beispielhaft zusammen: Die Tendenz zu Anonymität, Pseudonymen und Äquivalenz, zu einem Vielflieger-Internationalismus, der Produzenten aus Tokio, London, Köln, Berlin und Barcelona mit einschließt. Die Tendenz zu einem als Signatur dienenden Klang als Markenzeichen, zu einer Nüchternheit im Klang, zu einer visuellen Reduktion auch durch die konsequente Kleinschreibung. Und zur Klangkunst: Nicolai, Ikeda, Pan Sonic und Scanner produzieren alle auch Installationen für so prestigeträchtige Kunsträume wie das Centre Pompidou, das Hayward und das ICC.

Vor allem aber typifiziert *20' to 2000* die Tendenz zur Zusammenarbeit zwischen Label, Kurator/Produzenten und Designern, die ein Teil dieser post-medialen Praxis ist. Wie der Medientheoretiker Howard Slater argumentiert, entsteht ein postmedialer Operator

lations for prestigious art spaces such as the Pompidou, the Hayward and the ICC.

Above all, *20' to 2000* typifies the tendency towards collaboration between label, curator producer and designers that's part of post media practice. As argued by the media theorist Howard Slater, a post-media operator emerges wherever practices "escape the institutional control of the industry and the media" thereby eluding the "dominant repressive models of an inherited subjectivity."

Instead of choosing "competition, exposure and the labor of success," the composer operates "outside these monetary and conceptual constraints." Concurrently, a new experimental attitude emerges which signals "the end of the need to conform to what is expected and 'understood'." The result is a "renewed appreciation for the idiosyncrasies of sound and the transgression of perceptual habits they inspire." Tomorrow will be otherwise but for now, *20' to 2000* sums up the moment.

References

1 'perception of processes', Sascha Kosch, Sleevenotes to Clicks and Cuts , Mille Plateaux, 2000
2 'the digital routing ...' Sascha Kosch, Sleevenotes to Clicks and Cuts , Mille Plateaux 2000.
3 'the technical errors ...' Rob Young, Worship the Glitch, The Wire 190/191 NewYear 2000
4 'practices escape ...' Howard Slater, Post Media Operators p398-9 in Read Me Read Me Read Me (Autonomedia,1999)

immer dann, wenn die Praktiken „der institutionellen Kontrolle der Industrie und der Medien entfliehen" und damit sich den „dominanten repressiven Modellen einer ererbten Subjektivität" entziehen. Anstatt sich dem Wettbewerb und der Mühe des Erfolgs auszusetzen, operiert der Komponist „außerhalb dieser monetären und konzeptuellen Zwänge". Gleichzeitig entwickelt sich eine neue experimentelle Attitüde, die uns „das Ende der Notwendigkeit, dem Erwarteten und dem ‚Verstandenen' zu entsprechen," signalisiert. Das Ergebnis ist eine erneuerte „Wertschätzung der Idiosynkrasien des Klangs und die Überwindung der Wahrnehmungsgewohnheiten, die sie inspirieren."

Morgen wird es schon anders sein, aber *20' to 2000* fasst die Gegenwart zusammen.

Quellen

1 *perception of processes*, siehe Sascha Kosch, Covertext zu Clicks and Cuts, Mille Plateaux, 2000
2 *the digital routing* ..., siehe Sascha Kosch, Covertext zu Clicks and Cuts, Mille Plateaux 2000
3 *the technical errors* ... siehe Rob Young, Worship the Glitch, in: *The Wire 190/191*, NewYear 2000
4 *practices escape* ..., siehe Howard Slater, „Post Media Operators", S. 398f, in: *Read Me Read Me Read Me*, Autonomedia, 1999

20' to 2000
raster-noton / Carsten Nicolai

notorious for the close interaction between experimental forms of music, art and science, label noton announces a new series of 12 "20 minutes" cds. the artistic concept behind it, is the creation of a piece for the last 20 minutes of the year 1999 as a cutting edge of the century.

20' to 2000 is a monthly magazine-like release for the last year of this millennium. in the 12 issues 12 artists are invited to write down their ideas, possibly a manifest of the millennium; design for the series uses new technologies and the characteristic translucent materials of noton.

all issues are limited and released at the end of every month. A binding magnet construction-kit for the 12 cds will be designed and produced by "bless", a designer company based in paris and berlin.

timetable for 20' to 2000

01	komet	germany, rastermusic, member of signal, product
02	ilpo vaisanen	finland, sähkö, blast first, member of pan sonic
03	ryoji ikeda	japan, touch, cci, member of dumb type
04	coh	russia, rastermusic
05	beytone	germany, rastermusic, member of signal, product
06	senking	germany, karaoke kalk, work also as kandis
07	thomas brinkmann	germany, ernst label
08	scanner	united kingdom, subrosa
09	noto	germany, noton, rastermusic, member of product, signal
10	mika vainio	finland, sähkö, blast first, Ø, philus, member of pan sonic
11	wolfgang voigt	germany, profan, studio 1, mike ink, gas
12	elph	united kingdom, members of coil

issued monthly

das für die enge interaktion zwischen experimentellen musikformen, kunst und wissenschaft bekannte label noton kündigt eine neue serie von zwölf 20 minutes-CDs an. dahinter steht das künstlerische konzept, ein stück für die letzten zwanzig minuten des jahres 1999 zu schaffen – am cutting edge des jahrhunderts.

20' to 2000 ist eine monatliche magazin-ähnliche ausgabe für das letzte jahr des jahrtausends. in den zwölf ausgaben werden zwölf künstler eingeladen, ihre ideen niederzulegen, möglichst als manifest ihres millennium-designs, denn die serie verwendet neue technologien und das charakteristisch transluzente material von noton.

alle ausgaben sind limitiert und erscheinen jeweils zum monatsende. das magnetische binde-kit für die zwölf cds wurde von bless, einer designfirma mit sitz in paris und berlin, entworfen und konstruiert.

zeitplan für 20' to 2000

01	komet	deutschland, rastermusic, mitglied von signal und produkt
02	ilpo vaisanen	finnland, sähkö, blast first, mitglied von pansonic
03	ryoji ikeda	japan, touch, cci, mitglied von dumb type
04	coh (son)	russland, rastermusic
05	beylone	deutschland, rastermusic, mitglied von signal und produkt
06	sanking	deutschland, karaoke kalk, arbeiten auch als Kandis
07	thomas brinkmann	deutschland, ernst label
08	scanner	großbritannien, subrosa
09	noto	deutschland, noton, rastermusic, mitglied von produkt und signal
10	mika vaino	finnland, sähkö, blast first, ø, mitglied von pansonic
11	wolfgang voigt	deutschland, profan, studio 1, mike inke, gas
12	elph	großbritannien, mitglied von coil

erscheint monatlich

20' to 2000

RASTERMUSIC/NOTON IS A LABEL COOPERATION FOR ELECTRONIC MUSIC BASED IN GERMANY. IN 1999 BOTH ARTIST LABELS RASTERMUSIC AND NOTON.ARCHIV FÜR TON UND NICHTTON MERGED. RASTERMUSIC WAS FOUNDED BY OLAF BENDER AND FRANK BRETSCHNEIDER IN 1996. THEIR MAIN INTEREST WAS TO GIVE REPETITIVE MINIMAL-MUSIC FOLLOWING THE DIRECTION OF POP. NOTON.ARCHIV FUER TON UND NICHTTON WAS FOUNDED BY CARSTEN NICOLAI MORE AS A PLATFORM FOR CONCEPTIONAL AND EXPERIMENTAL-RELATED PROJECTS IN MUSIC, ART AND SCIENCE.

rastermusic/noton (D) ist eine label-cooperation für elektronische musik. 1999 erfolgte die verschmelzung der beiden künstler-labels rastermusic und noton.archiv für ton und nichtton. rastermusic wurde 1996 von olaf bender und frank bretschneider gegründet. ihr hauptinteresse lag in repetitiver minimal-music in richtung popmusik. noton.archiv für ton und nichtton wurde von carsten nicolai eher als plattform für konzeptionelle und experimentelle verwandte projekte in musik, kunst und wissenschaft gegründet.

MINIDISC
GESCOM

Following several short releases *minidisc* is Gescom's only full length release. It is a celebration/critique of the format both in terms of it's immediate aesthetic and the technical and mathematical limitations it imposes on its users.

Reviews

"Of course *Minidisc* is another attempt to exploit the market of digitalia, after CD, DAT, DCC, and of course it has added some gadgets: you can have many more track numbers on a disc, it's easier to use in loop mode, or shuffle the information. Therefore I doubt whether many people who actually own a minidisc player will have it hooked in their home stereo system, or packed next to their records for tonight's DJ set ... Nevertheless this is the first official independent minidisc only release (and I'm sure it will be bootlegged to CDR by those who think that the format is just a waste of money) (Yep, you're right there, Franky—MP), and it's not hand copied, but factory manufactured. Gescom, as we all know, is one of the alter ego's of those very intelligent technoids Autechre (my favorite in this field). So far I've heard the odd piece, and a dancy 12", but this is altogether something different. The 88 tracks are quite short, and very noisy. Occasionally it hints at something, that may be a beat ... but it's rare. Knowing that this is released by OR, we are not surprised it sounds like Farmers Manual or Pita—the Viennese abuser of digital media. Of course I tried looping some tracks (as the minidisc is better capable of performing that than the CD) and it worked fine. No doubt this minidisc will appeal to the adventurous DJ's (I'm not) and that bootleg copies will be made to CDR (the other exciting new medium) for home consuming (Right again,

Nach mehreren kürzeren Releases ist *minidisc* Gescoms einziges Release in voller Länge. Es feiert und kritisiert das Format zu gleicher Zeit, sowohl hinsichtlich seiner Ästhetik wie der technischen und mathematischen Beschränkungen, die es seinen Usern auferlegt.

Kritiken

„Natürlich ist *Minidisc* ein weiterer Versuch, den digitalen Musikmarkt auszubeuten, nach CD, DAT, DCC, und natürlich hat es einiges an Schnickschnack mitbekommen: Man kann viel mehr Track-Nummern auf einer Scheibe haben, es lässt sich leichter im Loop-Modus fahren und man kann leichter Information mischen. Deshalb frage ich mich, ob viele der Besitzer eines Minidisc-Players den tatsächlich an die Heim-Stereoanlage angehängt oder ob sie ihn nicht eher zu den Platten für die DJ-Session heute abend gepackt haben ... Jedenfalls ist dies die erste unabhängige Produktion ausschließlich auf Minidisc (und ich bin sicher, dass es bald auf CDR umkopiert wird von all denen, die das Format für reine Geldverschwendung halten) [Richtig, Frank! – M.P.], und sie ist nicht handkopiert, sondern Fabrikware. Wie wir alle wissen, ist Gescom ein Alter Ego der sehr intelligenten Technoiden von Autechre [meinen Favoriten auf diesem Gebiet]. Bisher habe ich das eine oder andere Stück gehört und eine tanzbare 12", aber dies hier ganz was anderes. Die 88 Tracks sind ziemlich kurz und sehr geräuschvoll. Manchmal erinnern sie an etwas, das ein Beat sein könnte – aber eben nur manchmal. Wer sich vor Augen hält, dass dies bei OR herausgekommen ist, ist nicht erstaunt, dass es wie Farmers Manual klingt oder wie Pita – der Wiener Missbraucher digitaler Medien. Natürlich habe ich versucht, einige Tracks als Schleife laufen zu lassen (die Minidisc ist dafür besser geeignet als die CD) – und es hat gut geklappt. Ohne Zweifel wird diese Minidisc allen abenteuerlustigen DJs gefallen [ich bin keiner], und Raubkopien auf CDR werden für den Hausgebrauch entstehen [Wieder richtig, Frank! – M.P.]. Wem die Shuffle-Modi früherer CDs von Jos Smolders, Farmers Manual oder Otomo Yoshide

Franky - MP). If the shuffle modes of previous CD's by Jos Smolders, Farmers Manual or Otomo Yoshide appealed to you, then this is another treasure-trove of sounds."
VITAl (The Netherlands)

"Autechre have a second album out this month, this time in their Gescom guise, and it's definitely the better release. As it's only available on MiniDisc, Autechre/Gescom completists will have to shell out for a MiniDisc player, too, if they want to hear it. Gescom are undoubtedly aware of the MiniDisc "wow" factors—like the title scrolling across the LED screen while the track is playing; what next, bouncing ball singalong texts?—but they've also investigated how the MD's digital compression of information might affect the sound. You don't need to know the psychoacoustics of the thing to enjoy it, though. If some of the disc's 45 tracks (spread over 88 cue points) are little more than spare parts from the Autechre toolkit, parts of MiniDisc are as good as anything they've done: the blue hour meditations of "Shoegazer" and "Dan Dan Dan"; the compacted perforations of "Is We" and "Vermin"; the envelope folds and sugar twists of "Wab Wat" and "Squashed To Pureness". A consistently astounding range of effects, treatments and transformations here give GESCOM a leg up into the realm of such studio alchemists as Coil or Luc Ferrari."
The Wire (UK)

gefallen haben, der hat hier eine weitere Schatzkiste an Klängen gefunden."
Vital (Niederlande)

„Autechre, die dieses Mal als GESCOM auftreten, haben dieses Monat ein zweites Album herausgebracht, und das ist zweifellos das Bessere der beiden. Da es nur auf Minidisc erhältlich ist, werden Sammler von Autechre/Gescom wohl für einen Minidisc-Player in die Tasche greifen müssen, wenn sie es hören wollen. Gescom ist sich der „Wow-Faktoren" der Minidisc wohl bewusst – etwa dass der Titel beim Abspielen eines Tracks über das LCD-Panel läuft. Was folgt als nächstes? Etwa ein im Takt des Textes mithüpfender Ball? –, aber sie haben auch untersucht, wie die digitale Informations-Kompression der Minidisc sich auf den Klang auswirkt. Man braucht aber von der Psychoakustik des Dings nichts zu verstehen, um es zu genießen. Wenn auch einige der 48 Tracks (die über 88 Cue-Points verteilt sind) wenig mehr als Ersatzteile aus dem Werkzeugkoffer von Autechre sind, so gehören Teile der Mindisc zum besten, was sie je gemacht haben: Die Blue-Hour-Meditationen von *Shoegazer* und *Dan Dan Dan*; die kompakten Perforationen von *Is We* und *Vermin*; die sich umschlaggleich faltenden und doch zuckerigen Verdrehungen von *Wab Wat* und *Squashed to Pureness*. Mit ihrer permanent faszinierenden Bandbreite von Effekten, Bearbeitungen und Umformungen stellen GESCOM einen Fuß in die Tür des Königreichs solcher Studio-Alchemisten wie Coil oder Luc Ferrari."
The Wire (Großbritannien)

THE GESCOM GROUP BEGAN ITS ACTIVITIES IN 1991. IT EXISTS AS A PLATFORM FOR SEVERAL ARTISTS TO WORK IN VARIOUS DIFFERENT COMBINATIONS, WHILST REMAINING—OTHERWISE—ANONYMOUS. WORKING MEMBERS FOR THE PROJECT *MINIDISC* WERE RUSSELL HASWELL, SEAN BOOTH AND ROB BROWN. ▬▬ Die Gescom-Gruppe begann ihre Aktivitäten 1991. Sie besteht als Plattform für mehrere Künstler, die in unterschiedlichen Kombinationen zusammenarbeiten, ansonsten aber anonym bleiben. Am Projekt *minidisc* haben

OUTSIDE THE CIRCLE OF FIRE

Chris Watson

Alternative Press (US):
"French conceptual artist Marcel Duchamp once predicted that the artist of the future would point at what already exists and it would become art. Many sound artists have taken a philosophically similar outlook, using found sound, plagiarism and media manipulation. Former Hafler Trio and Cabaret Voltaire collaborator Chris Watson takes a different approach to this aesthetic, preferring to let the world around us do its own talking.

Stepping is an engaging collection of field recordings made at exotic locales while Watson was doing location sound for various documentaries. He presents a rich variety of environments ranging from flies near the Mara River in Kenya, to nesting rooks in an old churchyard, to fishing bats in Venezuela.

Animals and the elements take center stage throughout the disc. It seems strange to have any human's name on the sleeve at all. Watson has wisely chosen to leave these sounds raw and realistic, making this about as close as most of us will get to a world tour. Use your ears and drift.
(Joseph Cross, Wired, UK)

Watson's lead instrument is the tape recorder. After working with Cabaret Voltaire and The Hafler Trio, he became sound recordist for the Royal Society for the

Der französische Konzeptkünstler Marcel Duchamp hat einst vorhergesagt, dass der Künstler der Zukunft auf das hinweisen würde, was bereits existiert, und dass es dadurch Kunst würde. Viele Klangkünstler haben einen ähnlichen philosophischen Ansatz gewählt, indem sie gefundene Klänge verwenden, plagiieren und Medien manipulieren. Chris Watson, früher Mitglied des Hafler Trio und bei Cabaret Voltaire, geht einen anderen Weg zu seiner Ästhetik: Er zieht es vor, die Welt um uns selbst sprechen zu lassen.

Stepping ist eine engagierte Sammlung von Vor-Ort-Aufnahmen an exotischen Orten, wo Watson Ton für mehrere Dokumentationen aufgenommen hat. Er bringt eine reichhaltige Sammlung an verschiedenen Umgebungen mit, von Fliegen nahe dem Mara-Fluss in Kenia über nistende Saatkrähen auf einem alten Friedhof bis hin zu fischenden Fledermäusen in Venezuela.

Tiere und die Elemente haben auf der ganzen Disk die tragenden Rollen inne. Es erscheint fast seltsam, überhaupt einen menschlichen Namen auf dem Umschlag zu sehen. Watson hat sich klugerweise entschlossen, die Klänge roh und realistisch zu lassen, und bringt uns damit so nahe an eine Reise um die Welt, wie das nur möglich ist. Setzt eure Ohren ein und gleitet dahin!
Joseph Cross, Wired (Großbritannien)

Watsons Lead-Instrument ist das Tonband. Nachdem er bei Cabaret Voltaire und beim Hafler Trio gearbeitet hatte, wurde Watson Tonaufnahmeleiter der Royal Society for the Protection of Birds. Er hat sich inzwischen einer Film- und Videoproduktionsgesellschaft angeschlossen, die für die BBC Naturdoku-

Protection of Birds. He has since joined a film and video production company, working for BBC wildlife documentaries and occasional feature films. "In recent years I have noticed that some of the particular locations I have visited had an overall characteristic—sparkling acoustics, a special timbre, sometimes rhythmic or transient animal sounds". Watson's interest goes beyond the brief of the programmes he works on: he takes the chance to explore "the intangible sense of being in a special place—somewhere that has a spirit—a place that has an atmosphere".

The 13 recordings on *Stepping into the Dark* contrast a windswept forest in Glen Cannich with the gathering conversations of rooks roosting in a churchyard in Northumberland. Other atmospheres include the heat and wall of sound found on the River Mara in Kenya, fishing bats on a mangrove pool in Venezuela, the ritual dance of snipe at dusk in the Northern Hebrides ... a hydrophone at 5m. depth in the Moray Firth captures the signature whistles and clicks of botlenose dolphins.

But it is not simply a question of capture: nor do the atmospheres settle softly into the genre of New Age-style environments. "These recordings avoid background noise, human disturbance and editing. They are made with sensitive microphones camouflaged and fixed in position well in advance of any recording or animal behaviour. The mics are cabled back on very long leads or radio-linked back to a hiding place or concealed recording point. Sites can be discovered by chance, by researching features on a map, in history through anecdote, and also in conversation with local people about their feelings (both for and against) particular places. Tom Lethbridge identified places for several spirits within the local topography of an area. I suspect this also includes flora and fauna, the time of day, the elements and the season."

mentarfilme und bisweilen Spielfilme produziert. „In den letzten Jahren habe ich festgestellt, dass einige der Orte, die ich bereist habe, ein gemeinsames Merkmal aufweisen – eine strahlende Akustik, ein spezielles Timbre, manchmal rhythmische oder transiente Tierklänge." Watsons Interesse geht über das reine Drehbuch der Programme, an denen er arbeitet, hinaus: Er nimmt die Chance wahr, „das ungreifbare Gefühl" zu erforschen, „an einem speziellen Ort zu sein – einem Ort, der Geist hat, einer Stelle mit Atmosphäre."

Die 13 Aufnahmen in *Stepping into the Dark* stellen einen winddurchrauschten Wald in Glen Cannich den Versammlungsrufen von horstenden Saatkrähen auf einem Friedhof in Northumbria gegenüber. Andere atmosphärische Orte umfassen die Wärme und das Wabern eines Klanges, den er am Mara-Fluß in Kenia gefunden hat, fischende Fledermäuse in einem Mangrovensumpf in Venezuela, den rituellen Tanz der Schnepfen in der Morgendämmerung auf den nördlichen Hebriden ... ein Unterwassermikrofon in fünf Metern Tiefe fängt die Signalpfiffe und -klicks von Großen Tümmlern im Morey Firth ein.

Aber es geht nicht nur um das Einfangen und auch die Atmosphäre lässt sich nicht einfach in das Genre der Umweltklänge im New-Age-Stil einreihen. „Diese Aufnahmen vermeiden Hintergrundgeräusche, menschliche Störungen und Schnitte. Sie werden mit empfindlichen Mikrofonen gemacht, die lange vor der eigentlichen Aufnahme irgendwelchen Tierverhaltens getarnt angebracht wurden. Die Mikros werden über lange Leitungen oder über kabellose Verbindungen an das ebenfalls getarnte bzw. versteckte Aufnahme-Equipment angeschlossen. Die Aufnahmeorte werden entweder zufällig entdeckt oder durch Recherchen auf Landkarten und in der Geschichte gefunden, über Anekdoten oder im Gespräch mit den Einheimischen über ihre (sowohl positiven wie negativen) Gefühle verschiedenen Orten gegenüber. Tom Lethbridge hat Orte für verschiedene Geister innerhalb der lokalen Geografie identifiziert und ich vermute, zu diesen gehören auch Flora und Fauna, Tageszeit, die Elemente und die Jahreszeiten."

Chris Watson (GB). 1972—1981: Cabaret Voltaire, 1981—1984: Tyne Tees television; sound recordist and founder member of The Hafler Trio. 1985-1987: Royal Society for the Protection of Birds; sound recordist and Hafler Trio projects. 1987—1993: Freelance sound recordist. 1994 to present: Partner with Hoi Polloi Film & Video; sound recordist and projects for Touch and Ash international. Current: Location and post-production of recordings to films by Sir David Attenborough. ▬ Chris Watson (GB), 1972–1981: Cabaret Voltaire; 1981–1984: Tyne Tees Television, Tontechniker und Gründungsmitglied des Hafler Trio. 1985–1987: Royal Society for the Protection of Birds, Tontechniker und Hafler-Trio-Projekte. 1987–1993: Freiberuflicher Aufnahmetechniker. Seit 1994: Partner von Hoi Polloi Film & Video; Aufnahmetechniker und Projekte für Touch und Ash International. Derzeit: Vor-Ort-Aufnahmen und Postproduk-

SOUND CHARACTERS FOR "THIRD EAR MUSIC"

Maryanne Amacher

Tracks 1, 4, 4 and 6 contain what I call "Third Ear Music", - when our ears act as instruments and emit sounds as well as receive them. "Third Ear Music" is composed to stimulate our ears to "sound" their own tones and melodic shapes. When played at the right sound level, which is quite high and exciting, the tones in this music will cause your ears to act as neurophonic instruments that emit sounds that will seem to be issuing directly from your head. In concert my audiences discover music streaming out from their head, popping out of their ears, growing inside of them and growing outside of them, meeting and converging with the tones in the room. They discover they are producing a tonal dimension of the music which interacts melodically, rhythmically, and spatially with the tones in the room. Tones "dance" in the immediate space of their body, around them like a sonic warp, cascade inside ears, and out to space in front of their eyes, mixing and converging with the sound in the room. Do not be alarmed! Your ears are not behaving strangely or being damaged! Nor are loudspeakers being damaged. These virtual tones are a natural and very real physical aspect of auditory perception, similar to the fusing of two images resulting in a third three dimensional image in binocular perception. Produced interaurally, these virtual sounds and melodic patterns originate in ears and neuroanatomy, not in your loudspeakers. I believe that such response tones exist in all music, where they are usually registered subliminally, and are certainly masked within more complex timbres. I want to *release* this music which is produced by the listener, bring it out of subliminal existence, make it an important sonic dimension in my music.

Die Tracks 1, 4, 5 und 6 enthalten etwas, das ich als „Musik fürs Dritte Ohr" bezeichnen möchte – wenn unsere Ohren als Instrumente fungieren und Klänge nicht nur empfangen, sondern auch aussenden. *Third Ear Music* ist komponiert, um unsere Ohren anzuregen, ihre eigenen Töne und melodischen Formen „klingen zu lassen". Wenn die Stücke beim richtigen – ziemlich hohen und recht aufregenden – Klanglevel abgespielt werden, lassen die Töne in dieser Musik die Ohren zu neurophonischen Instrumenten werden, die Klänge produzieren, welche direkt aus dem Kopf zu kommen scheinen. Bei Konzerten entdecken meine Hörer, wie Musik aus ihren Köpfen strömt, aus ihren Ohren quillt, innerhalb und außerhalb von ihnen wächst und mit der Klangwelt im Raum verschmilzt. Sie entdecken, dass sie eine tonale Dimension der Musik produzieren, die melodisch, rhythmisch und räumlich mit den Tönen im Raum interagiert. Die Töne „tanzen" im unmittelbaren Raum ihres Körpers, rund um sie wie eine Klanghülle, wie ein Wasserfall innerhalb der Ohren und hinaus in den Raum vor ihren Augen, vermischen sich und verschmelzen mit dem Klang im Raum. Erschrecken Sie nicht! Ihre Ohren verhalten sich nicht seltsam, sie erleiden keinen Schaden! Auch die Lautsprecher werden nicht beschädigt. Diese virtuellen Töne sind ein natürlicher und sehr realer physische Aspekt unserer Gehörswahrnehmung, ähnlich der Verschmelzung zweier Bilder zu einem dritten dreidimensionalen bei der beidäugigen Wahrnehmung. Interaural produziert, entstehen diese virtuellen Klänge und melodischen Muster in den Ohren und der Neuroanatomie, nicht in den Lautsprechern. Ich glaube, dass solche Response-Töne in jeder Musik existieren, aber normalerweise nur subliminal registriert und sicherlich von den komplexeren Timbres überlagert werden. Ich möchte diese vom Hörer erzeugte Musik *freisetzen*, aus ihrer Existenz unterhalb der Wahrnehmungsschwelle befreien und sie zu einer wichtigen klanglichen Dimension in meiner Musik machen.

MARYANNE AMACHER (USA) IS A COMPOSER AND SOUND INSTALLATION ARTIST. MUSICAL STUDIES IN SALZBURG, AUSTRIA AND DARTINGTON DEVON, ENGLAND AND COMPOSITION STUDIES WITH KARLHEINZ STOCKHAUSEN. STUDIES IN MUSIC AND COMPUTER SCIENCE AT THE UNIVERSITY OF PENNSYLVANIA. FELLOW AT THE CENTER FOR ADVANCED VISUAL STUDIES, MIT (1972—76)—COLLABORATION WITH THE VISUAL ARTISTS SCOTT FISHER, LUIS FRANGELLA, AND THE ARCHITECT JUAN NAVARRO BALDEWEG. CREATED WORKS WITH JOHN CAGE AND MERCE CUNNINGHAM. ■ Maryanne Amacher (USA) ist Komponistin und Klanginstallations-Künstlerin. Sie studierte Musik in Salzburg und Dartington (Devon, GB) und Komposition bei Karlheinz Stockhausen sowie Musik und Computerwissenschaften an der University of Pennsylvania. 1972–76 war sie Fellow am Center for Advanced Visual Studies des MIT. Zusammenarbeit mit den visuellen Künstlern Scott Fisher, Luis Frangella und mit dem Architekten Juan Navarro Baldeweg sowie mit John Cage und Merce Cunningham.

VILLIGER
Dat Politics

Quickly Dat Politics gained a name for themselves. With three people of Tone Rec plus one more member they have a split 12" on *Fat Cat*, a LP for their own label skipp and this CD for A-musik. The 12" and LP were nice already, but this CD blows my mind! Dat Politics are announced as a computer group, or laptop quartet, so I expected heavy improv doodlings like Rehberg & Bauer or the rest of the Mego posse, but how delighted I am to find something different. Dat Politics brings us 11 cuts of cute music—popmusic maybe. Each track—no titles—is well-balanced, structured around guiding themselves. Hardly a week brother in here. There are hints of hip hop and techno elements, but it avoids falling in these traps ...
VITAL (february 2000)

Dat Politics hat sich schnell einen Namen gemacht. Mit drei Leuten von Tone Rec und einem weiteren Mitglied haben sie eine Zwölf-Zöller für Fat Cat, eine LP für ihr eigenes Label Skipp und eben diese CD –*Villiger* – für A-Musik herausgebracht. Die 12"-Scheibe und die LP waren schon nett, aber diese CD haut mich um! Dat Politics werden als Computergruppe oder als Laptop-Quartett angekündigt, deshalb erwartete ich harte improvisierte Spielereien wie von Rehberg & Bauer oder dem Rest der Mego-Truppe, aber wie erfreut war ich, etwas völlig anderes zu finden! Dat Politics bringt uns elf Cuts einer hübschen Musik – vielleicht Popmusik. Jeder der Tracks – alle ohne Titel – ist gut ausgewogen und um führende Themen strukturiert. Kaum einer ist schwächer. Es gibt Andeutungen von Hip-hop und Techno-Elemente, aber die CD vermeidet sorgfältig, in diese Fallen zu gehen ...
(VITAL, Februar 2000)

FORMED IN OCTOBER 1998, DAT POLITICS ARE A 4-PIECE LAPTOP UNIT (EMERIC AELTERS, CLAUDE PAILLIOT, GAETAN COLLET, VINCENT THIERION) BASED IN THE CITY OF LILLE IN NORTHERN FRANCE. IN ONE YEAR THEY RELEASED A SPLIT 12" ON FAT CAT, VILLIGER ALBUM ON A-MUSIK, AND THEIR NEW *TRACTO FLIRT* ALBUM ON THEIR OWN LABEL SKIPP, WHICH THEY SET UP RECENTLY. ■■■ Die im Oktober 1998 gegründete Gruppe dat politics ist eine Vier-Laptop-Einheit (Emeric Aelters, Claude Pailliot, Gaetan Collet, Vincent Thierion) und in der nordfranzösischen Stadt Lille beheimatet. Innerhalb eines Jahres haben die vier eine 12"-Scheibe bei Fat Cat, das *Villiger*-Album bei A-musik und das *Tracto Flirt*-Album für ihr eigene kürzlich

THREE
Kevin Drumm

Three was assembled on a computer using a simple editing program. The sound sources used were mostly derived from prepared guitar, analog synthesizer and field recordings, some of which were treated with a modular synthesizer and/or computer sound synthesis program.

I had approximately five or six sections that I wanted to work with. Each section had its own character. There is no specific system of logic in the way I assembled the piece. I was merely guided by the "feel" of the sections, and everything seemed to fall into place the way it did without regard for any personal aesthetic considerations (with regards to form) that I might normally have.

Three was realized in fall and winter of 99. A slightly different version using a quadraphonic sound system was presented in Chicago in December 99, a revised version should appear on Perdition Plastics in late 00.

Three wurde auf einem Computer mit Hilfe eines simplen Editierprogramms zusammengesetzt. Als Klangquellen wurden überwiegend präparierte Gitarren, analoge Synthesizern und Vor-Ort-Aufnahmen eingesetzt, die anschließend mit einem modularen Synthesizer bzw. einem Computer-Klangsyntheseprogramm bearbeitet wurden.
Ich wollte mit fünf oder sechs Abschnitten arbeiten, von denen jeder seinen eigenen Charakter hat. Das Stück entstand so, wie es ist – ohne zu Grunde liegende Logik. Ich ließ mich einfach vom Feeling der einzelnen Abschnitte leiten, und alles schien so an Ort und Stelle zu rutschen, wie es jetzt ist, ohne Rücksicht auf irgendwelche persönlichen ästhetischen Überlegungen (hinsichtlich der Form), wie sie normalerweise zu haben pflege.
Three entstand im Herbst und Winter 1999. Eine leicht veränderte Version unter Verwendung eines quadrophonischen Klangsystems wurde im Dezember 1999 in Chicago vorgestellt, eine revidierte Variante sollte gegen Ende 2000 bei Perdition Plastics erscheinen.

KEVIN DRUMM (USA), AVANT-GARDE TABLETOP GUITARIST AND SYNTHESIZER PLAYER (FORMERLY OF BRISE-GLACE AND A NUMBER OF OTHER GROUPS) HAS BEEN WORKING FOR SEVERAL YEARS WITH OTHER MEMBERS OF CHICAGO'S IMPROV SCENE—MOST NOTABLY KEN VANDERMARK AND JIM O'ROURKE. HIS ALL-WHITE SELF-TITLED ALBUM ON *PERDITION PLASTICS* GIVES YOU A CHANCE TO CHECK OUT SOME OF HIS REALLY OUT THERE SOLO WORK. ■ Kevin Drumm (USA), der Avantgarde-Tabletop-Gitarrist und Synthesizer-Spieler (früher bei Brise-Glace und anderen Gruppen) arbeitet seit etlichen Jahren mit anderen Mitglieder der Improv-Szene von Chicago, u. a. mit Ken Vandermak und Jim O'Rourke. Sein ganz weißes, nach ihm selbst benanntes Album bei Perdition Plastics gibt Hörern die Möglichkeit einige seiner allerbesten Soloarbeiten zu

MAKESND CASSETTE

Mark Fell / Mat Steel

The apparent simplicity which unites those coming from "glitch" agendas with those coming from the darkest fringes of the dancefloor necessitates a drift around the fringes of the scene, skirting the least Euclidean sectors of the dancespaces and headspaces. But snd leave clues as to their seeing of a wider topology away from the "field studies" mentality of the minute glitch collectors, and *makesnd cassette* ensures that any schematics for a new digital masterplan are written with soluble ink on blotting paper. A mathematical analysis of this scene works to some degree, but obviously doesn't go far enough. Whence it has been suggested that this music operates "between 0 and 1"—focussing on infinity not as some techno or trance defined idea as progressing to an invisible point of ecstatic oblivion, but instead as a highly introspective and bifurcating process—this definition can play into the hands of those who wish to enforce this critical distance between source and process / product. Thus many tracks may exist as reconfigured fragments from something much greater, but this fetishises a process of piecing back together whilst the original dissection is left imagined as something akin to a laboratory technician's specimen tank. Listen again to *makesnd cassette* and you hear the swarming existence of various source sounds, not taken and stripped back a priori to reconfiguration, but taken and stripped back to model some process of natural decay glimpsed in both the swell of the speaker system and the rebelling headspace of the nurtured nightclubber. In some instances that's all there is, modelled decay processes left to stimulate the mind: music on the verge of unbecoming music, yet still retaining a vast grace and elegance. All the time collapsing and all the time rebuilding—not through some textbook process of cold, clinical modelling, but through radically changing topologies and geographies. (*I.M.Trowell*)

Die augenfällige Einfachheit, die jene, die von den „Glitch"-Agenden herkommen, mit den anderen vereint, die von den finstersten Rändern des Tanzbodens kommen, erfordert ein Driften entlang des Randes der Szene, das die am wenigsten euklidischen Sektoren der Tanzflächen wie der Kopfräume streift. Aber snd hinterlässt Hinweise darauf, dass hier eine weitere Topologie betrachtet wird als jene der „Feldstudien"-Mentality der kleinen Glitch-Sammler, und *makesnd cassette* stellt sicher, dass jedes Schema für einen neuen digitalen Masterplan mit löslicher Tinte auf Löschpapier geschrieben wird.
Eine mathematische Analyse dieser Szene funktioniert bis zu einem gewissen Grad, aber sie geht offensichtlich nicht weit genug. Deshalb wurde auch suggeriert, dass diese Musik „zwischen 0 und 1" operiert – konzentriert auf Unendlichkeit nicht als eine Techno- oder Trance-definierte Idee, die zu einem unsichtbaren Punkt ekstatischen Vergessens fortschreitet, sondern als ein höchst introspektiver und zwiespältiger Prozess. Diese Definition kann in die Hände jener spielen, die die kritische Distanz zwischen Quelle und Prozess/Produkt durchzusetzen versuchen. Und deshalb mögen viele Tracks als neu konfigurierte Fragmente von etwas viel Größerem existieren, aber dies macht einen Prozess des Wiederzusammenfügens zum Fetisch, während man sich die ursprüngliche Zerteilung als etwas vorstellt, das einer Probensammlung von Labortechnikern gleichzusetzen ist.
Man höre sich *makesnd cassette* nochmals an und man wird die schwärmende Existenz verschiedener Quellgeräusche hören, die nicht zur Rekonfiguration aufgenommen und a priori zerpflückt wurden, sondern aufgenommen und zerpflückt wurden, um einen Prozess des natürlichen Verfalls zu modellieren, der sowohl im Schweller des Lautsprechersystems wie auch im rebellierenden Kopfraum des Nachtclubgängers aufblitzt. In einigen Fällen ist da nicht mehr – modellierte Verfallsprozesse, um den Geist zu stimulieren: Musik an der Grenze zu einer sich auflösenden Musik und dennoch erfüllt mit einer weitläufigen Grazie und Eleganz. Stets in sich zusammenfallend und immer wieder aufbauend – nicht durch irgendeinen Textbuch-Prozess des kalten, klinischen Modellierens, sondern durch sich radikal verändernde Topologien und Geografien. (*I.M. Trowell*)

MAT STEEL (UK) AND MARK FELL (UK) FIRST MET IN THE EARLY NINETIES IN SHEFFIELD. AFTER COLLABORATING ON OTHER PROJECTS, THE TWO BEGAN WORK AS SND IN 1998. *MAKESND CASSETTE* WAS RECORDED BETWEEN JANUARY AND MARCH 1999 IN SHEFFIELD (UK). IT FOLLOWED THE RELEASE OF SEVERAL VINYL ONLY PROJECTS ON THEIR OWN LABEL, ALSO CALLED 'SND'. DESPITE HAVING DIVERGENT CREATIVE APPROACHES AND MUSICAL TASTES, THEIR AIM AS SND IS TO BRING TOGETHER COMMON INTERESTS IN BOTH VERY EARLY TECHNO AND THE NEW MINIMAL ELECTRONIC MUSIC SCENE. ■ Mat Steel (GB) und Mark Fell (GB) haben sich erstmals in den frühen neunziger Jahren in Sheffield getroffen. Nach einer Zusammenarbeit bei anderen Projekten begannen sie 1998 als snd zu arbeiten. Makesnd Cassette wurde zwischen Jänner und März 1999 in Sheffield eingespielt. Die Aufnahme folgte der Herausgabe mehrerer reiner Vinyl-Projekte auf ihrem eigenen – ebenfalls snd genannten – Label. Trotz unterschiedlichem kreativem Ansatz und verschiedenem musikalischem Geschmack ist ihr Ziel als snd, die gemeinsamen Interessen an frühem Techno und der neuen Minimal Electronic Music Szene zu

0°C
Ryoji Ikeda

_____ „Meine Musik ist ein Spiegel für den Höre

Wie der Titel bereits suggeriert, ist 0°C eine klangliche Erkundungsreise am untersten Grunde der
menschlichen Wahrnehmung – an der hauchdün-
nen Linie zwischen Leben und Tod, Licht und Dunkel,
Geräusch und Stille. Ein authentisches Sein kann niemals mit der kompletten und endgültigen Wahrnehmung des Universums erfolgen – der Mensch in seiner Beschränkung ist nur ein vernachlässigbares
Tröpfchen irgendwo zwischen Unendlichkeit und
dem Nichtseienden. Es gibt keinen Ansatzpunkt für
pretenziöse Wachheit und Frieden für den eigenen
arroganten Geist – Dunkelheit, Unsicherheit und
Zögerlichkeit sind die einzigen wahren Begleiter des
Menschen in seiner Lebensbahn. Das untere Limit
von 0°C kann niemals überquert werden, egal welches Etikett man daranhängt – Geräusch, Umwelt
oder Techno.
0°C ist eine Forschungsreise an der Schwelle der
Wahrnehmung – eine Erweiterung von +/- und
gleichzeitig der nächste Schritt.
(Calmant, Litauen)

"My music is a mirror for the listener"

As the title may suggest, 0°C is a sonic exploration at
the very depths of human perception—the fragile
line between life and death, dark and light, noise and
silence. The authentic being is never to be achieved
with complete and definitive perception of the universe—the limited human is just a negligible drop
somewhere between infinity and nonentity. There is
no fulcrum for pretentious awareness and peace for
one's arrogant mind—obscurity, uncertainty and hes-
itation are the only true man's satellites all along his
lifeway. The limit of minimal 0°C is never to be over-
stepped, no matter what labels might be attached to
it—noise, ambient or techno.
0°C is an exploration at the edge of one's perception—an extension of +/- and the next step.
(Calmant, Lithuania)

RYOJI IKEDA (J) STARTED IN 1985 WITH THE PRODUCTION OF "CHANNEL X", A TEN MINUTE COLLAGE-STYLE SEQUENCE OF FAST-SPLICED TV AND
RADIO SOUND SEGMENTS. IT WASN'T UNTIL TEN YEARS LATER THAT HE PUBLISHED A CD (1000 FRAGMENTS) ON HIS OWN CCI LABEL. MEANWHILE
IKEDA HAS BECOME A MEMBER OF THE JAPANESE PERFORMANCE GROUP DUMB TYPE THAT MEETS AT THE ART SCHOOL IN TOKYO. ■■■ Ryoji
Ikeda (J) beginnt 1985 mit der Produktion von „Channel X", einer knapp zehnminütigen, collagehaften Aneinanderreihung von rasant geschnitte-
nen TV/Radio-Klangfragmenten. Erst zehn Jahre später veröffentlicht er auf dem eigenen Label CCI Recordings eine CD (1000 Fragments). In der
Zwischenzeit ist Ikeda Teil der japanischen Performancegruppe Dumb Type, die sich an der Kunstschule in Tokio zusammenfindet.

THE WEST
Matmos (Andrew Daniel / M.C. Schmidt)

Five new tracks from this boundary exploring Bay Area electronic duo. Their third release finds the pair bringing in some pals to help them create a hybrid of digitally constructed acoustic music that melds aspects of rock, country and folk styles and transforms them into something entirely new. How very exploratory! Guests include folks from Acetone, Aerial M, Cars Get Crushed, Cul de Sac, Run On and more! Nice, beautifully packaged six panel chipboard digipak sleeve.

Fünf neue Tracks dieses Grenzen erforschenden Elektronik-Duo aus der Bay Area. Gemeinsam mit einigen Freunden schaffen sie in ihrer dritten Veröffentlichung ein Hybrid digital konstruierter akustischer Musik, die Rock, Country und Folk vermischt und in etwas vollkommen Neues transformiert. Echte Grenzgänger! Zu den Gästen zählen Mitglieder von Acetone, Aerial M. Cars Cet Crushed, Cul de Sac, Run On und andere. Das Ganze nett verpackt in einer sechsteiligen Digipak-Hülle.

MATMOS (USA) CONSISTS OF THE TWO MUSICIANS ANDREW DANIEL AND M.C. SCHMIDT. THE ENTHUSIASM WITH WHICH THE PRESS RECEIVED THEIR FIRST 2 CDS LED BJÖRK TO OFFER HER ASSISTANCE. SINCE 1998 THEY HAVE BEEN WORKING WITH MUSICIANS FROM BANDS SUCH AS AERIAL M, TORTOISE, ACETONE AND AMBER ASYLUM. AT THE MOMENT THEY ARE WORKING TOGETHER WITH BJÖRK ON A PROJECT USING NOISES ASSOCIATED WITH THE MEDICAL PROFESSION. ■■■ Matmos (USA) besteht aus den beiden Musikern Andrew Daniel und M.C. Schmidt. Die begeisterte Aufnahme ihrer ersten beiden CDs in der Presse veranlasste Björk dazu, ihnen ihre Zusammenarbeit anzubieten. Seit 1998 arbeiten sie fallweise mit Musikern von Aerial M. Tortoise, Acetone, Amber Asylum u. a. zusammen. Derzeit arbeiten sie gemeinsam mit Björk

CD CÉCILE
<u>Kaffe Matthews</u>

starting each show with no sound in her laptop, matthews samples from the crowd and the venue and the atmosphere, transforming, then occasionally weaving stabs and threads of mutilated violin into her crackling layers of processings as she goes.

this time it's a warehouse in london, a rock club in chicago and a converted spinning factory on a snowy sunday night with an audience of 400 in oslo. before that one, she had just seen the sea frozen for the first time.

the work then is entirely live and improvised and gig specific, pulling on the uniqueness of each event to make a new piece.

cd cécile pulls these events together into one object, with minimal editing, mixing and mastering done by km at annette works, londong. trackIDs within each piece are for the listener's use. they are not edit points.

Please refer to http://www.stalk.net/annetteworks for reviews, info on other releases and installations, calendar and pics.

technically

matthews uses a wee powermac 5300cs (48megRAM) running LiSa, a Peavey 1600 midi controller a Boss SE70FXunig a Behringer 8 channel 802 mixer. a violin with pickup and specially designed midi switches 2 x PZM microphones with very long cables.

Matthews beginnt ihre Show stets, ohne dass sie auf ihrem Laptop Klänge gespeichert hätte, sondern nimmt die Atmosphäre und Geräusche des Aufführungsortes und der Zuseher auf, bearbeitet sie und webt nach und nach gelegentliche kurze Passagen mit einer gedämpften Violine in ihre knackenden Ebenen von Be- und Verarbeitung ein.

Diesmal sind es ein Lagerhaus in London, ein Rock-Club in Chicago und eine umgebaute Spinnerei in Oslo mit 400 Zusehern an einem schneereichen Sonntagabend. Kurz vor dem Auftritt hatte sie zum ersten Mal das zugefrorene Meer gesehen.

Das Werk selbst ist komplett live und ortsspezifisch und benützt die Einzigartigkeit jedes Events, um ein neues Stück zu schaffen.

CD Cécile fasst diese Events in ein Objekt zusammen, mit einem Minimum an Schnitt, Mix und Mastering, das von KM bei Annette Works in London durchgeführt wurde. Die Track-IDs innerhalb der einzelnen Stücke dienen der Orientierung des Hörers und sind keine Edit-Punkte.

Technische Daten

Matthews verwendet einen kleinen PowerMac 5300 mit 48 MB RAM unter LiSa, einen Peavey 1600 MIDI-Controller, eine Boss SE70 FX-Unit, einen Boehringer 802 8-Kanal-Mixer, eine Violine mit Tonabnehmer und speziell entwickelten MIDI-Schaltern, zwei PZM-Mikrofone mit extra langen Kabeln.

TGII
Radian (Martin Brandlmayr/Stefan Nemeth/John Norman)

als prinzip das spannungsfeld zwischen elektronischen und nichtelektronischen instrumenten. daraus schält sich ein selbstdefinierter mikrokosmos, in dem minimale verschiebungen und modulationen innerhalb der struktur von ton und rhythmik die eigentliche dynamik bestimmen. die einzelnen ebenen schichten sich, ähnlich einem übereinander verschieden rauer oberflächen, zu einer einheitlichen geräusch-ton-textur.

as a principle, the charged field between electronic and non-electronic instruments. a self-defined microcosm emerges from it, in which minimal shifts and modulations within the tonal and rhythmic structure determine the actual dynamics. the individual levels are layered, similar to a layering of differently rough surfaces, to form a unified noise-tone texture.

"The trio Radian appear rather reserved and calm with just the right aural measure for temporal processes. As a band, which is rather unusual in the field of electronic music, using percussion, bass, analog synthesizer and sampler, Radian process more the inner charge of sounds, playing around them and linking them with new branches rediscovered in playing. Yet this also results in grooves that would invite dancing or at least swaying along, if the audience were brave enough for that. Perhaps, however, Radian is simply the first exemplary band, whose rhythmical structures are not thirsting for speed, but rather for tranquilizers."
(*Christof Kurzmann*)

„Ziemlich zurückhaltend und ruhig und mit dem nötigen Ohrenmaß für zeitliche Abläufe präsentiert sich das Trio Radian. Als Band, was im Umfeld elektronischer Musik ja eher seltenheitswert hat, an Schlagzeug, Bass, Analogsynthesizer und Sampler, bearbeiten Radian mehr das innere Spannungsfeld der Sounds, spielen um jene herum und verbinden sie mit aus dem Spiel heraus neuentdeckten Verzweigungsmöglichkeiten. So entstehen aber auch Grooves die, hätte das Publikum nur den Mut dazu, auch zu Tanz oder zumindest zu Mitswingen verleiten. Vielleicht ist Radian ja aber einfach die erste Vorzeigeband deren rhythmische Strukturen nicht nach Speed, sondern nach Tranquilizern dürsten."
(*Christof Kurzmann*)

RADIAN WAS FOUNDED SOMETIME IN 1996 BY MARTIN BRANDLMAYR (DRUMS, PROGRAMMING), STEFAN NEMETH (ANALOGUE & DIGITAL SYNTH) AND JOHN NORMAN (BASS) AND PERFORMED FOR THE FIRST TIME IN 1997 AT PORGY & BESS IN VIENNA BEFORE A CONCERT BY PETER REHBERG. PERFORMANCES SINCE THEN IN AUSTRIA, INCLUDING THE WIENER FESTWOCHEN, ARS ELECTRONICA, MUSIC UNLIMITED, STEIRISCHER HERBST, AND IN GERMANY, SWITZERLAND, FRANCE, PORTUGAL AND THE NETHERLANDS. ■ Radian wurde irgendwann 1996 von Martin Brandlmayr (drums, programming), Stefan Nemeth (analogue & digital synth) und John Norman (bass) gegründet und hatte seinen ersten Auftritt 1997 im Porgy & Bess in Wien vor einem Konzert von Peter Rehberg. Seither Auftritte in Österreich u. a. bei den Wiener Festwochen, Ars Electronica, Music Unlimited, Steirischer Herbst und in Deutschland, der Schw.. Frankreich, Portugal und in

SATOR ROTAS
Marcus Schmickler

Sator Rotas evolves like an organic process, like cell division. Small motifs consisting of sinus waves split up into smaller and smaller parts. A field is thus built up of extremely compact sounds, of noises—until maximum compactness is reached. Yet something else is repeatedly added: there is static.

The title stems from ancient number and letter mysticism. It involves a "magical square":

The title *Sator Rotas* is a play on the various approaches to the contradiction, to which we owe our existence. At the end of the last century, representatives from science and religions came closer together again. Especially because science is concerned with increasingly specialized problems, it runs into epistemological boundaries: what is perception? Will it never be possible to reconcile the different models for explaining reality? Representatives from different faiths seek proof of God, for example in the *Tora*, which is full of codes encrypted in a secret language. Similar to those of logic. What is peculiar about these squares is that regardless of how they are turned and twisted, they only refer to themselves. The world in ruins (cf. Dürer's *Melancolia*, among others). This one means: "The creator of the spheres holds the work in his hands." Yet this sentence in a magical square—a contradiction: for regardless of the interpretation of its contents, the square retains its logical regularity. Eco, Deleuze and Luhmann were not the first to do away with the notion of authorship.

The material for the piece was successively developed in different cities under quadrophonic conditions. The spatiality resulting from this process is an integral component.

s	a	t	o	r
a	r	e	p	o
t	e	n	e	t
o	p	e	r	a
r	o	t	a	s

Sator Rotas entwickelt sich wie ein organischer Prozess, wie Zellteilung. Kleine aus Sinuswellen bestehende Motive zerlegen sich in immer kleinere Teile. So baut sich ein Feld auf, aus extrem dichten Klängen, aus Geräuschen – bis ein Maximum an Dichte erreicht wird, doch es kommt wieder etwas hinzu: Es rauscht.

Der Titel stammt aus der antiken Zahlen- und Buchstabenmystik. Es handelt sich um ein „magisches Quadrat":

Der Titel Sator Rotas spielt mit den unterschiedlichen Herangehensweisen an jenen Widerspruch, dem wir unsere Existenz verdanken. Am Ende des letzten Jahrhunderts sind sich Vertreter der Wissenschaft und der Religionen wieder nähergekommen. Gerade weil die Wissenschaft sich mit immer spezialisierteren Problemen beschäftigt, stößt sie an erkenntnistheoretische Grenzen: Was ist Wahrnehmung? Werden sich die unterschiedlichen Modelle, Realität zu erklären, nie vereinbaren lassen? Vertreter der verschiedenen Glaubensrichtungen suchen nach einem Gottesbeweis, beispielsweise in der Thora, die voll ist von in Geheimsprache verschlüsselten Codes. Ähnlich derer der Logik.

Das Besondere daran, egal wie man diese Quadrate dreht und wendet, sie verweisen nur auf sich selbst. Die Welt als Scherbenhaufen (vgl. a. A. Dürers Melancolia). Dieses hier bedeutet: „Der Schöpfer der Sphären hält das Werk in den Händen". Doch dieser Satz in einem magischen Quadrat – ein Widerspruch: Denn das Quadrat wird unabhängig von einer inhaltlichen Auslegung seine logische Gesetzmäßigkeit behalten. Vorstellungen von Urheberschaft sind nicht erst seit Eco, Deleuze und Luhmann von den Tischen.

Das Material zum Stück wurde sukzessive in verschiedenen Städten unter quadrophonischen Bedingungen entwickelt. Die bei diesem Prozess entstandene Räumlichkeit ist integraler Bestandteil.

MARCUS SCHMICKLER (D), BORN 1968, GRADUATED FROM SECONDARY SCHOOL IN 1988 IN COLOGNE. STUDIED FROM 1993–1999 AT THE MUSIC COLLEGE, COLOGNE, WITH PROF. H. U. HUMPERT (ELECTRONIC COMPOSITION) AND WITH PROF. J. FRITSCH (COMPOSITION). ▬ Marcus Schmickler (D), geb. 1968. Abitur 1988 in Köln. Von 1993–1999 Studium an der Hochschule für Musik Köln bei Prof. H.U. Humpert (elektronische Komposition) und bei Prof. J. Fritsch (Komposition).

PRIX ARS ELECTRONICA 2000

.NET
INTERACTIVE ART
COMPUTER ANIMATION
VISUAL EFFECTS
DIGITAL MUSICS
U19/CYBERGENERATION

NO STRAW STARS
KEINE STROHSTERNE ...

A colleague at work received two disks as a Christmas present from his 10-year-old daughter. The contents: a web project, an assignment from school, especially prepared for "Daddy." In this case, Daddy, who had been expecting something more like a star made out of straw or a handmade picture frame, did not even have a web browser installed on his old PC. Some things never change, like making Christmas presents for parents at school, for instance. It seems that in the institutions of education here too, attempts are being made to prepare the younger generation for a digital society. This endeavor is strongly supported by the Prix Ars Electronica, at least since the beginning of the category Cybergeneration / U19 two years ago. Over 800 entries this year—many of them school-related— were welcomed by the Austrian Cultural Service (ÖKS) as well, as co-organizer of the competition. The U19 category is often regarded as the young talent competition of the Prix Ars Electronica. This is also correct. And yet here it is more than just a matter of familiarizing young people with new technologies and media. A subtitle of the category is "Freestyle Computing." To begin with, this is an invitation to submit everything that can be done with or on a computer, and this is exactly what happened: from school web pages to a self-made autonomous robot, from a six-year-old's paintbrush graphics to short animation films, self-coded software, games and music on MP3. Yet "Freestyle Computing" also stands for a very specific approach to computers and all things digital on the part of the "cybergeneration" (another subtitle of the category). For this generation, computers are as commonplace as refrigerators, so they use technology differently from the way adults do. The motif "future" has long been promising a "better tomorrow." The only prerequisite for us to enter into the digital wonderland is an unconditional faith in the optimism of technological progress. A fear that educational systems may not be compatible with the future may be found around the world. A generation that is unprepared for the grand project "Future" can only lose, according to the parent generation's greatest fear. Yet it may be that this younger

Ein Arbeitskollege bekam von seiner 10-jährigen Tochter zu Weihnachten zwei Disketten geschenkt. Der Inhalt: ein Web-Projekt, eine Arbeit aus der Schule, speziell gestaltet für „Papa". Dabei hat Papa, der sich eher Strohsterne oder einen selbstgebastelten Fotorahmen erwartet hat, gar keinen Web-Browser auf seinem alten PC installiert. Manche Dinge bleiben immer gleich. Das Basteln der Weihnachtsgeschenke für die Eltern in der Schule zum Beispiel. Anscheinend bemüht man sich auch in den Erziehungsanstalten hierzulande, den Nachwuchs für die digitale Gesellschaft vorzubereiten. Ein Vorhaben, das vom Prix Ars Electronica zumindest seit dem Start der Kategorie Cybergeneration / U19 vor zwei Jahren stark unterstützt wird. Über die 800 Einsendungen dieses Jahr – viele aus dem schulischen Bereich – freut sich auch der Österreichische Kulturservice (ÖKS) als Mitveranstalter des Bewerbs. Die Kategorie U19 wurde oft als der Nachwuchswettbewerb der Prix Ars Electronica verstanden. Das stimmt auch. Und trotzdem geht es hier um viel mehr als nur darum, junge Menschen an neue Technologien und Medien zu gewöhnen. Ein Untertitel der Kategorie lautet „Freestyle Computing". Zuerst einmal eine Aufforderung, alles einzuschicken, was mit oder am Computer gestaltet werden kann. Dies ist auch passiert: von der Schulwebpage bis zum selbergebastelten autonomen Roboter, von der Paintbrush-Grafik eines 6-Jährigen bis zum Animationskurzfilm, selbstgecodete Software, Spiele und Musik auf MP3. „Freestyle Computing" steht aber auch für den ganz besonderen Zugang der „Cybergeneration" (ein weiterer Untertitel) zu Computer und Digitalität. Eine Generation, für die der Computer so alltäglich ist wie der Kühlschrank, benützt Technologie ganz anders als die Erwachsenenwelt. Das Leitmotiv „Zukunft" verspricht seit langer Zeit das „Better Tomorrow". Die einzige

Bedingung , die man von uns für den Eintritt in das digitale Wunderland verlangt, ist der uneingeschränkte Glauben an den technologischen Fortschritt. Die Angst, dass die Erziehungssysteme nicht zukunftskompatibel sind, gibt es weltweit. Eine Generation, die nicht gut für das große Projekt „Zukunft" vorbereitet ist, kann nur verlieren, so lautet die Hauptangst der Elterngeneration. Doch vielleicht sieht es der Nachwuchs, für den diese Zukunft konstruiert wird, ganz anders.

Die 10-jährige Tochter mit dem Web-Projekt für Papa hätte wahrscheinlich genauso liebevoll die Weihnachtsstrohsterne gebastelt. Das Mädchen, das die Computergrafik *Die Nixe scheut das verschmutzte Wasser* (Anerkennung) eingesandt hat, hätte vielleicht das Bild genauso in der analogen „Wir-reißen-jetzt-Buntpapier-und-machen-damit-schöne-Bilder-Technik" gestaltet (die Optik ist zumindest sehr ähnlich). Der Computer ist Nebensache. Technologie steht nicht im Vordergrund. „Freestyle Computing" umschreibt die Möglichkeiten außerhalb eines erwachsenen Denken in technischen Standards. Es geht um das Schaffen, das außerhalb der üblichen Produktionsprozesse stattfindet.

Durch die große Anzahl der Einsendungen war die U19-Kategorie besonders gefordert, einen repräsentativen Überblick über die verschiedenen Welten von Kindern und Jugendlichen der „Cybergeneration" zusammenzustellen. Bei den Musikeinsendungen tauchte dieses Jahr natürlich das Thema MP3 auf. Das gesammelte Werk von Alexander Fischl bekam von der Jury gerade für seinen besonderen Umgang mit diesem Codierungsformat eine Anerkennung. Extrem kurze Stücke auf niedriger Samplingrate, die durch eine geschickte Bearbeitung trotzdem nicht allzu sehr an Qualität verlieren, sind geradezu ideal für die Verbreitung über Download. Gerade die Kleinteiligkeit von Fischls Selbstproduziertem zeigt einen interessanten Ansatz, wie die Musik für ein vernetztes Publikum bald klingen könnte. Die Auflösung des Werks an sich, kleine feine Stücke statt des digital behäbigen „Liedes", ideal für die Weiterverarbeitung durch ein Internet-Musikpublikum, das billige oder kostenlose Musikprogramme zu benutzen weiß. Eine Entwicklung, die viel interessanter ist als die große Diskussion um Copyright oder darüber, ob man in Zukunft die CD noch

generation, for whom this future is being constructed, views the matter differently.

The 10-year-old girl with her web project for Daddy would probably have made a straw Christmas star with just as much devotedness. The girl who entered the computer graphic *Die Nixe scheut das verschmutzte Wasser* (Honorary Mention) might have made the picture just the same with the analog technique "now we tear off strips of colored paper and make nice pictures with them" (there is a certain optical similarity). The computer is secondary. Technology is not in the foreground. "Freestyle Computing" describes possibilities for thinking outside the adult realm of technical standards. It is a mat-

ter of creativity that takes place outside the conventional production processes.

The large number of entries particularly challenged the U19 category to assemble a representative survey of the different worlds of the children and young people of the "cybergeneration." Naturally the topic of MP3 came up with the music entries. The jury awarded an Honorary Mention to Alexander Fischl for his collected work, particularly because of his special way of dealing with this coding format: extremely short pieces at a low sampling rate, which do not lose too much quality because of the clever way they are processed, are practically ideal for distribution via downloading. Especially the small scale of Fischl's self-produced pieces demonstrates an interesting approach to the way music for a networked audience may sound very soon. The resolution of the work itself, small fine pieces rather than the digitally ponderous "song", makes them ideal for further processing on the part of an Internet music audience that knows how to use cheap or free music programs. This development is much more interesting than the major discussions about copyright, or whether we will buy CD's in the future in a store or download them from the Internet.

Another example of using the Internet for one's own community is one of the two Awards of Distinction: the project *Cybervoting*. Politicians around the world have long paid lip-service to the topic of teledemocracy, but a group of Austrian school students has put this promise into practice. Transregional elections for school representatives can be conducted over a telematic distance using a web page. Here the management of the database, the treatment of the problem of data protection, the impeccably designed web page, and especially the aspect of a democratic self-initiative illustrate the possibilities of social cooperation through networks.

A completely different approach to networked society is evident in the other Award of Distinction. Gerhard Schwoiger is an excellent web designer and programmer. In addition, he is also interested in other people's rubbish. *Netdump* is the name of his web art project. A self-coded client links the desktop trashcan with a server. Registered users can dispose of their digital rubbish there. And naturally, anyone can recycle anything in the dump on the server. Gerhard

im Geschäft kauft oder vom Netz herunterlädt.

Ein anderes Beispiel für den Gebrauch vom Internet für die eigene Community ist eine der beiden Auszeichnungen: das Projekt *Cybervoting*. Das Thema Teledemokratie ist schon lange ein Lippenbekenntnis vieler Politiker weltweit. Eine österreichische Schülergruppe setzte das Versprechen in die Tat um. Über eine Webpage können überregionale Schülervertretungswahlen über telematische Distanzen abgehalten werden. Die Handhabung der Datenbank, der Umgang mit dem Problem des Datenschutzes, die einwandfreie Gestaltung der Webpage und vor allem der Aspekt einer demokratischen Selbstinitiative zeigen hier die Möglichkeiten gesellschaftlicher Kooperation über Netzwerke.

Einen ganz anderen Zugang zur vernetzten Gesellschaft zeigt die zweite Auszeichnung. Gerhard Schwoiger ist ein ausgezeichneter Webgestalter und Programmierer. Außerdem interessiert er sich für den Müll anderer Leute. *Netdump* heißt sein Webkunstprojekt. Ein selbst gecodeter Client verbindet den Desktop-Mistkübel mit einem Server. Angemeldete User können dort ihren digitalen Müll abladen. Und natürlich ist auf der Serverdeponie alles für jeden recyclebar. Gerhard Schwoiger entwirft hier eine Online-Community, die durch ihren gemeinsamen Abfall verbunden ist – die Privacy-Thematik aus der Sicht eines jungen Developer.

Auch dieses Jahr war die Anzahl der teilnehmenden Mädchen die Minderheit. Trotzdem war nicht eine qualitative Aufbesserung der Quote der Grund, warum die Goldene Nica an ein Projekt niederösterreichischer Schülerinnen ging. Ihr Projekt *Harvey* ist nach einem unsichtbaren Hasen aus einem alten Schwarzweißfilm benannt, der aus dem Nichts zu James Stewart spricht, und ist im Grunde eine Computeranwendung, die kein Mensch braucht. Doch Markttauglichkeit ist kein Kriterium in der U19-Kategorie. Die Gruppe 14-jähriger Mädchen überlegte sich im EDV-Unterricht, wie man den Computer zum Sprechen bringen könnte. Auf einen Bierdeckel wurde die Schaltung für eine Soundkarte gelötet, ein altes Transistorradio dient als Lautsprecher. Das dazugehörige Programm wurde selbst programmiert.

Schwoiger has created an online community that is linked through its common rubbish. This is the theme of privacy from the perspective of a young developer.

Once again this year, the participating girls were a minority. Yet the reason why the Golden Nica went to a group of girls from Lower Austria, was not to improve the quota. Their project *Harvey* is named after an invisible rabbit from an old black and white film, that speaks to James Stewart out of nowhere. On the whole, it is a computer application that no one needs. However, marketability is not a criterion in the U19 category. In their computer science class, this group of 14-year-old girls was thinking about how one could make the computer speak. The circuitry for a soundcard was soldered onto a beer mat, an old transistor radio served as a loudspeaker. They programmed the corresponding program themselves. Naturally the question arises as to why you would build something that you can buy in any computer shop. Yet perhaps an impossible question should be posed here: why buy something, if you can build it yourself? In a world where technology is sold in inviolable units, the girls practically broke a taboo. They opened up the casing the houses the sacrosanct technology and misused everyday objects for their purposes. Shamelessly uninhibited, they found an unconventional solution that defies all claims of standardization. The predetermined concept of hardware was redefined here in an unusual way. "The thing is charming," said one of the jury members, a trained technician himself. "If that's not freestyle computing, then what is?"

Natürlich stellt sich hier die Frage, wozu man etwas bauen soll, das man in jedem Computergeschäft kaufen kann. Vielleicht sollte aber hier aber eine unmögliche Frage gestellt werden: Wozu etwas kaufen, was selber gebaut werden kann? In einer Welt, in der Technik in unantastbaren Einheiten verkauft wird, begingen die Mädchen fast einen Tabubruch. Sie öffneten das Gehäuse, das das Heiligtum Technik verbirgt, und missbrauchten Alltagsgegenstände für ihre Zwecke. In einer schamlosen Unbekümmertheit fanden sie einen unkonventionellen Lösungsweg, der sich über jeden Standardisierungsanspruch hinwegsetzt. Auf eine besondere Weise wurde hier der festgesetzte Begriff der Hardware neu definiert. „Das Ding hat Charme", meinte eines der Jurymitglieder, selbst ein gelernter Techniker, „wenn das nicht Freestyle Computing ist, was dann?"

HARVEY

Verena Riedl, Michaela Hermann

Our project is concerned with the output of a stored text in spoken form. It is intended to be a reading aid for people, who are either blind or cannot read a written text for other reasons. The program Harvey makes it possible to have a text read out loud that is stored on a data medium, such as a disk.
A handmade soundcard on a beer mat was developed for this. The quality of the voice output is limited naturally, in keeping

with this simple program, however it is sufficiently intelligible. IBM-ASCII files can be used, also text files (.TXT) created with the Windows 3.1 text editor. Even Write files (.WRI) (without graphics) can be made audible, if one doesn't mind background noise at the beginning and the end. The program runs best under MS-DOS, but can also be run under Windows 3.1. Windows 95 has not been tested. It was originally developed on an Intel 486 processor with 33 Mhz, but it should also be possible to use it on a Pentium with a higher clock rate. However, this has not been tested.

Unser Projekt beschäftigt sich mit der Ausgabe eines gespeicherten Textes in gesprochener Form. Es ist gedacht als Lesehilfe für Menschen, die entweder blind sind oder aus anderen Gründen einen geschriebenen Text nicht lesen können. Das Programm Harvey ermöglicht es, sich einen Text, der auf einem Datenträger, z. B. einer Diskette, gespeichert ist, vorlesen zu lassen. Dazu wurde eine Selbstbau-Soundkarte entwickelt, die auf einem Bierdeckel aufgebaut wird. Die Qualität der Sprachausgabe ist natürlich nur einem einfachen Programm entsprechend. Trotzdem ist die Verständlichkeit ausreichend. Es können IBM-ASCII-Dateien verwendet werden, aber auch vom Windows 3.1 Texteditor erstellte Textdateien (.TXT). Wenn man einige Nebengeräusche am Anfang und Ende hinnimmt, kann man sogar Write-Dateien (.WRI) (ohne Grafik) hörbar machen. Das Programm läuft am besten unter MS-DOS, ist aber auch unter Windows 3.1. lauffähig. Windows 95 wurde nicht getestet. Ursprünglich auf einem Intel-486-Prozessor mit 33Mhz entwickelt, sollte es auch auf einem Pentium mit höherer Taktfrequenz einsetzbar sein. Das wurde aber nicht überprüft.

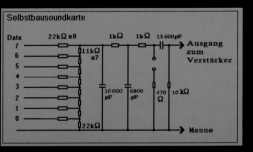

Verena Riedl (17) was born in Gmünd (Lower Austria) and currently attends the secondary school BG Waidhofen/ Thaya. She took computer science as an elective class in the 7th and 8th grade and did a training course with a computer company during school holidays. Michaela Hermann (17) attends the secondary school BRG Waidhofen/ Thaya and has taken computer science as an elective class for three years.

Verena Riedl (17) wurde in Gmünd (NÖ) geboren und besucht derzeit das BG Waidhofen/Thaya. Sie belegte in der 7. und 8. Schulstufe das Freifach „Informatik" und absolvierte 1999 ein Ferialpraktikum in einer Computerfirma. Michaela Hermann (17) besucht das BRG Waidhofen/ Thaya und belegt seit drei Jahren das Fach Informatik.

GARBAGE™
NETDUMP

Gerhard Schwoiger

This project is intended to call attention to the growing
problem of garbage. A shock is needed to convey the
desolate situation. Yet it should also provide information
about all the themes and statistics related to garbage.
These two aspects together are not intended to
represent an information platform in the conventional
sense, but should instead be regarded as an art
experiment. The project attempts to shift a media
design, which has only been familiar previously from

music broadcasters like MTV, to the on-line level.
However, this already poses certain system requirements as a prerequisite to the user. The absolute
minimum requirements for problem-free viewing of
gARbage TM Netdump are a picture resolution of 1024 x
768, Internet Explorer version 5.x (Netscape is only
partially supported), a soundcard and a processor with
at least 200 Mhz.

Gerhard Schwoiger (19) lives in Vienna, has already carried out numerous projects in the field of media, for which he has also received a number of prizes (press button award 96, Prix Multimedia 97, rheingold award 97, soundtrack competition 97).

Gerhard Schwoiger (19) lebt in Wien, hat schon zahlreiche Projekte im Medienbereich durchgeführt und dafür auch schon eine Reihe von Preisen erhalten (press button award 96, Prix Multimedia 97, rheingold award 97, soundtrack competition 97).

Das Projekt *gARbage™ (Netdump)* soll einerseits auf das permanent wachsende Müllproblem aufmerksam machen. Ein Schock zur Vermittlung der tristen Situation ist erwünscht. Andererseits soll es auch Auskunft über sämtliche mit Müll zusammenhängenden Themen und Statistiken geben. Beides gemeinsam soll nicht eine Informationsplattform im herkömmlichen Sinn darstellen, sondern vielmehr als Kunstexperiment betrachtet werden. Dieses Projekt versucht auch, ein bisher nur von Musiksendern wie MTV bekanntes Mediendesign auf die Online-Ebene zu verlagern. Als Voraussetzung dafür werden hiermit aber schon einige Systemanforderungen an den Benutzer gestellt. So sollte eine Bildschirmauflösung von 1024 x 768, der Internet-Explorer Version 5.x (Netscape wird nur zum Teil unterstützt), eine Soundkarte und ein Prozessor mit mindestens 200 Mhz als absolute Mindestvoraussetzung zum reibungslosen Betrachten von *gARbage™ (Netdump)* dienen.

CYBERVOTING

Erich Hanschitz / 7B BG/BRG Völkermarkt

As we discovered for ourselves this year, the election of a school representative is an extremely complicated matter. From the preparation of the ballots to the tedious counting, the organization is nothing but an ordeal. This led us to the decision to completely automate school representative elections. With databases, tables and countless working hours, we created a system that turns conducting school representative elections into a brief process: from entering the candidates via the virtual ballot to the output of all the results, the entire election process is displayed and conducted in a client-server environment on the Internet.

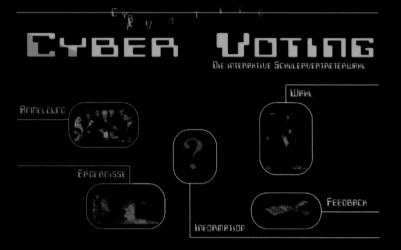

The technical demands that we faced with enthusiasm were over a thousand lines of code in a language (PHP) new to us, in conjunction with a high performance database link with fantastic possibilities, and learning animated web design with flash.
The high intensity of the work and the huge investment of time have already proven worthwhile, and our efforts have been rewarded with the complete functionality of the election program and the page itself.

Erich Hanschitz (18) attends the 7th year class of the secondary school BG Völkermarkt. For over five years he has been taking part in the school experiment "computer science." When he finishes school he wants to study computer science and then work in this field with an emphasis on technology and implementing his own ideas.

Erich Hanschitz (18) besucht die 7. Klasse des BG Völkermarkt. Seit nun mehr fünf Jahren nimmt er am Schulversuch "Informatik" teil. Berufswunsch: Nach dem Informatik-Studium möchte er in jedem Fall in diesem Bereich Fuß fassen, technologieorientiert arbeiten und seine Ideen einbringen.

Wie wir im heurigen Jahr selbst erfahren mussten, ist die Wahl des Schulsprechers eine äußerst aufwändige Angelegenheit. Von der Erstellung der Stimmzettel bis hin zur mühevollen Auswertung ist die Organisation eine einzige Tortur. Deshalb fassten wir kurzerhand den Entschluss, die Schulsprecherwahl komplett zu automatisieren. Mit Datenbanken, Tabellen und unzähligen Überstunden schufen wir ein System, das die Prozedur der Schulsprecherwahl zu einem kurzlebigen Unterfangen macht: Von der Eingabe der Kandidaten über den virtuellen Stimmzettel bis hin zur Ausgabe aller Ergebnisse wird der gesamte Ablauf im Internet in einer Server-Client-Umgebung abgebildet und durchgeführt. Weit über tausend Zeilen Code in einer für uns ganz neuen Sprache (PHP) in Verbindung mit einer leistungsfähigen Datenbankanbindung mit fantastischen Möglichkeiten und die Einarbeitung in animiertes Webdesign mit Flash waren die technischen Anforderungen, denen wir uns mit Freude stellten. Die hohe Arbeitsintensität sowie die rigorose Zeitinvestition trugen bereits ihre Früchte und entschädigten unseren Aufwand durch die vollständige Funktionalität sowohl des Wahlprogramms als auch der Seite an sich.

es geht auch anders

UNENDLICHE KUNST

Marlene Maier

My possibilities are very limited, because I do not have an Internet connection, although I keep asking my parents to get it, so I had the idea of drawing something with the computer. Although this is easy to do, by using many variations of colors and forms I succeeded in creating several different graphics, making them as diverse as possible and not monotonous. On the whole, I had no trouble carrying out my ideas, although some pictures look a bit boring, because I only had a certain number of colors to choose from. Otherwise it was fun and easy to make these kinds of graphics.

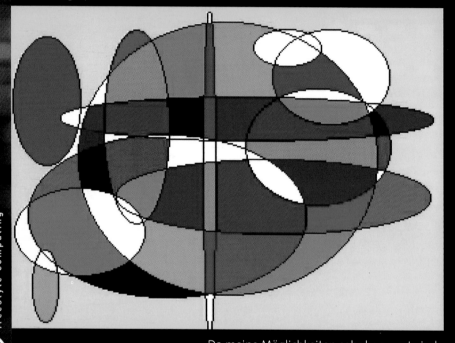

Marlene Maier (14) lives in Linz and attends the college preparatory school Petrinum.

Marlene Maier (14) lebt in Linz und besucht das Gymnasium Petrinum.

Da meine Möglichkeiten sehr begrenzt sind, weil ich über keinen Internetanschluss verfüge, obwohl ich meine Eltern schon ständig darum bitte, kam mir die Idee, mit dem Computer etwas zu zeichnen. Diese Idee ist zwar leicht auszuführen, doch durch vielerlei Variationen von Farben und Formen ist es mir gelungen, mehrere verschiedene Grafiken zu gestalten und sie möglichst vielfältig, nicht eintönig wirken zu lassen. Schwierigkeiten bei der Umsetzung meiner Idee hatte ich eigentlich nicht, doch manche Bilder sehen vielleicht etwas langweilig aus, weil mir nur eine bestimmte Menge an Farben zur Auswahl stand. Ansonsten war es lustig und leicht, solche Grafiken zu gestalten.

u19

DIE NIXE MEIDET VERSCHMUTZTES WASSER

Lisa Hofstadler

The mermaid cannot swim in dirty water. She is very sad. The crow and the butterflies cannot drink the contaminated water. They have to fly far away.

Die Nixe kann im verschmutzten Wasser nicht baden. Sie ist sehr traurig. Der Rabe und die Schmetter-linge können das verseuchte Wasser nicht trinken. Sie müssen weit fort-fliegen.

Lisa Hofstadler (8) lives on a farm in Staudach near Hartberg (Styria) and attends the second year of grammar school.

Lisa Hofstadler (8) lebt auf einem Bauernhof in Staudach bei Hartberg (Stmk) und besucht die zweite Klasse Volksschule.

DIE TOTENINSEL
<u>Markus Zwickl</u>

It all started last fall, when we learned about symbolism and about the picture "Die Toteninsel" ("Island of the Dead") by Arnold Böcklin in drawing class. Since I had been busy with 3D design for several months and was interested in building something larger, the very expressive painting provided me with the model I had been looking for. I started to work on it the same day and constructed the first parts of the temple that is located on the Island of the Dead.

Begonnnen hat alles im letzten Herbst, als wir im Zeichenunterricht über den Symbolismus und über das Bild *Die Toteninsel* von Arnold Böcklin lernten.
Da ich mich schon seit einigen Monaten mit 3D-Design beschäftigt hatte und

Lust hatte, etwas Größeres zu bauen, fand ich in dem ausdrucksstarken Gemälde endlich die Vorlage, die ich gesucht hatte. Noch am selben Tag ging ich ans Werk und konstruierte die ersten Teile des Tempels, der sich auf der Toteninsel befindet.

Markus Zwickl (18) lives in Weiz (Styria) and attends college preparatory school there. He is involved in 3D LAN games and spends his free time creating virtual worlds with the help of 3D design programs.

Markus Zwickl (18) lebt in Weiz (Stmk) und besucht dort das Gymnasium. Er beschäftigt sich in seiner Freizeit mit 3D-LAN-Spielen und mit der Gestaltung von virtuellen Welten mit Hilfe von 3D-Designprogrammen.

DANCING
Lisa Ratzenböck

I am currently taking part in a computer course at the department for arts education. Here I became acquainted with the program "Flash 4." Since dancing is my greatest hobby and I also like to work with the computer, I chose the theme "dancing" for the course. At first I only had a solo dancer. But to expand the animation a bit and because the dancer was lonely, I added a second dancer. Then it became a dance duet.

Lisa Ratzenböck (11) lives in Walding (Upper Austria) and currently attends the second year class of the college preparatory school Petrinum. She created this animation in conjunction with a course at the University for Industrial and Artistic Design in Linz.

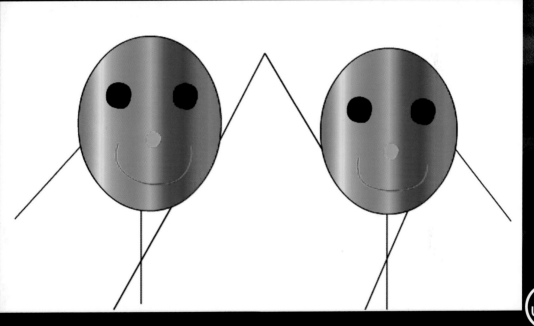

Ich besuche derzeit einen Computerkurs an der Lehrkanzel für bildnerische Erziehung. Hier lernte ich das Programm Flash 4 kennen. Da mein größtes Hobby Tanzen ist und ich mich auch sehr gerne mit dem Computer beschäftige, habe ich beim Kurs das Thema „Dancing" gewählt. Anfangs hatte ich nur einen Solotänzer. Um aber die Animation noch ein wenig auszubauen und weil sich der Tänzer so einsam fühlte, habe ich noch einen zweiten Tänzer hinzugefügt. Daraus wurde jetzt ein Tanzduett.

Lisa Ratzenböck (11) lebt in Walding (OÖ) und besucht derzeit die 2. Klasse des Gymnasiums Peuerbachstraße in Linz. Die Animation entstand im Rahmen eines Kurses an der Universität für industrielle und künstlerische Gestaltung in Linz.

ZIRKUS/AFFENBAUM
Sebastian Endt

I saw a picture of a monkey in my guitar music book and I wanted to copy it on the computer. First I built the monkey's body, although I had a lot of trouble with the stomach. The hands worked out right away. It took me about one day to make this picture. Then I wanted to make a little man for fun. I sat down and made a body.

Ich habe eine Affenzeichnung in meinem Gitarre-Notenheft gesehen und wollte diese am Computer nachzeichnen. Zuerst baute ich den Affenkörper, wobei der Bauch für mich sehr schwierig zu machen war. Die Hände gelangen mir gleich auf Anhieb. Für dieses Bild brauchte ich ungefähr einen Tag.
Dann wollte ich zum Spaß ein Manderl zeichnen. Ich setzte mich hin und gestaltete einen Körper. Dann dachte ich, das könnte ein Clown werden und so kam ich auf die Idee, einen Zirkus zu machen.

Sebastian Endt (11) lives in Wels and also goes to school there. For his picture *"Schweineherde"* he received an Honorary Mention in Cybergeneration /U19 Freestyle Computing last year.

Sebastian Endt (11) lebt in Wels und besucht dort auch die Schule. Er hat im Vorjahr für sein Bild *Schweineherde* eine Anerkennung bei Cybergeneration /U19 Freestylecomputing erhalten.

SPOOKY
Kevin Ku

With friends I saw some homepages that I liked very much. That gave me the idea of making my own homepage. First I started experimenting with Frontpage. Soon that wasn't enough for me, though, so I started making a site myself with HTML. But soon that bored me, too, so I started adding JavaScripts and Applets.

Then I started wanting more and more, so I added categories like awards, forum, etc. To make it more exciting, I added games, too. So my homepage became more and more elaborate, and it is a lot of fun to keep expanding it.

Ich habe bei Freunden Home-pages gesehen, die mir gut gefielen. Das gab mir den Anstoß, selbst eine eigene Homepage zu erstellen. Zuerst begann ich mit Frontpage zu experimentieren. Da mir das aber nach einiger Zeit zu wenig war, begann ich mit HTML selbst eine Seite zu basteln. Doch auch dies war mir bald zu langweilig und ich begann JavaScripts und Applets einzubauen. Dann wurde die Lust auf Mehr immer größer und ich fügte Kategorien wie Awards, Forum usw. ein. Damit es spannender wurde, kamen auch noch Spiele dazu. So wurde meine Homepage immer umfangreicher und es macht großen Spaß, sie weiter auszubauen.

Kevin Ku (actually Kevin M) (13) lives in Vienna and attends the fourth year class at college preparatory school. In addition to his hobbies snake-boarding, cinema and TV, he devotes intensive attention to his web page *Spooky*.

Kevin Ku (eigentlich Kevin M) (13) lebt in Wien und besucht die vierte Klasse Gymnasium. Neben seinen Hobbies Snakeboarden, Kino und Fernsehen widmet er sich intensiv seiner Web-page *Spooky*.

MMH-BROWSER
Mario Meir-Huber

I started with the program at the end of 1999. Originally it was just intended to be a browser for the school that is not displayed in the task bar. Some time later I continued developing the browser and it started to take shape. Addresses were added. For this, I took the addresses most familiar to me and those that I thought were most important. Then I realized, though, that every person has different preferences. So I simply added a subcategory "user defined" to the menu item addresses. That way, someone else can put their favorite address into a menu. The browser differs especially from the norm in the help files. You don't even need to look closely to see that this is not a conventional kind of help. There are three categories that are intended to simplify things. For instance, the menus are described separately, as well as the individual functions that the browser has.

Mario Meir-Huber (16) lives in Munderfing (Upper Austria) and currently attends the business academy in Braunau/Inn. In addition to playing tennis, he spends his time programming at the computer.

Mario Meir-Huber (16) lebt in Munderfing (OÖ) und besucht gegenwärtig die Handelsakademie Braunau/Inn. Neben dem Tennisspielen beschäftigt er sich mit der Programmierung am Computer.

Begonnen mit dem Programm habe ich gegen Ende 1999. Ursprünglich war es nur als Browser für die Schule gedacht, der nicht in der Taskleiste angezeigt wird. Einige Zeit später entwickelte ich den Browser weiter und er nahm allmählich Gestalt an. So bekam er Adressen dazu. Hier nahm ich diejenigen, die mir am geläufigsten waren und von denen ich dachte, dass sie am wichtigsten seien. Aber dann wurde mir klar, dass jeder Mensch verschiedene Vorlieben hat. So fügte ich dem Menüpunkt "Adressen" kurzerhand eine Unterkategorie „benutzerdefiniert" hinzu. Damit kann man seine Lieblingsadressen in einem Menü unterbringen. Besonders bei den Hilfedateien weicht der Browser stark von der Norm ab. Man muss nicht einmal besonders genau hinsehen, um festzustellen, dass es sich hier nicht um eine gewöhnliche Hilfe handelt. Sie besitzt nämlich drei Kategorien, die einiges vereinfachen sollen. So sind etwa die Menüs extra beschrieben und auch die einzelnen Funktionen, die der Browser besitzt.

ZEIT.LOS

David Feiler / Projektgruppe Brandfall

A school corridor is used as a virtual starting environment for exploring the project. The individual doors that can be freely chosen form portals to eleven audio and video installations. This enabled us to fulfill a wish we had had for a long time, of being able to arrange information and contents in three-dimensional space. In this project, as in reality, it is necessary to move in order to attain information.

During the development of the project, we ran into many difficulties, problems and obstacles. There were difficulties obtaining utensils, with filming and taking photographs, and finally, we seriously underestimated how much time we would need for the programming. For the interactive music installation we developed an environment based on Flash, in which the user can choose from eighteen sounds to produce their own compositions.

The group "Brandfall" is a project group consisting of 11 school students from a 7th year class at the Erich Fried School in Vienna

Die Gruppe „Brandfall" ist eine Projektgruppe, die aus elf Schülerinnen und Schülern aus einer 7. Klasse des Erich-Fried-Gymnasiums Wien Glasergasse besteht.

Ein Schulgang wird als virtuelle Ausgangsumgebung genützt, über die man das Projekt erforschen kann. Die einzelnen, frei ansteuerbaren Türen bilden Portale zu elf Audio- und Videoinstallationen. Damit wurde unser lang gehegter Wunsch erfüllt, Information und Inhalt im dreidimensionalen Raum anzuordnen. In diesem Projekt ist wie auch in der Realität Bewegung erforderlich, um zu Information zu gelangen. Bei der Entwicklung des Projekts stießen wir auf viele Schwierigkeiten, Probleme und Hindernisse: beim Beschaffen der Utensilien, beim Filmen und Fotografieren. Und schließlich haben wir auch die für die Programmierung erforderliche Arbeitszeit stark unterschätzt. Für die interaktive Musikinstallation haben wir eine Umgebung auf Flash-Basis entwickelt, in der dem User 18 ausgewählte Sounds zur Produktion einer eigenen Komposition zur Verfügung stehen.

CHAT
Gottfried Haider

Gottfried Haider
(15) lives in Vienna
and attends the
Benedictine college
preparatory school
there. He received
his first PC for his
10th birthday.

Chat, as the name indicates, is a program that you can use for conversation via the TCP protocol. This also makes it possible to use the program over the Internet. The program was programmed in Microsoft Visual Basic. All the settings are stored in the Windows registration.

At the moment I am working on a new version, with which more than two users can talk. Unfortunately, this version still has so many errors that I can't present it yet.

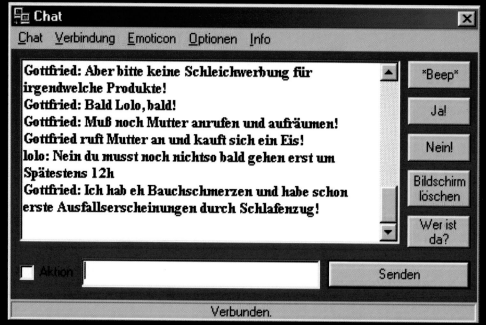

Chat

Chat Verbindung Emoticon Optionen Info

Gottfried: Aber bitte keine Schleichwerbung für irgendwelche Produkte!
Gottfried: Bald Lolo, bald!
Gottfried: Muß noch Mutter anrufen und aufräumen!
Gottfried ruft Mutter an und kauft sich ein Eis!
lolo: Nein du musst noch nichtso bald gehen erst um Spätestens 12h
Gottfried: Ich hab eh Bauchschmerzen und habe schon erste Ausfallserscheinungen durch Schlafenzug!

Beep

Ja!

Nein!

Bildschirm löschen

Wer ist da?

Aktion

Senden

Verbunden.

Gottfried Haider (15)
lebt in Wien und
besucht dort das
Gymnasium der
Benediktiner. Zu
seinem zehnten
Geburtstag erhielt
er seinen ersten PC.

Chat ist, wie der Name schon sagt, ein Programm, mit dem man sich über das TCP-Protokoll unterhalten kann. Somit ist es auch möglich, das Programm über das Internet zu nutzen. Das Programm wurde in Microsoft Visual Basic programmiert. Sämtliche Einstellungen werden in der Windows-Registrierung gespeichert.

Im Moment arbeite ich an einer neuen Version, mit der sich mehr als zwei Benutzer unterhalten können. Leider hat diese Version noch so viele Fehler, dass ich sie einstweilen noch nicht vorstellen kann.

HTL-ADVENTURE / TEILCHEN-BESCHLEUNIGER DES GRAUENS

Lukas Fichtinger / Thomas Köckerbauer

The game that we have created takes place at the technical secondary school in Braunau, which has been left at the start of the game to go shopping. Meanwhile, a particle accelerator experiment takes place and goes wrong. After returning and finding that there is no one left at school, one has to solve the riddle and bring back all the lost students. The game itself was created using a self-implemented interpreter system that is based on the SDL library, which allows a script-based implementation of multimedia presentations, adventure games, etc.

Lukas Fichtinger and Thomas Köckerbauer both attend the fifth year class of the technical secondary school in Braunau. In addition to their computer science class, working with the computer is one of their favorite free-time activities, too.

Das von uns erstellte Spiel handelt in der HTL-Braunau, welche als Spielannahme zum Einkaufen verlassen wurde. Währenddessen findet ein Teilchenbeschleuniger-Experi-ment statt, das fehlschlägt. Wenn man zurückkommt, entdeckt man, dass sich niemand mehr in der Schule befindet: Man muss die Lösung des Rätsels finden und die verloren gegangenen Schüler wieder zurückholen.
Das Spiel selbst wurde mittels eines selbst implementierten Interpretersystems erstellt, das auf der SDL-Library basiert und das scriptbasierte Implemen-tieren von Multimediapräsenta-tionen, Adventure-Spielen etc. erlaubt.

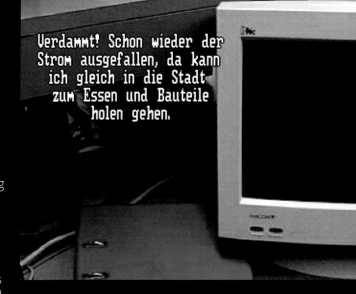

Verdammt! Schon wieder der Strom ausgefallen, da kann ich gleich in die Stadt zum Essen und Bauteile holen gehen.

Lukas Fichtinger und Thomas Köckerbauer besuchen beide die fünfte Klasse der HTL Braunau. Das Arbeiten mit dem Computer auch außerhalb des EDV-Unterrichts gehört zu den bevorzugten Freizeit-beschäftigungen der beiden.

TNSP–THE NEWTRON SOUND PROJECT

<u>Alexander Fischl</u>

The idea behind *TNSP* is to generate sounds or music for pure pleasure and to distribute these via the medium of the Internet for non-commercial purposes. TNSP is the art of creating art at no cost.

"Draft 5" that was particularly distinguished by the jury was not actually a "serious" production, but was just intended to prove to a friend how fast chart-oriented standard techno (aka "Uz-Uz") can be generated. The result was five minutes of drumbox playing in combination with a speech sample from a children's cartoon series. Then somehow the idea came up of combining this with a completely different style. The final result was ultimately 180bpm Jungle/Hardcore combined with a classical, somewhat re-transposed orchestra sample. Effect: "That sounds more interesting now."

Die Idee hinter *TNSP* ist es, Sounds bzw. Musik rein aus Vergnügen zu erzeugen und diese über das Medium Internet zu nicht kommerziellen Zwecken zu verbreiten. *TNSP* ist die Kunst, kostenlose Kunst zu erzeugen. Der von der Jury im Speziellen ausgezeichnete *Draft 5* wurde eigentlich nicht „ernst" produziert, sondern sollte einem Freund beweisen, wie schnell man chart-orientiertes Standard-Techno (aka „Uz-Uz") generieren kann. Das Resultat waren fünf Minuten Drumbox-Spielerei kombiniert mit einem Speech-Sample aus einer Kinder-Zeichentrickserie. Dann kam aber irgendwie die Idee auf, dies mit einem komplett anderen Stil zu kombinieren. Das Endergebnis war schließlich 180 bpm Jungle/Hardcore kombiniert mit einem klassischen, etwas herumtransponierten Orchestersample. Effekt „Klingt schon interessanter".

Alexander Fischl was born in Vienna and currently attends the fifth year class of the technical secondary school Wexstrasse. Together with a friend last year, he received a U19 Award of Distinction for the work "Von Ignoranten, Betriebssystemen und Atomraketen."

Alexander Fischl wurde in Wien geboren und besucht dort gegenwärtig die 5. Klasse der HTL Wexstraße. Gemeinsam mit einem Freund erhielt er bei U19 im Vorjahr eine Auszeichnung für die Arbeit *Von Ignoranten, Betriebssystemen und Atomraketen.*

CYBERTEC
Lukas Pilat

My robot may best be described as a self-steering device. It is constructed so that it can move around a defined area or find a defined goal completely independently in any space, while avoiding obstacles and seeking the best route. To do so, the robot has a microprocessor as a "brain." This receives the information about possible obstacles from several sensors, processes this information and steers the drive accordingly.

I am continuously improving my robot with regard to the quality of the sensors. Soon it should be able to perform a simple task, vacuum cleaning in my case, in permanent operation with no maintenance. In addition, I want to enhance my robot with an automatic charging station, so that it can charge its own batteries.

Lukas Pilat
(16) lives in Vienna and attends the Lycée francais de Vienne there. When he finishes school, he wants to study experimental physics. He received an award at the "International Science & Engineering Fair 1998" for a PC weather station that he built himself.

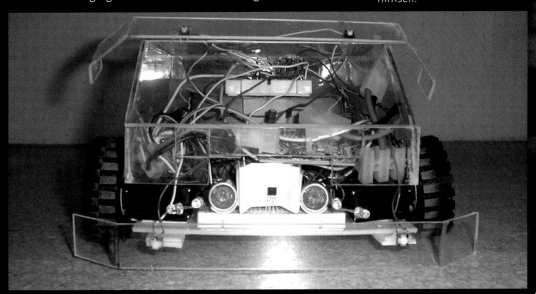

Mein Roboter lässt sich am besten als ein sich selbst steuerndes Gerät beschreiben. Er ist so konstruiert, dass er vollkommen selbstständig in jedem beliebigen Raum entweder einen definierten Bereich abfährt oder ein definiertes Ziel findet, wobei er jeweils Hindernissen ausweicht und den besten Weg sucht. Dazu hat der Roboter einen Mikroprozessor als „Gehirn". Dieser empfängt die Informationen über mögliche Hindernisse von mehreren Sensoren, verarbeitet diese und steuert den Antrieb dementsprechend.

In punkto Qualität der Sensoren verbessere ich meinen Roboter kontinuierlich, damit er demnächst völlig wartungsfrei eine einfache Aufgabe, in meinem Fall Staubsaugen, im Dauerbetrieb erledigen kann. Zusätzlich möchte ich den Roboter mit einer automatischen Ladestation ergänzen, bei der er seine Akkus selbstständig aufladen kann.

Lukas Pilat (16) lebt in Wien und besucht dort das Lycée français de Vienne. Sein Studienwunsch nach der Matura ist das Fach Experimentalphysik. Für eine selbstgebaute PC-Wetterstation erhielt er eine Auszeichnung bei der "International Sience & Engineering Fair 1998".

.NET

Brian Blau (USA)

recently joined Silicon Graphics as the Graphics API Evangelist for

Fahrenheit and OpenGL API's. Before joining SGI Brian was co-founder and Vice President at Intervista Software, an Internet startup building VRML and 3D web technologies and products. There he served in many roles, including software developer and manager of day-to-day engineering operations. Brian also worked at Autodesk Multimedia and the Institute for Simulation and Training, where he worked on desktop visual simulation systems and networked virtual realities. Brian also actively volunteers for SIGGRAPH. In 1996 he co-produced the acclaimed Digital Bayou interactive exhibition and produced the 1999 Computer Animation Festival and Electronic Theater.

Claudia Giannetti (E)

Claudia Giannetti (E) studied art history, specializing in media art.

Since 1993 co-founder and director of the L'Angelot Association for Contemporary Culture in Barcelona, the first space in Spain specialized in Electronic Art, and since 1998 Director of MECAD—Media Centre of Art and Design and Director of Electronic Art and Digital Design at ESDI—Escuela Superior de Diseño, Sabadell-Barcelona, Spain. Writer and curator of several cultural events and media art exhibitions in different countries. She edited, among others, the books *Media Culture* (1995), *Art in the Electronic Age—Perspectives of a New Aesthetic* (1997), *ARS TELEMATICA—Telecommunication, Internet and Cyberspace* (1998), *Arte Facto &*

Ciencia (1999). She is director of the editorial board of the on-line magazine about art, science and technology "MECAD e-journal". (http://www.mecad.org/e-journal).

Lisa Goldman (USA)

is a consultant who works at the intersections of cutting-edge technology, media, and culture. From 1995—2000, she was President and CEO of Construct, www.construct.net, an internet design company known for experimental web development. Lynn Hershman and Construct were awarded a Golden Nica in

Interactive Art in 1999 for *The Difference Engine*. Prior to co-founding Construct, Lisa was the Director of the Interactive Media Festival in Los Angeles. This international competition and exhibition reflected the range of creative accomplishment going on in the emerging digital society in 1994 and 1995. Lisa serves on the boards of the Russian American Foundation and the Art Technology Foundation. She holds a degree in Fine Arts from Rice University, and also studied at the Université de Paris IV, Sorbonne.

John Markoff (USA)

is based in San Francisco as West Coast Correspondent for the *New York Times* where he covers Silicon Valley, computers and information technologies. Before coming to the *Times* in 1988 he covered Silicon Valley for the *San Francisco Examiner* beginning in 1985. He has also been a writer at *Infoworld* and in 1984 he was West Coast

Technical editor for *Byte Magazine*. He is the co-author with Katie Hafner of *Cyberpunk: Outlaws and Hackers on the Computer Frontier* (1991) and with

Lenny Siegel of the *The High Cost of High Tech* (1985). In January of 1996 Hyperion published *Takedown: The Pursuit and Capture of America's Most Wanted Computer Outlaw*, which he co-authored with Tsutomu Shimomura. He was named as one of *Upside Magazine*'s Digital Elite 100 in both 1996 and 1997.

Joichi Ito (J/USA)

developer and producer in the areas of virtual reality and multimedia. Japan correspondent for Mondo 2000, Wired and others. Numerous publications, particularly on networks.

Interactive Art

Masaki Fujihata (J)

Masaki Fujihata (J), born 1956; BA in 1979 and MA in 1981 at Tokyo University of Arts / design course; 1982—1990 business career, board member of Japan Animation Film Association, since 1987 member of

ASIFA, since 1990 Associate Professor, Faculty of Environmental Information at Keio University. Since 1999 Professor at National University of Fine Art and Music, Inter Media Art course. Masaki Fujihata was awarded among others a Golden Nica by the Prix Ars Electronica jury for his entry *Global Interior Project* in the category of Interactive Art (1996).

Peter Higgins (GB)

trained at the Architectural Association, and worked as a scenographic designer in West End theatre and for BBC TV. In 1992 he was co founder of Land Design Studio who are interpretive

designers working with lottery based projects and museums throughout the UK. Most recently they have been responsible for The Playzone at The Millennium Dome London, which publicly demonstrates the extraordinary potential of Interactive digital art.

Joachim Sauter (D)
born 1959, MA in communication design at the University of the Arts, Berlin, further studies at the German Academy for Film and Television, Berlin. He has been using computers from the early stages of his work as a designer and filmmaker. Fueled by his interest, he founded Art+Com, an

independent design and research group in 1988 together with other designers, scientists and technicians. Today he is head of Art+Com e.V. He has been awarded a Distinction in the Prix Ars Electronica 92 and 97 (Interactive Art), as well as the Los Angeles Interactive Media Festival Impact Award in 1995 and the Prix Pixel INA in 1997.

Yukiko Shikata (J)
Self-educated in art, media and theory, working as independent curator and critic of media and contemporary art since late 80s. Since 1990, as co-curator of Canon ARTLAB, she has produced many media art works, which were shown also in Europe and the U.S. Other works include the curator of

Shiseido CyGnet, *Sound Garden* (Christian Möller, Spiral, Tokyo 1997), *Power of Codes* (Mischa Kuball, Tokyo National Museum, 1999), etc.

Jon Snoddy (USA)
is the S.V.P. of Design for Game-Works and CEO of MyTelescope.com. In 1993, Snoddy founded the Walt Disney Virtual Reality Studio. He also led the conceptual development and design of the ride system for the Indiana Jones attraction. Prior to joining Disney, Snoddy was with Lucasfilm's THX division where he was instrumental in transitioning the THX sound system from an industry studio mix product to a worldwide

consumer product. He began his carrier as a technical director working for Mational Public Radio in Washington with the program "All Things Considered." Snoddy has degrees in journalism and electronics from the University of South Carolina and holds several design patents.

Computer Animation / Visual Effects

Mark Dippé (USA)
has worked in feature film for many years, specializing in digital

filmmaking. He was a visual effects director on several films such as *Abyss*, *Terminator 2*, and *Jurassic Park* and was the director of *Spawn*. He was the recipient of the Prix Ars Electronica for *Terminator 2* and *Jurassic Park* and was a past member of the Ars Electronica Jury in 1991.

James Duesing (USA)
is an Associate Professor at Carnegie Mellon University. His concentration is in computer animation. His work has been exhibited widely, in hundreds of

international festivals and exhibitions and nationally televised in the United States, through out Europe, Asia and Australia. His work has received much recognition including: Grants from the National Endowment for the Arts, an American Film Institute Fellowship, an Emmy Award, the Deutscher Videokunstpreis, and a CINE Golden Eagle. His work is held in numerous public collections including: the Museum of Modern Art, New York; the Goethe Memorial Museum, Tokyo; the UCLA Film Archive, and the Israel Museum.

Ines Hardtke (CDN)
is Head of Digital Imaging in the ACI east and Animation/Jeunesse

Studios at the National Film Board of Canada in Montreal. She obtained a M.Math (Computer Science—Computer Graphics Lab) from the University of Waterloo in Waterloo, Ontario. She has been Computer Animation Festival Chair at SIGGRAPH 98.

Barbara Robertson (USA)
has been the West Coast Senior Editor for Computer Graphics World since 1985 and in that capacity has written many award-winning articles on computer animation, visual effects, and graphics technology as she watched the evolution of computer graphics art and technology. Prior to this work,

she was the Editor and Researcher for the *Whole Earth Software Catalog*, West Coast Bureau Chief for Popular Computing, and West Coast Editor for *Byte Magazine*.

Digital Musics

Kodwo Eshun (GB)
At 17, Kodwo Eshun won an Open Scholarship to read Law at University College, Oxford. After 8 days he switched to Literary Theory, magazine journalism and running clubs. He is not a cultural critic or cultural commentator so much as a concept engineer, an imaginist at

the millennium's end writing on electronic music, science fiction, technoculture, architecture, design, post war movies and post war art for *I-D, The Wire, The Face, Arena, Süddeutsche Zeitung, Die Zeit, Melody Maker, The Times* and *The Guardian*. He has published in *The Celluloid Jukebox*, the British Film Institute's critical anthology on the soundtrack and is the youngest writer in Jon Savage and Hanif Kureishi's *The Faber Book of Pop*.

Naut Humon (USA)
conducts, curates and performs his and outside works for Sound Traffic

Control, an omniphonic orchestral "dub dashboard" network which remorphs sonic spatial objects from live instrumentalists, audio sculptures, and multiple DJ/VJ configurations for the RECOMBINANT diffusion jockey summits. He also operates as producer and creative director of the Asphodel/Sombient labels from San Francisco to New York.

Zeena Parkins (USA)
studied dance, classical piano and harp, graduating from Bard College with a degree in Fine Arts. Arriving in New York City in 1984, she worked with and recorded in John Zorn's "Cobra" and Butch Morris' "Conduction ensemble". She joined "Skeleton Crew" with Tom Cora and

Fred Frith with whom she made the record *The Country of Blinds* and toured extensively in Europe, the States, Iceland, and Japan. The duo OWT with David Linton and the quintet NO SAFETY with Chris Cochrane were both projects that she co-founded, toured and recorded'with. From 1986 to the present, Zeena Parkins has written numerous scores for dance, theater, film and video, receiving commissions from Jennifer Monson, Neil Greenberg, Cydney Wilkes, Ana Maria Simo, The Kitchen, Performance Space 122, and Roulette.

Robin Rimbaud (GB)
With his work as Scanner, Robin Rimbaud implicates himself in processes of surveillance, engen-

dering access to both technology and language and the power games of voyeurism. Dubbed a "telephone terrorist", Rimbaud is a techno-data pirate whose scavenging of the electronic communications highways provides the raw materials for his aural collages of electronic music and 'found' conversations. Musician, writer, media critic, cultural engineer, and host of the monthly digital club the Electronic Lounge at the ICA since 1994, he is currently at work on a variety of projects. Scanner—A fearless exponent of ambient electronic soundscapes, Scanner is widely recognised as one of the most exciting artist-producers working in instrumental music. His compositions are absorbing, multi-layered soundscapes that twist state-of-the-art technology in gloriously unconventional ways. He has collaborated and spoken at conferences with Brian Laswell, Brian Eno, Peter Gabriel, Bjork, Derek Jarman and Neville Brot.

Peter Rehberg (aka Pita, A)
DJing at various locations in Viennese underground for almost a decade. Started the use of Pita moniker in early nineties as DJ playing ambient/experimental sounds in so called chill out rooms. Although vastly interested in DJ culture prefers to bring in the non

dance elements he learned at school (if the sound fits the hole let it live). His period of social activity bore the fruits which led to the setting up of the Mego label, and the serious start to recording own music.

U19/ Cybergeneration

Sirikit M. Amann (A)
studied political science, theater arts and economics in Austria, Germany and the USA. Since 1987 she has worked for the ÖKS—Austrian Culture Service—where she is responsible for the concept devel-

opment and implementation of multimedia projects in schools. Austrian Culture Service school projects: 1994 "Computer and Games" and 1995 "Speed" (both at the Ars Electronica Festival).

Norman Filz (A)
has worked for seven years for Austria's biggest youth and music magazine, the RENNBAHN EXPRESS, as a film journalist and is

also head of the counseling section for young people there. In addition, Filz works as a musician and freelance writer; his most recent publication was a

book of advice about love for young people.

Florian Hecker (A)

is an independent composer in the fields of computer music and digital production. Involved in

 various sonic projects simultaneously, such as recording for labels like Mego and Or, his current activities also include audio publishing and interconnection in new formats, e.g. mp3 via the *fals.ch* web project, cd_slopper, and visual abstraction with SKOT.

Horst Hörtner (A)

studied telematics at the technical University of Graz and worked as a freelance developer of realtime

 control systems as well as working for art projects. Co-founder of the group "x-space". He has worked for EXPO Sevilla, documenta IX, austromir, etc. Since 1995 he is technical director of the Ars Electronica Center in Linz and director of the AEC FutureLab.

Hans Wu (A)

 born 1969 in Vienna. ORF employee and freelance journalist since 1991. Writes about computer games, new media, youth culture, cyber culture etc. Responsible for electronic music and new media at FM4 since 1994.

PARTICIPANTS

80/81
Via Stampatori 12
10122 Torino
I
island@8081.com

Ichiro Aikawa
1535 Bellevue Ave # 201
Seattle, WA 98122
USA
Iordia@usa.net

Yae Akaiwa
Exonemo
3-15-15-301 Minami-cho
Kichijoji
Musashino-city Tokyo
180-0003
J
mail@exonemo.com

Jim Armstrong
2112 F/x
2801 Denton
Tap Rd # 835
Lewisville, TX 75067
USA
info@2112fx.com

Roy Ascott
64 Upper Cheltenham
Place
Bristol BS6 5HR
GB
roy_ascott@compuserve.
com

Angeliki Avgitidou
102 Wilberforce Rd
London N4 2SR
GB
aavg@yahoo.com

Marketa Bankova
Smilovskeho 11
120 00 Prague
CZ
marketa@avu.cz

Scott Becker
PO Box 578956
Chicago, IL 60657-8956
USA
artscb@interaccess.com

Peter Beckmann
Gilsingstr. 32
44789 Bochum
D
kunstbepe@yahoo.de

Wolfgang Beinert
Kaulbachstr. 92
80802 München
D
atelier@beinert.net

Thomas Berdel
Schießstätte 6
6800 Feldkirch
A
ttl@cyberdude.com

Joshua Berman
College Of Computing
Georgia Institute Of
Technology
Atlanta, GA 30332-0280
USA
berman@cc.gatech.edu

Luzius Bernhard
Ubermorgen.com
Kurrentg. 10/22
1010 Wien
A
hans@ubermorgen.com

Alberto Blumenschein
Luzcom Interactive
Rua Mourato Coelho,
1046
05417-001 São Paulo
BR
alblum@ruralsp.com.br

Mary-Anne [mez] Breeze
Point In The Fluid
Productions
25 Corrimal St
Tarrawanna 2518
AUS
mezandwalt@wollongong.
starway.net.au

Colin Bright
10/115 Pacific Parade
Dee Why
Sydney NSW 2099
AUS
colbright@netsydney.com.
au

Benjamin Britton
University Of Cincinnati
Ml # 16 College Of Daap
Cincinnati,
OH 45221-0016
USA
benb@cerhas.uc.edu

Paul Brown
PO Box 3603
South Brisbane QLD 4101
AUS
paul@paul-brown.com

Holger Bruckschweiger
Ottstorf 18
4600 Wels
A
holger@bruckschweiger.at

Niels Buenemann
Kavalleriestr. 13
33602 Bielefeld
D
buenemann@agentx.de

Brian Carroll
1336 Park St. # 214
Alameda, CA 94501
USA
human@architexturez.com

Paul Chan
National Philistine
113 W. 60th St., Rm 422
New York, NY 10023
USA
manwichartist@yahoo.
com

Dalia Chauveau
UQUAM
3591 Benny Avenue
Montreal, QC H4B 2S1
CDN
d.chauveau@videotron.ca

Motoshi Chikamori
1-31-1 Katsuradai-nishi
247-0035 Sakae-ku,
Yokohama
J
motoc@attglobal.net

Helen Cho
Quiet-time
5661 Estates Dr.
Oakland, CA 94618
USA
ifrancis@dnai.com

Thanasis Chondros
Konst Melenikou 34
546 35 Thessaloniki
GR
allipoli@magnet.gr

Alberto Cidraes
Kanazawa International
Design Institute
Wakakusa Machi - 4 - 27
921-8111 Kanazawa
J
alberto@kidi.hokuriku.
ne.jp

Victor Cintron
30th Century Productions
PO Box 541982
Houston, TX 77254
USA
sintron@30th.com

Jean-François Colonna
Cmap Ecole Poly-
technique
Route de Palaiseau
91128 Palaiseau Cedex
F
colonna@cmap.polytechni
que.fr

Tom Corby
Block H, Flat 8, Rodney
Rd, Peabody Estate
London SE17 1BN
UK
tom@athomas.demon.co.
uk

Jim Costanzo
Repohistory
350 Bleecker St. # 4p
New York, NY 10014
USA
costanzo@thing.net

David Crawford
168 1st Ave, 2nd Fl
New York, NY 10009
USA
crawford@lightofspeed.
com

Franco Dagani
0100101110101101.org
Via Avesella 1
40121 Bologna
I
propaganda@01001011101
01101.org

Douglas Davis
80 Wooster St.
New York, NY 10012-4347
USA
dd@nyworld.com

Etienne De Bary
21, Rue du Pdt Wilson
94250 Gentilly
F
edebary@club-internet.fr

Andy Deck
Artcontext
257 7th Ave
New York, NY 10001
USA
andy@andyland.net

Sharon Denning
8 Mercer St., 3rd Fl
Jersey City, NJ 07302
USA
sharon@denning.net

Dimiter Dimitrov
Mega Art
78, Samokov Blvd.,
bl. 305, ap. 61
1113 Sofia
BG
megaart@otel.net

Gisela Domschke
Ntl Interactive
18a Albert St.
London NW1 7NZ
UK
gisela.domschke@ntl.com

Sher Doruff
Society For Old And New
Media
Nieuwmarkt 4
1012 CR Amsterdam
NL
sher@waag.org

Stefan Asmus
Uni Wuppertal
Ludwigstr. 49
42105 Wuppertal
D
asmus@uni-wuppertal.de

Reynald Drouhin
31 Rue de Paris
94190 Villeneuve
St. Georges
F
reynald@ensba.fr

William Duckworth
6109 Boulevard East
West New York, NJ 07093
USA
wdckwrth@aol.com

Ursula Endlicher
224 Metropolitan Ave
Apt.19
Brooklyn, NY 11211
USA
ursula@thing.net

Daniel Englert
250 Washington Ave (4d)
Brooklyn, NY 11205
USA
denglert@blue.com

Viktoria Fabsits
HAK Floridsdorf;
Projektgruppe Organon
Schenkendorfg. 11/20
1210 Wien
A
vicky.fabsits@surfeu.at

Marton Fernezelyi
C3
Orszaghaz U. 9
1017 Budapest
H
demedusator@c3.hu

Skawennati Fragnito
Nation To Nation
102 St. Eugene
Chateauguay, QC J6K 1Y6
CDN
skawennati@yahoo.com

Tiziano Fratus
Lines
Dante Alighieri 36
10090 Trana (TO)
I
you@lines.8m.com

Jakob Freitag
Evolution-one
Rotherstr.17
10245 Berlin
D
jakob@evolution-one.de

Alvar Freude
Assoziations-blaster-team
/ Merz Akademie
Im Fuchsrain 44
70186 Stuttgart
D
info@assoziations-
blaster.de

Angela Fung
@radical.media, Inc.
435 Hudson Street
New York, NY 10014
USA
angela@radicalmedia.com

Colette Gaiter
Technographica
1342 Simpson St.
St. Paul, MN 55108
USA
colette_gaiter@mcad.edu

Martin Gantman
719b N. Fairfax Ave
Los Angeles, CA 90046
USA
gantman@pacbell.net

Caroling Geary
Wholeo Online
1031 Crestview Dr. # 318
Mountain View,
CA 94040-3445
USA
caroling@earthlink.net

Jie Geng
4310 Wells Dr.
Parlin, NJ 08859
USA
jie@sva.edu

Valery Grancher
53 Rue de Seine
75006 Paris
F
vgrancher@nomemory.org

John Grech
28 Lansdowne St., Surry
Hills
2010 Sydney
AUS
007grech@ion.apana.org.
au

Reinhold Grether
D
reinhold.grether@uni-
konstanz.de

David Guez
13, Rue Blondel
75002 Paris
F
dguez@tv-art.net

Karen Harman
Oyster Partners
4 Holford Yard,
Cruikshank St
London WC1X 9HD
GB
karen@oyster.co.uk

Isabelle Hayeur
Perte De Signal
2310 Holt
Montreal,
Quebec H2G 1Y4
CDN
isa@elfe.com

Jörg Heinrich
Die Vision Digital Ag
Mittererstr. 9
80336 München
D
jh@dvdigital.de

Lynn Hershman
1201 California St.
San Francisco, CA 94109
USA
lynn2@well.com

Paul Higham
Hfrl Virtual Reality Lab
141 Mariucci Arena
Operations,
1901 4th St. S.e.
Minneapolis, MN 55414
USA
highaoo1@tc.umn.edu

Leslie Huppert
Adalbertstr. 22
10997 Berlin
D
leslie@internett.de

David Hutchison
Eyedoll
35/2 Viewcraig Gdns
Edinburgh EH8 9UN
UK
deitch@eyedoll.com

Ron Hutt
School Of The Art
Institute Of Chicago
2742 N. Hampden Ct.
3c
Chicago, IL 60614
USA
rhutt@21stcentury.net

Lisa Hutton
1000 Hilltop Circle Fa111
Baltimore, MD 21250
USA
hutton@umbc.edu

Fransje Jepkes
Anjeliersstraat 19
1015 PT Amsterdam
NL
fransjej@xs4all.nl

Tiia Johannson
Mahtra 21 - 48
13811 Tallinn
EST
xtiiax@hotmail.com

Olgalyn Jolly
63 Greene St. Suite 204
New York, NY 10012
USA
jolly@somestrange.com

Wolf Kahlen
Ehrenbergstr. 11
14195 Berlin
D
wolf.kahlen@tu-berlin.de

Paras Kaul
Paras West Productions
PO Box 5275
39762 Ms State, MS
USA
paras@ra.msstate.edu

Andruid Kerne
Nyu Media Research Lab
+ Creating Media
719 Broadway, 12th Floor
New York, NY 10003
USA
andruid@mrl.nyu.edu

Olga Kisseleva
30 Rue de L'Echiquier
75010 Paris
F
kisselev@cnam.fr

Nicolaus Klinger
Daponteg. 11/12
1030 Wien
A
klinger@chello.at

Maria Klonaris
A.s.t.a.r.t.i.
71 Rue Leblanc
75015 Paris
F
klon.thom.astar@wanado
o.fr

Jeff Knowlton
25057 Chestnut St.
Newhall, CA 91321
USA
knowlton@calarts.edu

Alexander Kolev
Objectart Media Art
Group
Mladost 1 Bl. 36 Ent. 14
Apt. 15
1000 Sofia
BG
objectart@nat.bg

Fvezi Konuk
Hilschbacherstr. 31
66292 Riegelsberg
D
info@hyper-eden.com

Fabio Lauria
Timecode Records
Postlagernd
4018 Basel 18
CH
timecode@gmx.ch

Bing Lee
Bing Lee Studio
474 Greenwich St.
New York, NY 10013
USA
bing@pictodiary.com

Marketta Leino
Valiaitankatu 9 E 39
40320 Jyvaskyla
SF
marketta@art.net

Rudy Lemcke
540 Alabama
San Francisco, CA 94110
USA
rudy@sirius.com

Richard Lerman
13230 N 3rd Way
Phoenix, AZ 85022
USA
sonicjourneys@yahoo.
com

Jennifer Ley
Riding The Meridian
7 Ernst Ave
Bloomfield, NJ 07003
USA
anemone@sprynet.com

Patrick Lichty
8211 E. Wadora Nw
N. Canton, OH 44720
USA
voyd@raex.com

Simon Lucas
Amateur Enterprises
5 The Barons
Twickenham TW1 2An
GB
simon@othermedia.com

Ernest Lucha
RTMark
PO Box 14872
San Francisco, CA 94114
USA
ernest@rtmark.com

Stephen Mamber
UCLA, Dept. of Film/TV
East Melnitz Hall
Los Angeles, CA 90095
USA
smamber@ucla.edu

Fumio Matsumoto
Plannet Architectures
2-11-27-302 Yakumo
Meguro-ku
152-0023 Tokyo
J
matsumoto@plannet-
arch.com

Cezary Mazurek
Poznan Supercomputing
And Networking Center
Noskowskiego 10
61704 Poznan
PL
mazurek@man.poznan.pl

Jennifer McCoy
210 Congress St. # 2c
Brooklyn, NY 11201
USA
mccoy@earthlink.net

Conor McGarrigle
1 Annamount,
Mulgrave St.
Dun Laoghaire, Co Dublin
IRL
conor@stunned.org

Daniel Micaelli
Sarmiento 1411, 2º 6
1042 Ciudad De
Buenos Aires
RA
d_micaelli@hotmail.com

Bonnie Mitchell
Bowling Green State
University
Rm 1000 Fine Arts Center
Bowling Green,
OH 43403
USA
bonniem@creativity.bgsu.
edu

Enrico Mitrovich
Via Milano 53
36100 Vicenza
I
emitrov@goldnet.it

Rodolfo Muchela
C&c
C.c. 169 Suc 1b
1401 Buenos Aires
RA
muchela@bicentenario201
o.com

Christina Mueller
10 West 19th St.
New York, NY 1080
USA
christina@deijncallot.com

Kazushi Mukaiyama
2-45-2-8-302,
Shirakawadai, Suma
Kobe, Hyogo 654 - 0103
J
kazu@arizona.ne.jp

Petra Müller
Kunstkomm.t!
Waltgeristr. 50
32049 Herford
D
post@kunstkommt.de

Domenico Nadile
Via S. Antonio 6
37122 Verona
I
domeniconadile@libero.it

Mark Napier
451 East 14 St. # 12h
New York, NY 10009
USA
napier@interport.net

Dariusz Nowak-Nova
PO Box 1742
40-874 Katowice 22
PL
nova@entropia.com.pl

Matteo Pasquinelli
Net_institute
Via Zamboni, 59
40126 Bologna
I
net-i@zkm.de

Timo Piatkowski
Hfg Karlsruhe
Klauprechtstr. 23
76137 Karlsruhe
D
timopiat@hfg-
karlsruhe.de

Gerhard Pichler
Spielboden Kv GmbH
Färberg. 15
6850 Dornbirn
A
pichlerg@luxmate.co.at

Nina Pope
94 Hind House
London N7 7NB
GB
artists@somewhere.
org.uk

Stuart Pound
34 Blondin Ave
London W5 4UP
GB
bocadillo@cwcom.net

Psykinetica / Centre For
Metahuman Exploration
Studio For Creative
Inquiry, Carnegie Mellon
University
5000 Forbes Ave
Pittsburgh, PA 15213
USA
projectclub@onelist.com

Joseph Rabie
Magelis Sa
129 Rue Gaston
Doumergue
31170 Tournefeuille
F
joe@magelis.com

Melinda Rackham
Subtle.net
387 Riley St.
Surry Hills, Sydney 2010
AUS
melinda@subtle.net

Bruno Randolf
Subnet
Haydnstr 1/12
5020 Salzburg
A
br1@subnet.at

Simone Ricci
Via Miglietti 10
10144 Torino
I
kung@iol.it

Martin Rille
Hasenleiteng. 100
1110 Wien
A
animator@gmx.net

Gregory Royer
1 Égal 2
11, Av. De La Capelette
13010 Marseille
F
groyer@1egal2.com

Jacopo Ruggiero
Graceland 2000
V. Del Casalaccio 99
00046 Grottaferrata
(Roma)
I
j.ruggiero@tiscalinet.it

Frieder Rusmann
Arndtstr. 35
70193 Stuttgart
D
rusmann@das-deutsche-
handwerk.de

Elizabeth Santeix
Vaguepolitix
40 West 29th St.
New York, NY 10001
USA
esanteix@vaguepolitix.co
m

Herbert Schager
Mozartstr. 56
A-4020 Linz
A
maikel@fan.priv.at

Birgit Schmidt
Hochschule für
Gestaltung Karlsruhe
Adlerstr. 46
76133 Karlsruhe
D
bschmidt@teco.edu

Michele Schnabel
Artformation.com
Rathausg.1/1
4020 Linz
A
schnabel@artformation.
com

Manfred Seifert
B@d@rt
Stadtring 142
64720 Michelstadt
D
mansei@t-online.de

Yukiko Shikata
Canon Artlab
Dk Bldg. 5f, 7 - 18 - 23
Roppongi, Minato-ku
106-0032 Tokyo
J
yshikata@crpg.canon.
co.jp

Tomoo Shimomura
2-2, Nanryo-cho 2 Cho
Sakai-shi
Osaka 590-0811
J
tomoo@osk.3web.ne.jp

Alexei Shulgin
Easylife.org
Polotskaya 29-1-39
121351 Moscow
RUS
alexei@easylife.org

Deanne Sokolin
School Of Visual Arts,
NYC
148 Franklin St.
Brooklyn, NY 11222
USA
ds@interport.net

Vanessa Sowerwine
9 Garfield St.
Fitzroy 3065
AUS
van@netspace.net.au

David Steiner
Taunusstr. 9
14193 Berlin
D
steiner@hdk-berlin.de

Nicole Stenger
220 Bay St. # a
Santa Monica , CA 90405
USA
nicole@netgate.net

Igor Stepancic
Blueprint
Zarkovacka 18
11030 Beograd
YU
igor@blueprintit.com

Igor Stromajer
Intima Virtual Base /
Net Art Lab
Tabor 7
1000 Ljubljana
SLO
atom@intima.org

Gerd Struwe
Heinrichstr. 45
50676 Köln
D
struwe@biogenart.de

Joseph Tabbi
Electronic Book Review
1459 W. Cortez St.
Chicago, IL 60622
USA
ebr@uic.edu

Ray Thomas
RTMark
PO Box 14872
San Francisco, CA 94114
USA
ray@rtmark.com

Paul Thomas
29 Elizabeth St.
North Perth
Perth, WA 6006
AUS
prt@cyllene.uwa.edu.au

Henning Timcke
Ideen Werft 22 GmbH
Stadtturmstr. 5
5400 Baden
CH
henning.timcke@kunst.ch

Brad Todd
1125 Ave. Lajoie Apt. #10
Montreal, QC H2V 1N7
CDN
bt@mobilegaze.com

Hisayoshi Tohsaki
IMGSRC
Shinsentyo 5-2
Shioirikouji # 101
Shibuyaku,
Tokyo 150-0045
J
tohsaki@imgsrc.co.jp

David Tomas
527 Lansdowne Ave
Westmount, H3Y 2V4
CDN
auterima@internet.uqam.
ca

Bill Tomlinson
MIT
E15 - 320g /
20 Ames Street
Cambridge, MA 02139
USA
wireguy@media.mit.edu

Stephane Tougas
4098 Debullion
Montreal, QC H2W 2E5
CDN
toug@broue.com

Nina Velikova
Nova Art Ltd.
5 Alfred Novel, Block
125a
1113 Sofia
BG
nikke@omega.bg

Gary Venter
1001/12 Ithaca Road,
Elizabeth Bay
Sydney NSW 2011
AUS
gary@gva.net.au

Liz Vlx
Ubermorgen.com
Kurrentg. 10/22
1010 Wien
A
liz@ubermorgen.org

Marek Walczak
172 East 4th St. # 7c
New York, NY 10009
USA
marek@interport.net

Marilyn Waligore
6477 Fisher Rd
Dallas, TX 75214
USA
sorcery@flash.net

Timothy Weaver
P.o. Box 7048, Crescent
Branch, Coal Creek
Canyon
Golden, CO 80403-0100
USA
tweaver@artswire.org

Annette Weintraub
City College Of New York
C/o 2 Bond Street
New York, NY 10012
USA
anwcc@cunyvm.cuny.edu

Alex Weyers
Arts Alliance Labs
126 South Park
San Francisco, CA 94107
USA
weyers@artsalliance.co.uk

Zelko Wiener
Laimgrubeng. 19/7
1060 Wien
A
office@zeitgenossen.com

Maciej Wisniewski
53 Withers St.
Brooklyn, NY 11211
USA
mzw@interport.net

Matthias Wölfel
Wilhelm-Roether-Str. 12
76307 Karlsbad
D
matthias@wolfel.de

Nick Wray
27 Princes Ave
London N3 2DA
GB
nick@paperwork.demon.c
o.uk

Judson Wright
Plasma Studii
223 E 10 St., # 7
New York, NY10003
USA
judson@judson.net

Moritz Wurnig
Sechskrügelg. 2/14
1030 Wien
A
moritz@chello.at

Jaka Zeleznikar
Gornji Trg 22
1000 Ljubljana
SLO
jaka@kid.kibla.org

Jody Zellen
843 Bay St. # 11
Santa Monica, CA 90405
USA
jodyzel@aol.com

Marina Zerbarini
Teodoro Garcia 1939 P. 9
Dto.c
1426 Buenos Aires
RA
marinazerbarini@arnet.
com.ar

80/81
Via Stampatori 12
10122 Torino
I
island@8081.com

Romy Achituv
258 Union St. # 3
Brooklyn, NY 11231
USA
raa9624@is.nyu.edu

Marina Aina
Pohlinova 14
1000 Ljubljana
SLO
margrz@alpha.zrc-sazu.si

Montxo Algora
Art Futura
Plaza Doctor Laguna 12
28009 Madrid
E
earthdance7@yahoo.com

Ekkehard Altenburger
Ninja Biskit
20 B Sears St.
London SE5 7 JL
GB
ekkealt@britishlibrary.net

Mauro Annunziato
Plancton
Lungotevere Degli
Artigiani 28
00152 Roma
I
plancton@plancton.com

Josephine Anstey
EVL, University Of Illinois
At Chicago
851 S. Morgan, Rm 1120
Chicago, IL 60607
USA
anstey@evl.uic.edu

David Anthony
Metropolis Dvd
1790 Broadway, 9th Floor
New York, NY 10019
USA
david@metropolisdvd.
com

Kohei Asano
Tokyo Institute Of
Polytechnics
3-45-4-404 Higashi-
funabashi
Funabashi-shi
Chiba 273-0002
J
hey@a.office.ne.jp

Nicolas Baginsky
Admiralitätsstr. 74
20459 Hamburg
D
nab@provi.de

Annette Barbier
Northwestern University
1905 N. Sheridan Rd
Evanston, IL 60208
USA
abarbier@nwu.edu

Laura Beloff
Pengerkatu 4 A 11
00530 Helsinki
SF
laura@fl.aec.at

Timothy Benedict
8101 Richard Rd
Broadview Heights,
OH 44147
USA
tbenedi@hotmail.com

Erich Berger
Hauptstr. 2–4
4040 Linz
A
info@aec.at

Luzius Bernhard
Ubermorgen.com
Kurrentg. 10/22
1010 Vienna
A
hans@ubermorgen.com

Rodney Berry
ATR Media Integration
And Communications
Laboratories
2-2-2 Hikaridai,
Seika-cho, Soraku-gun
Kyoto, 619-0288
J
rodney@mic.atr.co.jp

Natalie Bewernitz
Luxemburger Str. 297
50939 Köln
D
n.bewernitz@hbks.uni-sb.de

Michael Bielicky
Kubelikova 38
13000 Praha 3
CZ
x@avu.cz

Roland Blach
Möhringerstr. 33
70199 Stuttgart
D
roland.blach@iao.fhg.de

Matt Black
Ninja Tune
90 Kennington Lane
London SE11 4XD
GB
mattb@ninjatune.net

Robert Brecevic
Phoenix Ab
Gervalds Hogrän 1:24
621 96 Visby
S
brecevic@algonet.se

Holger Bruckschweiger
Ottstorf 18
4600 Wels
A
holger@bruckschweiger.at

Jim Campbell
1161 De Haro St.
San Francisco, CA 94107
USA
jcam@sirius.com

Veronique Caraux
62 Rue Fondary
75015 Paris
F
verokaro@club-internet.fr

Bryan Carter
EVL
University Of Illinois at
Chicago, 851 S. Morgan
St. Rm 1120 Seo
Chicago, IL 60607-7053
USA
engbryan@showme.missouri.edu

Paul Chan
National Philistine
113 W. 60th St., Rm 422
New York, NY 10023
USA
manwichartist@yahoo.com

Isabelle Choiniere
Le Corps Indice
588 Bennett
Montreal, QC H1V 2S3
CDN
ddd@odyssee.net

Leon Cmielewski
4/49 Glenayr Avenue,
Bondi
2026 Sydney
AUS
leon@autonomous.org

Fred Collopy
Case Western Reserve
University
10900 Euclid Avenue
Cleveland,
OH 44106-7235
USA
flc2@po.cwru.edu

Maureen Connor
10 Leonard St. 1s
New York, NY 10013
USA
connor@thing.net

Shane Cooper
ZKM
Lorenzstr. 19
76135 Karlsruhe
D
shanecooper@shanecooper.com

Brigid Costello
323 Annandale St
Annandale
2038 Sydney
AUS
bm.costello@unsw.edu.au

Laurent Costes
Ecole d'Art
Rue Emile Tavan
13100 Aix-en-Provence
F
souc@wanadoo.fr

Gregory Cowley
Thetestproject
29a Hoff St.
San Francisco, CA 94110
USA
gregory@testsite.org

Till Cremer
Laboratorium
Körnerstr. 29 Hinterhaus
38102 Braunschweig
D
laboratorium@t-online.de

Peter D'Agostino
523 Shoemaker Rd
Elkins Park, PA 19027
USA
pda@astro.temple.edu

Calin Dan
Kinema Ikon /
Museum Arad
Enescu 1
2900 Arad
RO
atelier@kinema-ikon.sorostm.ro

Andrea Davidson
47 Avenue de Saint Ouen
75017 Paris
F
davidson@club-internet.fr

Cida De Aragao
Philippistr. 14
14059 Berlin
D
cida@slab-berlin.de

Diolio
Simonstr. 2
3012 Bern
CH

Ivor Diosi
Palarikova 4
811 05 Bratislava
SK
diosi@luna.sk

Brendan Earley
The Firestation Artists
Studios
9-11 Buckingham St.
Dublin 1
IRL
brendanearley@ireland.com

Jan Edler
[kunst Und Technik] E.v.
Leipziger Str. 50
10117 Berlin
D
play@berlin.heimat.de

Eike Eike
Lövölde Tér 3.
1068 Budapest
H
eike@c3.hu

Heather Elliott
Bowling Green State
University
1000 Fine Arts Center
Bowling Green,
OH 43403
USA
helliott@christo.bgsu.edu

Onno Ernst
Zuidlarenstraat 209
2545 VT Den Haag
NL
onno.ernst@arts-and-bits.org

Gregor Fischer
Cybereye
Landstraßer Hauptstr.
88/12
1030 Wien
A
obiwan@cybereye.at

Franz Fischnaller
Fabricators
Via Fratelli Bronzetti 6
20129 Milano
I
fabricat@galactica.it

Mary Flanagan
140 Linwood Ave Ste. E2
Buffalo, NY 14209
USA
marydot@attglobal.net

Joachim Fleischer
Hohenheimerstr.38
70184 Stuttgart
D
fleischer_kunst@hotmail.com

Alberto Frigo
Cdddi
Via Zanella, 212
36010 Mont. C. Otto (VI)
I
artdavi@hotmail.com

Noriyuki Fujimura
Keio Univ Graduate
School Of Media And
Governance
#7 Higashi Sou
251-0032
Katase,fujisawa,kanagawa
J
noriyuki@sfc.keio.ac.jp

Takaumi Furuhashi
Reinhold-Frank-Str.12
76133 Karlsruhe
D
taka@okay.net

Beate Garmer
Rotenbergstr. 24
66111 Saarbrücken
D
b.garmer@hbks.uni-sb.de

Dario Gibellini
Dario Gibellini Projects
Keplerstr. 21
22763 Hamburg
D
gibellini@aol.com

Piero Gilardi
Arslab Committee
Corso Casale 121
10132 Torino
I
piero.gilardi@tin.it

Jean-Pierre Giovanelli
Archimediart
Les Flouvières
06640 St . Jeannet
F
archimediart@nicematin.fr

Suguru Goto
82, Rue Charles Nodier
93500 Pantin
F
sgoto@ircam.fr

Christian Graupner
Human Magnetics
Cooperation
Leuschnerdamm 19
10999 Berlin
D
voov@voov.de

Peter Grzybowski
80 Varick Suite 2 D
New York, NY 10013
USA
grzybo@earthlink.net

Ale Guzzetti
V.le Rimembranze, 6
21047 Saronno (VA)
I
virale@pn.itnet.it

Martijn Hage
Emmastraat 5
3314 ZL Dordrecht
NL
m.hage@quicknet.nl

James Hancock
University Of Technology,
7 Stack St. Balmain
Sydney NSW 2041
AUS
hancockjames@hotmail.com

Janine Handler
Rechte Wienzeile 5/4/10
1040 Wien
A
janine.handler@amnis.at

Terrence Handscomb
Nil
15 Salisbury Garden
Court
Wellington 6001
NZ
mail@terrence.org

Laurent Hart
22 Rue Biot
75017 Paris
F
closer@club-internet.fr

Mongrel Harwood
Mongrel
16a Flodden Rd
London SE5 9LH
GB
harwood@mongrel.org.uk

Kelly Heaton
MIT Media Lab
20 Ames St. E15 - 413
Cambridge, MA 02139
USA
kelly@media.mit.edu

Aaron Hertzmann
NYU Media Research Lab
719 Broadway, 12th Floor
New York, NY 10003
USA
hertzman@mrl.nyu.edu

Jiro Hirano
Iamas
503-0014 Ryouke,Ogaki-City,Gifu
J
jiro@iamas.ac.jp

Rania Ho
Interactive
Telecommunications
Program, New York
University
721 Broadway, 4th Floor
New York, NY 10003
USA
r.ho.1@alumni.nyu.edu

Perry Hoberman
167 North 9th St.
Brooklyn, NY 11211
USA
hoberman@bway.net

Ulrich Höhne
Rudolfstr. 26
D 76131 Karlsruhe
D
ea02@rz.uni-karlsruhe.de

Tiffany Holmes
University Of Michigan
2055 Commerce Blvd. #
221
Ann Arbor, MI 48103
USA
tgholmes@umich.edu

Martin Huemer
C/o Bg Xix,
Gymnasiumstr. 83
1190 Wien
A
ps@to.or.at

Elvira Hufschmid
Hbk Saar, Saarbrücken
Lessingstr.45
66121 Saarbrücken
D
e.hufschmid@hbks.uni-sb.de

Ivan Iliev
Alxingerg. 5/43
1100 Wien
A
emanuel.wenger@oeaw.ac.at

Hiroshi Ishii
MIT Media Lab
20 Ames St.
Cambridge, MA 02139
USA
ishii@media.mit.edu

Haruo Ishii
Aichi Prefectural Art
University
30-1 Ishibata Narumi-cho
Midori-ku Nagoya-shi
458-0801
J
mxco0275@nifty.ne.jp

Naoko Ishizaki
Iamas
154-101
503-0031 Makino-cho,
Ogaki-City, Gifu-ken
J
poko98@iamas.ac.jp

Eiji Ito
Keio University
4-8-13-102
252-0804 Shonandai,
Fujisawa-shi
Kanagawa-ken, J
ito@sfc.keio.ac.jp

Teiji Iwata
Dept. Of Aeronautics And
Astronauts, University Of
Tokyo
2-6-13, Akitsu-town
Higashimurayama-city,
189-0001
J
teiwa@mbd.sphere.ne.jp

Tiia Johannson
Mahtra 21-48
13811 Tallinn
EST
xtiiax@hotmail.com

Art Jones
PO Box 357 Cooper
Station
New York, NY 10276
USA
artjones@earthlink.net

Istvan Kantor
1160 A Dundas St. East
Toronto, ONT, M4M 1S1
CDN
amen@interlog.com

Tetsuji Katsuda
Tsurumai Nisi-cho 2 - 15
409
631-0022 Nara City
J
katsuda@net.email.ne.jp

Karen Kelly
Dia Center For The Arts
542 West 22nd St.
New York, NY 10011
USA
kkelly@diacenter.org

Raivo Kelomees
Estonian Academy Of
Arts
Tartu Str. 1
10145 Tallinn
EST
offline@online.ee

Josh Kimberg
444 Park Ave South 1201
New York, NY 10016
USA

Ronit Kirchman
4211 Laurel Canyon Blvd
220
Studio City, CA 91604
USA
freeflow@mindspring.com

Ivika Kivi
Estonian Academy Of Art,
E-Media Center
Tartu Str.1.
10145 Tallinn
EST
ivika@artun.ee

John Klima
130 Broadway
Brooklyn, NY 11211
USA
klima@echonyc.com

Selene Kolman
Kkep
Tollensstraat 60
1053 RW Amsterdam
NL
selene@kkep.com

Kenji Komoto
Osaka University Of Arts
1-1-18-812 Tatuminisi,
Ikunoku
544-0012 Osaka
J
komoto@osk3.3web.ne.jp

Motoki Kouketsu
Iamas
503-0014 Ryoke Ogaki
City Gifu Pref.
J
ketsu99@iamas.ac.jp

Stacy Koumbis
Nearlife, Inc
147 Sherman St.
Cambridge, MA 02140
USA
stacy@nearlife.com

Reed Kram
Gotlandsgatan 53, 3tr.
11665 Stockholm
S
kram@kramdesign.com

Kritz Kratz-Institut
Hans-Sachs-Str. 19
50931 Köln
D
info@kritzkratz.de

Henrike Kreck
Hochschule Der
Bildenden Künste
Alvenslebenstr. 2
66117 Saarbrücken
D
h.kreck@hbks.uni-sb.de

Orit Kruglanski
Calle Nou De San
Francesc 17 Pral
08082 Barcelona
E
krugie@bigfoot.com

Olga Kumeger
Theremin Center
Bolshaja Filevskaja 45-1-
16
121433 Moscow
RUS
kumeger@hotmail.com

Felipe Lara
Walt Disney Imagineering
5158 1/2 Franklin Ave
Los Angeles, CA 90027
USA
felipe.lara@disney.com

George Legrady
Merz Akademie
Teckstr. 58
70190 Stuttgart
D
george.legrady@merz-
akademie.de

Golan Levin
MIT Media Lab
20 Ames Street, E15-447
Cambridge, MA 02139
USA
golan@media.mit.edu

Jason E. Lewis
Arts Alliance Laboratory
126 South Park
San Francisco, CA 94107
USA
ars@thethoughtshop.com

Patrick Lichty
8211 E. Wadora Nw
N. Canton, OH 44720
USA
voyd@raex.com

Jarryd Lowder
41 E. 28th St. # 3a
New York, NY 10016
USA
jarryd@sva.edu

Rafael Lozano-Hemmer
Ciudad Rodrigo 2, 4-izq
28012 Madrid
E
rafael@csi.com

Michael Lutz
Vor Den Pferdeweiden 16
27726 Worpswede
D

Desbazeille Magali
9 Rue Brûle Maison
59000 Lille
F
mdesbazeille@hotmail.
com

Aliyah Marr
119 Hamilton Ave
Fairfield, NJ 07022
USA
amarr@newmediaforge.
com

Massimo Mascheroni
Via Alla Boschina 13
24039 Sotto Il Monte
(BG)
I
onderozze@hotmail.com

Hiroshi Matoba
Human Media
Laboratories, Nec
4-1-1 Miyazaki
216 Kawasaki, Kanagawa
J
matoba@mxb.mesh.ne.jp

Nobuyuki Matsushita
Sony Csl Intaraction
Lab Inc
3-14-13, Higashigotannda,
Shinagawa-ku
Tokyo 141-0022
J
tota@csl.sony.co.jp

Cat Mazza
25-46 44th St., Apt. 3r
Astoria, NY 11103
USA
cath_siena@yahoo.com

Yiannis Melanitis
Digital Art Lab
Peanieon 6, Metaxourgio
10436 Athens
GR
melanitis@hotmail.com

Laurent Mignonneau
ATR Media Integration
And Communications
Research Lab
2-2 2-chome, Hikaridai,
Seika-cho, Soraku-gun
61902 Kyoto
J
laurent@mic.atr.co.jp

Steve Miller
48 Gold Street
New York, NY 10038
USA
steve@stevemiller.com

Enzo Minarelli
Via Cremonino 14
C.p. 152
44042 Cento
I
3vitre@iii.it

Norikazu Mitani
Keio Univ. Graduate
School Of Media And
Governance
Kanagawa Daimachi 2-5
602
221-0834 Yokohama
J
norikazu@sfc.keio.ac.jp

Christian Möller
Meiseng. 28
60313 Frankfurt am Main
D
moelchr@ipf.de

L Moren
Umbc
1000 Hilltop Circle
Baltimore, MD 21250
USA
lmoren@umbc.edu

Peter Morse
39 Green St.
Richmond 3121
AUS
marionh@netspace.net.au

Peter Muehlfriedel
Skop
Horandweg 27
13465 Berlin
D
skop@skop.com

Hisao Nagahama
Ghetto Graphics
3 - 5 - 11 Fukushima
Fukushimaku
553-0003 Osaka City
J
aag82300@popo2.odn.ne
.jp

Yoichi Nagashima
Art & Science Laboratory
10-12-301,
Sumiyoshi-5
430-0906 Hamamatsu,
Shizuoka
J
nagasm@kobe-
yamate.ac.jp

Ryo Nakatsu
Atr Mic Labs
Hikari-dai 2-2,
Seika-cho Soraku-gun
619-0288 Kyoto
J
nakatsu@mic.atr.co.jp

Anne Niemetz
Leopoldstr. 3
76133 Karlsruhe
D
a@adime.de

Haruki Nishijima
Iamas
Fujire-cho 1-1-7 Rist202
Ogaki-City Gifu 503-0893
J
haruki99@iamas.ac.jp

Georg Nussbaumer
Ottensheimer Str. 42
4111 Walding
A
g.nuss@aon.at

Karen O'Rourke
10, Rue Véronèse
75013 Paris
F
orourke@univ-paris1.fr

Ben Owen-Jones
13 Mertyr Ave
Portsmouth, NH PO6 2AR
USA
boj72@hotmail.com

Ursula Palla
Wickenweg 42
8048 Zürich
CH
palla@nabor.net

Tony Patrickson
Farravaun
Glann, Oughterard,
Co. Galway
IRL
asp@iol.ie

Dennis Paul
Colabart
532 1/2 N. Spaulding Ave
Los Angeles, CA 90036
USA
colabart@ix.netcom.com

Richard Pell
College Of Fine Arts
Rm 111
Carnegie Mellon
Pittsburgh, PA 15213
USA
iaa@appliedautonomy.co
m

Olof Persson
Äkps
Björkuddsgatan 17
412 62 Göteborg
S
artsystem@telia.com

Kjell Yngve Petersen
Bjørnsonsvej 85
2500 Valby
DK
kjell.yngve@teliamail.dk

Frank Petschull
Keplerstr.1
D-76351 Linkenheim
D
frank petschull@hfg-
karlsruhe.de

Norbert Pfaffenbichler
Degeng. 67/17
1160 Wien
A
norbert@monoscope.co.at

Liz Phillips
Parabola Arts Foundation
39 - 39 45th St.
Sunnyside, NY 11104-2103
USA
sculpsound@unidial.com

Andrea Polli
Columbia College
600 S. Michigan
Chicago, IL 60605-1996
USA
apolli@interaccess.com

Nina Pope
94 Hind House
London N7 7NB
GB
artists@somewhere.
org.uk

Andreas Reisewitz
Fachhochschule Augsburg
Talstr. 12
85250 Wollomoos
D
travel@compudrom.de

Don Ritter
204 15 St.
Brooklyn, NY 11215
USA
ritter@interport.net

Keith Roberson
Florida State University
2105 High Rd
Tallahassee, FL 32303
USA
cyanmom@mindspring.
com

Mike Roche
11 Great Russell
Mansions, 60 Great
Russell St.
London WC1B 3BE
GB
emmakaufmann@hotmail.
com

Heiko Rometsch
Lontelstr. 22
70839 Gerlingen
D
heiko@erdbeerhund.com

Danilo Rometsch
Erdbeerhund
Mediengestaltung
Lontelstr. 22
70839 Gerlingen
D
danilo@erdbeerhund.com

Caroline Ross
Les Productions
Recto-verso Québec
650, Côte d'Abraham
Québec, G1R 1A1
CDN
recto-verso@meduse.org

Teri Rueb
University Of Maryland,
Baltimore County
1000 Hilltop Circle,
Rm. Fa111
Baltimore, MD 21250
USA
rueb@umbc.edu

Manfred W. Rupp
Max-Planck-Str. 16 A
85435 Erding
D
manfred.rupp@mail.fracta
l-art.com

Marie Ruprecht
Kauri Medienproduktion
Volksgartenstr. 7
4020 Linz
A
marie@fl.aec.at

Shoken Sakai
Iamas
#202 The 3rd Tahara
Haimu
610-1106 13-272
Kutsukake-cho Ohe
Nishikyo-ku Kyoto
J
sakai98@iamas.ac.jp

Philip Samartzis
62 Clauscen St.
3068 North Fitzroy
AUS
p.samartzis@rmit.edu.au

Luz Maria Sanchez
Radio Udg
Av. San Francisco 2170-a
44500 Guadalajara
MEX
mariasc@radio.udg.mx

Carmine Arturo
Sangiovanni
Daygo Multimídia
Rua Francisco Tapajos,
513 Ap. 11
04153-001 Sao Paulo–Sp
BR
daygo@zaz.com.br

Lisa Schmitz
Eisenacher Str. 73
10823 Berlin
D
lisa.schmitz@snafu.de

Gebhard Sengmüller
Margaretenstr. 106/17
1050 Wien
A
gebseng@thing.at

Kohji Setoh
Keio University
1470 Dai, Kamakura
Kanagawa 247-0061
J
setoh@sfc.keio.ac.jp

Agueda Simó
Avda. Madariaga N. 18,
3a
48014 Bilbao
E
asimo@jet.es

Christa Sommerer
ATR Media Intergration
And Communications
Research Lab
2-2 2-chome Hikaridai,
Seika-cho, Soraku-gun
61902 Kyoto
J
christa@mic.atr.co.jp

Jörg Sonntag
Bautzner Str. 82
01099 Dresden
D
salich@gmx.de

Stenslie Stahl
Sverdrupsgate 4
0559 Oslo
N
stahl@sirene.nta.no

Arthur Stammet
29, Rue Léon Metz
4238 Esch-sur-Alzette
L
arthur.stammet@ci.
educ.lu

Douglas Edric Stanley
Chemin de Cassade
13122 Ventabren
F
destanley@teaser.fr

Birte Steffan
Hochschule Der Künste
Berlin
Manteuffelstr. 54
10999 Berlin
D
birte@hdk-berlin.de

Jennifer Steinkamp
Art Center College Of
Design
12029 Marine St.
Los Angeles, CA 90066
USA
jennifer@artcenter.edu

Andrew Stern
Pf Magic / Mindscape
1783 8th Ave
San Francisco, CA 94122
USA
apstern@ix.netcom.com

Yuri Sunahara
250 Mercer St. 605b
New York, NY 10012
USA
ys265@is7.nyu.edu

Jun Tani
Sony Csl
3-14-13, Higashi-gotanda,
Shinagawa
141 Tokyo
J
tani@csl.sony.co.jp

Joseph Tasnádi
Nyár U. 32 Iv.em
1072 Budapest
H
joseph@revolution.hu

Tug Of War Team
Hauptstr. 2–4
4040 Linz
A
info@aec.at

Sergey Teterin
Korolyova St. 12–28
614061 Perm
R
tet@raid.ru

Blast Theory
Toynbee Studios, 28
Commercial St
London E1 6LS
GB
blasttheory@easynet.
co.uk

Tamiko Thiel
Baaderstr. 64
80469 München
D
tamiko@alum.mit.edu

Henning Timcke
Ideen Werft 22 Gmbh
Stadtturmstr. 5
5400 Baden
CH
henning.timcke@kunst.ch

Takuji Tokiwa
Keio University, Shonan
Fujisawa Campus
6-41-9-102,shonanndai,
Fujisawashi
Kanagawaken 252-0804
J
tokiwa@sfc.keio.ac.jp

Naoko Tosa
ATR Mic Labs
Hikaridai 2 - 2, Seika-cho
Soraku-gun
Kyoto, 619-0288
J
tosa@mic.atr.co.jp

Scott Townsend
North Carolina State
University/School Of
Design
200 Pullen Rd. Box 7701
Raleigh, NC 27695-7701
USA
sttwn@unity.ncsu.edu

Christine Treguier
Les Virtualistes
90 Avenue de Paris
92320 Châtillon
F
lagadu@altern.org

Suzanne Treister
35 Harris St.
5019 Exeter
AUS
suzyrb@va.com.au

Tomoko Ueyama
International Academy Of
Media Arts And Sciences
2 - 1 - 9, Dainichigaoka-
cho, Nagata-ku
653-0872 Kobe City
J
ueyama98@iamas.ac.jp

Time's Up
Industriezeile 33b
4020 Linz
A
info@timesup.org

Camille Utterback
C/o Romy Achituv, 258
Union St # 3
Brooklyn, NY 11231
USA
raa9624@is.nyu.edu

Paul Vanouse
University At Buffalo, Art
Department
202 Center For The Arts
Buffalo, NY 14260-6010
USA
vanouse@buffalo.edu

Oliver Waechter
Abraxas Medien
Otto-Gildemeister-Str. 43
28209 Bremen
D
hamm@abraxas-
medien.de

Fabian Wagmister
Hypermedia Studio -
UCLA
330 De Neve Dr.
Los Angeles, CA 90024
USA
fabian@ucla.edu

Maciej Walczak
Chryzantem 7
91-817 Lodz
PL
walczaki@yahoo.com

Margaret Watson
PO Box 9627
Mississippi State,
MS 39762-9627
USA
watson@erc.msstate.edu

André Werner
Föhrerstr. 7
13353 Berlin
D
werner.19@t-online.de,

Alex Weyers
Arts Alliance Labs
126 South Park
San Francisco, CA 94107
USA
weyers@artsalliance.co.uk

Axel Wirths
Am Kölner Brett 6
50825 Köln
D
mail@235media.com

Kirk Woolford
32 1st St
Brooklyn, NY 11231
USA
junkmail@bomb-ny.com

Judson Wright
Plasma Studii
223 E 10 St, #7
New York, NY 10003
USA
judson@judson.net

Alexa Wright
Ucl
13a Woodstock Rd
London N4 3ET
GB
alexa@dircon.co.uk

Chris Yoculan
753 Oak St.
Columbus, OH 43205
USA
partnersms@aol.com

Jakov Zaper
6/14 Normanby Ave
Thornbury Vic 3071
AUS
zapj@ihug.com.au

Andrea Zapp/
Paul Sermon
Cuvrystr. 33
10997 Berlin
D
zapp@berlin.snafu.de

Romy Achituv
258 Union St. # 3
Brooklyn, NY 11231
USA
raa9624@is.nyu.edu

Miguel Almiron
Université Paris 8
2 Rue De La Liberté
93526 Saint Denis
F
almiron@wanadoo.fr

Le Meur Anne Sarah
Université Paris 8
257 Rue Du Faubourg
Saint Martin
75010 Paris
F
aslemeur@free.fr

Ronald Anzenberger
Inlume – Cg Workgroup
Rautenkranzg. 34/4
A-1210 Wien
A
anzenberger@inlume.com

Masatsugu Arakawa
Production I.g., Llc
10983 Bluffside Dr.,
Suite 6214
Studio City, CA 91604
USA
prodig@ibm.net

Candice Argall
The Video Lab
Eileen Rd
2123 Johannesburg
ZA
candicea@videolab.co.za

Pako Bagur
University Of Balearic
Islands
Carretera de Valldemossa
Km 7.5
07071 Palma de Mallorca
E
info@studio1.uib.es

Zach Bell
Savannah College Of Art
And Design
106 East Oglethorpe Lane
Savannah, GA 31401
USA
zjbell@msn.com

Timothy Benedict
8101 Richard Rd
Broadview Heights,
OH 44147
USA
tbenedi@hotmail.com

Nicolas Billiotel
C.n.b.d.i.
121, Rue De Bordeaux
16000 Angoulême
F
nbillio@caramail.com

Denis Bivour
Invalidenstr. 145
10115 Berlin
D
dbiv@gmx.de

Hisham Bizri
MIT - Cavs
265 Mass Ave, N52-390
Cambridge, MA 01239
USA
bizri@mit.edu

Albert Piella Blanch
Via Laietana 48 A
08004 Barcelona
E
mcanovas@mail.bcn.es

Cheyenne Bloomfield
35 McCaul Street,
Suite 200
Toronto, Ont., M5T1V7
CDN
cheyenne@topix.com

Michael Blum
Pipsqueak Films
5711 Vesper Ave
Van Nuys, CA 91411
USA
mike@pipsqueakfilms.
com

Walter Bohatsch
Mariahilfer Str. 57–59/8c
1060 Wien
A
wbohatsch@graphics.at

Sophie Bordone
22, Rue Hégésippe
Moreau
75018 Paris
F
sbordone@exmachina.fr

Jean-François Bourrel
Paf Production
2, Bvd De Strasbourg
75010 Paris
F
jfboo@worldonline.fr

Gary Breslin
Panoptic
83 Canal St. # 409
New York City, NY 10002
USA
gary@panoptic.org

Michèle Bret
Université Paris 8
2 Rue De La Liberté
93526 Saint Denis
F
mh.tramus@univ-paris8.fr

Rebecca Brown-Dana
Digital Domain
300 Rose Ave
Venice, CA 90291
USA
sunnybrk@d2.com

Holger Bruckschweiger
Ottstorf 18
4600 Wels
A
holger@bruckschweiger.at

Edgar Brueck
FH Wiesbaden
Kurt-Schumacher-Ring 18
65197 Wiesbaden
D
ebrueck@rz.fh-
wiesbaden.de

Michael Buchwald
Pixfixion
19 Heisesgade
2100 Ø Kopenhagen
DK
michbuch@centrum.dk

Pierre Buffin
Buf Compagnie
3 Rue Roquepine
75008 Paris
F
contact@buf.fr

Jose Carlos Casado
School Of Visual Arts
677 Metropolitan Ave,
Apt 7d
Brooklyn, NY 11211
USA
josecasado@usa.net

Ching Clara Chan
Texas A&m University
Visualization Lab
805a Oran Circle
Bryan, TX 77801
USA
clara@viz.tamu.edu

Sam Chen
Pixel Playground
Animations
950 High School
Way # 3328
Mountain View, CA 94041
USA
sambochen@yahoo.com

Se-lien Chuang
Atelier Avant
Grabenstr. 17
8010 Graz
A
cse-lien@sime.com

Fred Collopy
Case Western Reserve
University
10900 Euclid Ave
Cleveland,
OH 44106-7235
USA
flc2@po.cwru.edu

Jorge Cosmen
Goldenedge
Urb. Puebla Tranquila,115
29650 Mijas (Málaga)
E
jorge@goldenedge.com

Paul Debevec
University Of California At
Berkeley
545 Soda Hall
Berkeley, CA 94720
USA
debevec@cs.berkeley.edu

Markus Degen
Piaristeng. 56/16
1080 Wien
A
markusdegen@hotmail.
com

Luis Antonio Delgado
C. Marina, Nº 92, 4º - 3ª
08018 Barcelona
E
ldelgado@eic.ictnet.es

André-Marc Delocque-
Fourcaud
121 Rue De Bordeaux
16000 Angoulême
F
din@cnbdi.fr

Georg Eckmayr
Einar
Rögerg. 16–15
1090 Wien
A
cbar@menonthemoon.
com

Sergey Elizarov
Gragdansky Pr. D.115 K.1
Kv.16
195299 St. Petersburg
R
s_elizarov@mail.ru

Licio Esposito
Cactus Film Produzioni
Via Lago Laceno, 8
84098 Pontecagnano
(SA)
I
cactusfilm@freemail.it

Heure Exquise!
B.p. 113
59370 Mons-en-Baroeul
F
exquise@nordnet.fr

Kota Ezawa
876 Vallejo St.
San Francisco, CA 94133
USA
litky@earthlink.net

Sebastian Faber
Medienallee 7
85774 Unterföhring
D
sebastian.faber@szm.de

Michael Faulkner
Dfuse
3rd Floor, 36 Greville St.
London EC1N 8TB
GB
claire@sulphur.demon.co.
uk

Jeff Faustman
PO Box 22253
Santa Barbara, CA 93121
USA
faustracks@earthlink.net

Michael Garon
122 Hartford St.
San Francisco, CA 94114
USA
salaryman@pacbell.net

Yanick Gaudreau
Nad Centre
335, De Maisonneuve
East Blvd, Suite 300
Montreal, Quebec,
H2X 1K1
CDN
yanickg@buzzimage.com

James Gibson
Animation-art
3126 Bell Dr.
Boulder, CO 80301
USA
jamesgibson@pcisys.net

Felix Goennert
Hff Konrad Wolf
Düsseldorfer Str. 67
10719 Berlin
D
goennert@hff-potsdam.de

Andreas Graichen
Medienallee 7
85774 Unterföhring
D
andreas.graichen@szm.de

Maia Gusberti
Notdef
Zieglerg. 69/15
1070 Vienna
A
not.def@re-p.at

Miya Han
School Of Visual Arts
3703 Buckingham Circle
Middletown, NJ 07748
USA
miya@sva.edu

Niklas Hanberger
Ericsson Media Lab
Årstaängsv 5b
117 43 Stockholm
S
niklas@medialab.ericsson
.se

David Haxton
2036 Sharon Rd
Winter Park, FL 32789
USA
haxtond@aol.com

Claudia Herbst
Pratt Institute, Arc Ll-11c
200 Willoughby Ave
Brooklyn, NY 11205
USA
cherbst@pratt.edu

Rolf Herken
Mental Images
Fasanenstr. 81
10623 Berlin
D
office@mental.com

Aaron Hertzmann
NYU Media Research Lab
719 Broadway, 12th Floor
New York, NY 10003
USA
hertzman@mrl.nyu.edu

Martin Hirsch
Interactive Systems
Scheppe-Gewisse-G. 28
35039 Marburg
D
martin.hirsch@brainmedia
.de

Tina Hochkogler
Favoritenstr. 17/10
1040 Wien
A
tinhoko@chello.at

Lars Magnus Holmgren
(dr. Frankenskippy)
The Moving Picture
Company
127 Wardour St.
London W1V 4NL
GB
frankenskippy@moving-
picture.co.uk

Kenta Hoshi
3-11-20 Kandaiji
221-0801 Kanagawa-ku
Yokohama City
J
s-raja@super.win.ne.jp

Klaus Hu
Hu Production
Wildenbruch 77
12045 Berlin
D
klaushu@hotmail.com

Ivan Iliev
Alxingergasse 5/43
1100 Wien
A
emanuel.wenger@oeaw.
ac.at

Iska Jehl
Schlierseestr. 61
81539 München
D
jehl@adbk.mhn.de

Toshiyuki Kamei
Midorigaoka3 - 1 - 10
594-1155 Osaka-hu Izumi-
shi
J
swimiepixel@mail.goo.ne.
jp

Ayman Saad Kamel
14 Moustafa Abu Heif -
Bab El Look
Cairo
ET
mgallery@ritsec2.com.eg

Paras Kaul
Paras West Productions
PO Box 5275
Ms State, MS 39762
USA
paras@ra.msstate.edu

Toshifumi Kawahara
Polygon Pictures
Ariake Frontier Buinding
Tower
Tokyo 135-0063
J
contact@ppi.co.jp

Ioan Kirilov
Baba Vida 17 Et.2 Ap.6
3700 Vidin
BG
kioan@hotmail.com

Martin Koch
304 Bedford Ave,
3rd Floor
Brooklyn, NY 11211
USA
sukkoch@concentric.net

Wobbe Koning
Accad / The Ohio State
University
1224 Kinnear Rd
Columbus,
OH 43212-1154
USA
wkoning@cgrg.ohio-
state.edu

Ralf Kopp
ln.for.fe.mel
Erbacherstr. 113
64287 Darmstadt
D
in.for.fe.mel@pepup.net

Andrew Kotting
52 Bence House Pepys
Estate
Deptford London SE8
5RU
UK

Gesine Krätzner
5536 NE 27th
Portland, OR 97211
USA
gesine@mykle.com

Annja Krautgasser
Sixpack Film
Neubaug. 36
1070 Wien
A
sixpack@t0.or.at

Amy Krider
Pdi/dreamworks
3101 Park Boulevard
Palo Alto, CA 94306
USA
amyk@pdi.com

Dariusz Krzeczek
Sixpack Film
Neubaug. 36
1071 Wien
A
sixpack@t0.or.at

Anne Marie Kürstein
Danish Film Institute
Gothersgade 55
1123 K Copenhagen
DK
kurstein@dfi.dk

Alen Lai
School Of Visual Arts
677, Metropolitan
Ave # 6d
Brooklyn, NY 11211
USA
alen_lai@hotmail.com

Guy Lampron
Pygmee Productions
1290, Rue St. Denis,
9th Floor
Montreal, Quebec
H2X 3j7
CDN
p_abarca@hotmail.com

John Lasseter
Pixar Animation Studios
1001 West Cutting Blvd.
Richmond, CA 94804
USA
hart@pixar.com

Fabio Lauria
Timecode Records
Postlagernd
4018 Basel 18
CH
timecode@gmx.ch

Felix Lazo
Lazo Inc.
Los Jesuitas 769
Santiago
RCH
flx@netexpress.cl

Christopher Leone
Digital Filmworks
3330 Cahuenga Blvd.
West # 300
Los Angeles, CA 90068
USA
leone@dfw-la.com

Patrick Lichty
8211 E. Wadora Nw
N. Canton, OH 44720
USA
voyd@raex.com

Chia-horng Lin
379 Washington Ave 5f
Brooklyn, NY 11238
USA
clin@pratt.edu

Rüdiger Mach
Art & Engineering
Sternbergstr. 3
76131 Karlsruhe
D
r.mach@t-online.de

Josselin Mahot
Pitch Inc
304 Hudson St.
(6th Floor)
New York, NY 10013
USA
josselin@pitchinc.com

Margit M. Marnul
Reichsratsstr. 13
1010 Wien
A
office@marnul.at

Nuno Maya
Rua Joao Dias Nº 11
1400 Lisboa
P
nuno.maya@clix.pt

Robert McKeever
Robert E. McKeever
Studios
46 Elton Ave
Yardville, 08620
USA
evermax@nerc.com

Cameron McNall
University Of California,
Los Angeles
12034 Navy St.
Los Angeles, CA 90066
USA
cmcnall@ucla.edu

Phil Captain 3d McNally
142b Grosvenor Terrace
London SE5 0NL
UK
captain3d@pump-
action.co.uk

André Metello
Rua Antonio Parreiras
24210-320 Niteroi
BR
metello@microlink.com.br

Gabrielle Mitchell-Marell
Curious Pictures
440 Lafayette St.,
6th Floor
New York, NY 10003
USA

Masahiko Mitsunami
Tokyo Institute Of
Polytechnics
2-9-5, Honcho,
Nakano-ku
Tokyo 164-8678
J
miyatak@acm.org

Peter Moehlmann
Virchowstr. 27
88048 Friedrichshafen
D
ray47peter@aol.com

Juan Montes De Oca
University Of Balearic
Islands
Carretera de Valldemossa
Km 7.5
07071 Palma de Mallorca
E
info@studio1.uib.es

Hiromitsu Murakami
International Academy Of
Media Arts And Sciences
1-21-18, Utsukushigaoka-
nishi, Aoba-ku
Yokohama-city,
Kanagawa 225-0001
J
mura98@iamas.ac.jp

Masayo Nishimura
350 West 88th St.,
Apt. 302
New York, NY 10024
USA
mniart@aol.com

Koichi Noguchi
301-9-19-5 Minai-oi
Shinagawa-ku, 140-0013
J
q-minao@d3.dion.ne.jp

Carlos Nogueira
Universidade De São
Paulo
Rua Prof Luciano
Gualberto, Trav. 3 , 158
05508-900 São Paulo
BR
educar@usp.br

Renate Oblak
Leitnerg. 7/5
8010 Graz
A
remi@algo.mur.at

James O'Brien
Georgia Institute Of
Technology,
College Of Computing
801 Atlantic Dr.
Atlanta, GA 30332-0280
USA
obrienj@cc.gatech.edu

Yasuo Ohba
Namco Limited
1-1-32 Shin-urashima-cho
Kanagawa-ku
Yokohama 221-0031
J
ohba@rd.namco.co.jp

Timm Osterhold
Fiftyeight3d Animation &
Digital Effects Gmbh
Gustav-Freytag-Str. 18
65189 Wiesbaden
D
siham@fiftyeight.com

Muki Pakesch
Mukivision
Glaserg. 4/7
1090 Wien
A
mpakesch@to.or.at

Min Su Park
Sungbukgu Jungrung 2
Dong 168-24
136-102 Seoul
ROK
park_minsu@yahoo.com

Van Phan
University Of Southern
California
4644 W 137th St. A
Hawthorne, CA 90250
USA
vphan@usc.edu

Mary Phillipuk
Pratt Institute
369 Washington
Ave # 3a
Brooklyn, NY 11238
USA
mphillip@pratt.edu

Jakub Pistecky
Jp Animation
3319 W. 3rd Ave
Vancouver, BC, V6R 1L3
CDN
jpistecky@ea.com

Ben Pointeker
Schiffamtsg. 11/14
1020 Wien
A
pointecre@hotmail.com

Stuart Pound
34 Blondin Ave
London W5 4UZP
GB
bocadillo@cwcom.net

Sebastian Rätsch
Medienallee 7
85774 Unterföhring
D
gerald.gutberlet@szm.de

Marcel Ritter
Kärntnerstr. 64/7/60
6020 Innsbruck
A
csab7885@uibk.ac.at

Xavier Roig
University Of Balearic
Islands
Carretera de Valldemossa
Km 7.5
07071 Palma de Mallorca
E
info@studio1.uib.es

Peter Rubin
Max-a-vision!
Transvaalkade 133 3hg
1091 LT Amsterdam
NL
maxavision@hotmail.com

John Ryan
Click3x
345 Peachtree Hills Ave
Atlanta, GA 30305
USA
jryan@click3x.com

Semi Ryu
4628 Bayard St.,
Apt. # 413
Pittsburgh, PA 15213
USA
sryu@andrew.cmu.edu

Leigh Santilli
32 Atlantic Ave.
Toronto,
Ontario M6K 1X8
CDN
mikef@nelvana.com

Mauricio Santos
Mansão De Artes
Rua Da Mansão 205
90050-030 Porto Alegre
BR
mauricions@vanet.com.br

Minoru Sasaki
Technonet Co., Ltd.
6-8-8, Matsubara Bldg 2f
107-0052 Minato-
ku,akasaka
Tokyo, J
sasaki-m@technonet.co.jp

Ruth Saunders
Australian Film Tv &
Radio School
Balaclava & Epping Rds
North Ryde Nsw 2113
AUS
ruths@aftrs.edu.au

Sajan Skaria
Texas A&m University
216 Langford Architecture
Center
College Station,
TX 77843-3137
USA
sajan@viz.tamu.edu

Lisa Slates
Xaos, Inc.
444 De Haro St.,
Suite 211
San Francisco, CA 94107
USA
lisa@xaos.com

Julio Soto
30549 San Martinez Rd
Castaic, 91384
USA
jsotogur@yahoo.com

Jack Stenner
Texas A & M University
4004 Oaklawn St.
Bryan, TX 77801
USA
stenner@pursebuilding.
com

George Stennis
Mississippi State
University
102 Freeman Hall,
Barr Ave
Mississippi State,
MS 39762
USA
gls3@ra.msstate.edu

Oliver Stephan
Mariannenstr. 43
47799 Krefeld
D
stolooo1@fh-
niederrhein.de

Thomas Sternagel
Cuvilliestr. 1a
81679 München
D
sternagel@maspex.de

Douglas Struble
15455 Court Village Lane,
Bldg # 4
Taylor, MI 48180
USA
dugstruble@ili.net

Makoto Sugawara
Production I.g
10983 Bluffside Dr.,
Suite 6214
Studio City, CA 91604
USA
prodig@ibm.net

Dennis Summers
Quantum Dance Works
3927 Parkview Dr.
Royal Oak, MI 48073
USA
dennisqdw@home.com

Maarit Suomi
Tallbergink. 1 E 76
00180 Helsinki
SF
maarit.suomi@av-arkki.fi

Alexander Szadeczky
Nofrontiere Design
Zinckg. 20–22
1150 Wien
A
s.loeschner@nofrontiere.
com

Istvan Szakats
Vlahuta 59–61, N3, Ap 5
3400 Cluj
RO
altart@mail.dntcj.ro

Nobuo Takahashi
Namco Limited, Vs
Development Department
1-1-32 Shin-urashima-cho,
Kanagawa-ku
Yokohama, Kanagawa
221-0031
J
nobuo@vs.namco.co.jp

Atsushi Takeuchi
Production I.g., Llc
10983 Bluffside Dr.,
Suite 6214
Studio City, CA 91604
USA
prodig@ibm.net

Ying Tan
Department Of Fine Arts,
University Of Oregon
5232 University Of
Oregon
Eugene, OR 97403
USA
tanying@darkwing.uorego
n.edu

Helmut Telefont
Schwarzwiesenstr.3
3340 Waidhofen
A
helmut.telefont@bene.
com

Laurence Trouve
C.n.b.d.i.
26 Bis Bd Chabasse
16000 Angoulême
F
l_trouve@hotmail.com

Christian Tyroller
Aberlestr. 18
81371 München
D
ctyroller@t-online.de

Martin Venema
Happy Ship
Mispellaan 11
9741 GJ Groningen
NL
tv@inn.nl

Christian Volckman
6, Boulevard De
Strasbourg
75010 Paris
F
method@cybercable.fr

Arisa Wakami
Iamas
Aichi Okazaki Syoujida
444-0821
J
arisa98@iamas.ac.jp

D'amir Wanguard
Tellurian Cgi
Petrova 113
10000 Zagreb
HR
tellurian@tellurian.hr

Max Weintraub
California Institute Of The
Arts
24700 McBean Parkway
Valencia, NM 91355
USA
maxweintraub@hotmail.c
om

Geri Wilhelm
Oddworld Inhabitants
869 Monterey St.
San Luis Obispo,
CA 93401
USA
geri@oddworld.com

Hanne Winarsky
Blue Sky Studios
1 South Rd
Harrison, NY 10528
USA
hanne@blueskystudios.co
m

Judson Wright
Plasma Studii
223 E 10 St, #7
New York, NY 10003
USA
judson@judson.net

Chi-min Yang
School Of Visual Arts
507 E 5th St. # 6
New York, NY 10009
USA
chimin_y@hotmail.com

Jörg Zimmer
Kappenberger Str. 6
44339 Dortmund
D

Alessandro Amaducci
Via Muriaglio 1
10141 Torino
I
a.amaducci@tiscalinet.it

Candice Argall
The Video Lab
Eileen Rd
2123 Johannesburg
ZA
candicea@videolab.co.za

Pablo Bach
Medienallee 7
85774 Unterföhring
D
pablo.bach@szm.de

Sophie Bordone
22, Rue Hégésippe
Moreau
75018 Paris
F
sbordone@exmachina.fr

Rebecca Brown-Dana
Digital Domain
300 Rose Ave.
Venice, CA 90291
USA
sunnybrk@d2.com

Pierre Buffin
Buf Compagnie
3 Rue Roquepine
75008 Paris
F
contact@buf.fr

Arnd Buss von Kuk
Velvet Mediendesign
Osterwaldstr. 10
80805 München
D
busskuk@wtal.de

Se-lien Chuang
Atelier Avant
Grabenstr. 17
8010 Graz
A
cse-lien@sime.com

Cid Collins Walker
Bet On Jazz/bet
International
2000 W. Place,
Ne/studio 2
Washington, DC 20018
USA
dsngideas@smart.net

Michael Faulkner
Dfuse
3rd Floor, 36 Greville St.
London EC1N 8TB
GB
claire@sulphur.demon.
co.uk

Jessie Fischer
Sihlquai 346
8005 Zürich
CH
jezone@hotmail.com

Akemi Fujita
New England
Conservatory
13 Symphony Rd, Apt 3
Boston, MA 02115
USA
poesia@netscape.net

Philipp Geist
Schönholzer Str. 6
10115 Berlin
D
p-geist@gmx.de

Mathias Gmachl
Franzensg. 6/1
1050 Wien
A
hiaz@live.fm

Ruth Greenberg
The Computer Film
Company
19 - 23 Wells St.
London W1P 3FP
GB
ruth@cfc.co.uk

Simone Haberland
Velvet Mediendesign
Osterwaldstr. 10
80805 München
D
simone@velvet.de

Horst Hadler
Medienallee 7
85774 Unterföhring
D
horst.hadler@szm.de

Rob Heydon
Proteus
199 Cranbrooke Ave
Toronto, ON M1W 1P1
CDN
robheydon@hotmail.com

Greg Kam
The Greg Kam Company
17 Westwood Close, Little
Heath
Potters Bar EN6 1LH
GB
redyquip@netcapeonline.
co.uk

Kohske Kawase
2-22-14 Sakuradai
176-0002 Nerima-ku
Tokyo
J
neocube@intacc.ne.jp

Robert Kent
Link 2 Media
63 Pinemeadow Blvd.
Toronto, ON M1W 1P1
CDN
rob@link-2.com

Rolf-Jürgen Kirsch
Expimat
Luetzowstr. 23
50674 Köln
D
r.j.kirsch@netcologne.de

Alen Lai
School Of Visual Arts
677, Metropolitan Ave, #
6d
Brooklyn, NY 11211
USA
alen_lai@hotmail.com

Neil Landstrumm
Scandinavia Animation
214 Smith St. # 3
New York, NY 11201
USA
scandinavianyc@worldnet
.att.net

Manfred Laumer
Velvet Mediendesign
Osterwaldstr. 10
80805 München
D
m.laumer@t-online.de

Traude Pichler
Wolffstr. 3
22525 Hamburg
D
g.pichler@hanse-net.de

Stuart Pound
34 Blondin Ave
London W5 4UP
GB
bocadillo@cwcom.net

Claudius Schulz
Ehrengutstr. 16
80469 München
D
superschulz@t-online.de

Mark Simon
Aargh! Animation
8137 Lake Crowell Circle
Orlando, FL 32836
USA
mark@storyboards-
east.com

Leslie Streit
New Name Performance
33 Jennings Court
San Francisco, CA 94124
USA
leslie@new-
performance.org

Mika Taanila
Stenbäckinkatu 1 A 5
00250 Helsinki
SF
mtaanila@kuva.fi

Tippett Studio
2741 Tenth St.
Berkeley, CA 94710
USA
lisa@tippett.com

Cornelia Unger
Velvet Mediendesign
Osterwaldstr.10
80805 München
D
conny@velvet.de

Peter Wandmose
Echo Film
Mikkel Bryggers Gade 11
1460 K Copenhagen
DK
danwan@teliamail.dk

Sebastian Weidner
Medienallee 7
85774 Unterföhring
D
sebastian.weidner@szm.
de

Chi-min Yang
School Of Visual Arts
507 E 5th St. # 6
New York, NY 10009
USA
chimin_y@hotmail.com

Matthias Zentner
Velvet Mediendesign
Osterwaldstr. 10
80805 München
D
matthias@velvet.de

Paul Abad
Syntactic
8 Egbert St.
West End 4101
AUS
paulabad@powerup.com.
au

MasayGBi Akamatsu
Iamas
3-95, RyoGBe-cho
503-0014 Ogaki, Gifu
J
aka@iamas.ac.jp

Bill Alves
1286 Reims St.
Claremont, CA 91711
USA
alves@hmc.edu

Maryanne Amacher
Digital Musics
80 Marius St.
Kingston, NY 12401
USA
citylinks@aol.com

Peter Androsch
Domg. 5/3
4020 Linz
A

Anonymous Anonymous
Too
Igloo_records(c)
3, Rue Eugène Varlin
75010 Paris
F
oolgi_rec@hotmail.com

Brett Aplin
352 Barkly St, Brunswick
Melbourne 3056
AUS
b.aplin1@pgrad.unimelb.e
du.au

Phil Archer
6 Bury St.
Norwich NR2 2DN
GB
o_k_boy@hotmail.com

Artemiy Artemiev
Ul. Krilatskaya 31-1-321
121614 Moscow
RUS
eshock@cityline.ru

Daniel Asia
University Of Arizona
5230 N. Apache Hills
Trail
Tucson, AZ 85750
USA
asia@u.arizona.edu

Bjorn Askefoss
Bjorn Askefoss Atelier
PO Box 469
3193 Horten
N
askefoss@notam.uio.no

Miguel Azguime
Rua Do Douro 92
Rebelva
2775-318 Parede
P
misomusic@misomusic.co
m

Curtis Bahn
Rensselaer Polytechnic
Institute
298 New Boston Rd
Sturbridge, MA 01566
USA
crb@rpi.edu

Steven Ball
88 Victoria St.
3011 Footscray
AUS
sball@starnet.com.au

Christian Banasik
Querstr. 16
40227 Düsseldorf
D
c.banasik@t-online.de

Alberto Bario
V. Emilia Ponente 211
40133 Bologna
I
whazz@libero.it

Plinio Barraza
Adsmenu
Diag 74 # 6-28 Apt 703
00000 Bogotá
CO
plinio@orientation.com

Natasha Barrett
Nota Bene (n.b.)
Music Production
Blaasbortvn. 10
0873 Oslo
N
natashab@notam.uio.no

Ed Baxter
PO Box 556
London SE5 0RW
GB
ed@lmcltd.demon.co.GB

Shawn Bell
Polycarpe Studios
1567 Ch. St. philippe
St. Polycarpe, QC J0P 1X0
CDN
shawn@polycarpe.qc.CA

Sonja Bender
U.m.a. + Ginseng
Schönhauser Allee 177
10119 Berlin
D
ginseng@hdk-berlin.de

Douglas Benford
63 Windmill Rd
Brentford,
Middlesex TW8 0QQ
GB
douglas@benfo.demon.
co.GB

Rodney Berry
ATR Media Integration
And Communications
Laboratories
2-2-2 Hikaridai,
Seika-cho, Soraku-gun
Kyoto, 619-0288
J
rodney@mic.atr.co.jp

Laura Bianchini
CRM – Centro Ricerche
Musicali
Via Lamarmora 18
00185 Roma
I
laura.bianchini@usa.net

Matt Black
Ninja Tune
90 Kennington Lane
London SE11 4XD
GB
mattb@ninjatune.net

Markus Bless
Ledererg. 9
4861 Schörfling
A
markus.bless@ufg.ac.at

Edward C. Bobb
sony_mao@hotmail.com

DGBe Bojadziev
150 Massachusets Ave
Student Mailbox # 150
Boston, MA 02115
USA
dGBe.b@usa.net

Drazen Bosnjak
Tomandandy
89 Greene St., 2nd Floor
New York, NY 10012
USA
www.walcroface@earth
link.net

Glen Braun
Grand Valley State
University
3374 Byron Center
Apt.c-311
Wyoming, MI 49509
USA
braung@river.it.gvsu.edu

Colin Bright
10/115 Pacific Parade Dee
Why
Sydney NSW 2099
AUS
colbright@netsydney.com.
au

Holger Bruckschweiger
Ottstorf 18
4600 Wels
A
holger@bruckschweiger.at

Brigid Burke-Rigo
PO Box 315
Elsternwick, Victoria 3185
AUS
alexr@tbm.com.au

Nicola Buso
Via Montegrappa 36
31020 San Sisto di
Villorba (TV)
I
busoni@libero.it

Christoph Camenzind
Rue De Lausanne 70
1700 Fribourg
CH
chcd@bluewin.ch

Joseph Cancellaro
Poznan Supercomputing
And Networking Center
Ul. Noskowskiego 10
61-704 Poznan
PL
joec@man.poznan.pl

Kim Cascone
Anechoicmedia.com
748 Edgemar Ave
Pacifica, CA 94044
USA
kim@anechoicmedia.com

Tim Catlin
73 Little Charles St.
Abbotsford, Victoria 3067
AUS
timc@micronica.com.au

Dario Cavada
Via Bolzano, 20
38030 Molina Di Fiemme
(TN)
I
info@dchs-music.com

Victor Cerullo
Via Caneve, 77
30173 Mestre (VE)
I
moog@libero.it

Paulo César Chagas
Klangstudio C
Rolandstr. 47
50677 Köln
D
nc-
chagaspa@netcologne.de

Paul Chan
National Philistine
113 W. 60th St, Rm 422
New York, NY 10023
USA
manwichartist@yahoo.
com

Christophe Charles
Taitoku
Yanaka 3-21-5-303
Tokyo 110-0001
J
charles@tkd.att.ne.jp

Amar Chaudhary
1765 Oxford St. Apt #8
Berkeley, CA 94709
USA
amar@cnmat.berkeley.
edu

Insook Choi
University Of Illinois
405 N. Mathews Dr.
Urbana, IL 61801
USA
ichoi@ncsa.uiuc.edu

Martin Christel
Saarbrückenstr. 36
24114 Kiel
D
mchristel@t-online.de

Paul Clouvel
6 Rue du Charrier
18000 Bourges
F
paulclouvel@laposte.net

Michael Cohen
Granular
6, Petite Rue d'Austerlitz
67000 Strasbourg
F
homemade@club-
internet.fr

Maya Consuelo Sternel
Die Patinnen Teil II
Neuer Pferdemarkt 21
20359 Hamburg
D
die_patinnen@arcor
mail.de

Frank Corcoran
Hochschule Für Musik
Lenharzstr. 8
20249 Hamburg
D
volkerbartsch@hotmail.
com

Gregory Cowley
Thetestproject
29a Hoff St.
San Francisco, CA 94110
USA
gregory@testsite.org

Andrew Daniel
Vague Terrain
800 Hampshire St.
San Francisco, CA 94110
USA
mcess@pop.slip.net

Robin Davies
4209 Drolet
Montreal QC H2W 2l7
CDN
robin@music.mcgill.ca

Kelly Davis
1-3-5- #202
152-0001 Chu-o-cho,
Meguro-ku
J
tokyoaudioarts@hotmail.c
om

Ignacio De Campos
Rua Constancio Alves
N222
13080090 Campinas Sp
BR
ignaciocampos@fcmail.co
m

Peter De Lorenzo
PO Box 138
2577 Robertson
AUS
pdls@ozemail.com.au

Roderik De Man
1e Tuindwarsstraat 3
1015RT Amsterdam
NL
a.deman@cva.ahk.nl

Romulo Del Castillo
so@schematic.net

Christopher Delaurenti
PO Box 45655
Seattle, WA 98145-0655
USA
composer@oz.net

Richard Deuine
rdeuine@mindspring.com

Paul Dibley
12 Teal Close
Oxford OX4 7GU
GB
pdibley@brookes.ac.GB

Patricia Dirks
#52-21 Holborn Dr.
Kitchener, ON N2A 2E1
CDN
p.dirks@sympatico.CA

Roberto Doati
Via Giorgione 66
35020 Albignasegno (PD)
I
r.doati@flashnet.it

Paul Douglas
Sigma Editions
2/155 Karangahape Rd
Auckland Central 1001
NZ
rparlane@hotmail.com

Roger Doyle
Silverdoor
Rynville Mews
Killarney Rd Bray, Co
Wicklow
IRL
rogerd@eircom.net

Hugo Victor Druetta
Francia 1374 Dpto 2
S3000FBJ Santa Fe
Pcia De Santa Fe
RA
hdruetta@unl.edu.ar

Kevin Drumm
1523 N Wicker Park Ave,
2nd Floor
Chicago, IL 60622
USA
lefondst@ripco.com

Curd Duca
Blechturmg. 13/4/17
1050 Wien
A
cduca@t0.or.at

Michael Edgerton
The University Of
California At San Diego
3237 92nd St.
Sturtevant, WI 53177
USA
edgertonmichael@netscap
e.net

William Eldridge
241 S. Laurel St.
Richmond, VA 23220
USA
eldridge@jps.net

Angie Eng
307 Mott St. 2d
New York, NY 10012
USA
angieeng@hotmail.com

Rolf Enström
Helgestavägen 127
125 41 Älvsjö
S
enstroem@kacor.kth.se

Gino Esposto
Micromusic.net
Langstr. 197
8005 Zürich
CH
carl@micromusic.net

Jan Faroe
Andreas Bjorns Gade 28,
4. Th
1428 Copenhagen
DK
janfaroe@hotmail.com

Suzanne Farrin
Yale University
28 High St. N6
New Haven, CT 06510
USA
suzanne.farrin@yale.edu

Jeff Faustman
PO Box 22253
Santa Barbara, CA 93121
USA
faustracks@earthlink.net

Gino Favotti
Péroux
26750 Châtillon St Jean
F
ginofavotti@fnac.net

Annamaria Federici
Via Michelangelo 630
41056 Savignano Sul
Panaro (MO)
I

Mark Fell
Snd
12 Vere Rd
Sheffield S6 1SB
GB
info@sndloc.force9.co.GB

Antonio Ferreira
Ap. Corte Real, Av.
Gaspar Corte Real N18 4e
2750-164 Cascais
P
nop35624@mail.telepac.
pt

Boris Firquet
406 Richelieu
QC
CDN
eltractor@videotron.CA

Fon
Werkzeug
PO Box 5
3482 Gösing/Wagram
A
fon@redirect.to

Carlo Forlivesi
Via Fratelli Bandiera, 9
40026 Imola (BO)
I
carlo_forlivesi@hotmail.
com

Katharina Franck
Sans Soleil
Prinz Albert Str. 65
53037 Bonn
D
pociao@t-online.de

Andi Freeman
Deepdisc Collective
67 Millbrook Rd
Brixton SW9 7JD
GB
andi@deepdisc.com

Günther Friebis
Sauerbruchstr. 2
67063 Ludwigshafen
D
friebis.nimbostratus.recor
ds@t-online.de

Lawrence Fritts
University of Iowa
1006 Vmb
Iowa City, IA 52242
USA
lawrence-
fritts@uiowa.edu

Funkstörung
Max-Josef-Platz 6
83022 Rosenheim
D
mas@ovb.net

David Gamper
491 Broadway #3
New York, NY 10012
USA
dgamper@earthlink.net

Marie-Luise Goerke
Seelingstr. 28
14059 Berlin
D
ise@is.in-berlin.de

Annie Gosfield
301 East 12 St # 3d
New York, NY 10003
USA
agosfield@aol.com

Paul Gough
34 Camellia St.
Greystanes, NSW 2145
AUS
pimmon@bRdcast.net

Oliver Peter Graber
Josef-Weber-G. 11
2700 Wiener Neustadt
A
spitzentanz@utanet.at

Christian Graupner
Human Magnetics
Cooperation
Leuschnerdamm 19
10999 Berlin
D
voov@voov.de

Reinhold Grether
D
reinhold.grether@uni-
konstanz.de

Brent Gutzeit
Boxmedia
1806 W. Rice, Apt 3r
Chicago, IL 60622
USA
bgutzeit@boxmedia.com

Chris Halliwell
PO Box 24886
London E1W 3GP
GB
sousrature@angelfire.com

Heath Hanlin
633 Village Blvd North
Baldwinsville, CA 13027
USA
hahanlin@syr.edu

Michael Harding
Ash International [r.i.p.]
13 Osward Rd
London SW17 7SS
GB
msc@touch.demon.co.GB

Msc Harding
13 Osward Rd
London SW17 7SS
GB
touch@touch.demon.
co.GB

Hanna Hartman
C/o Chamier
Kyffhäuserstr. 10
10781 Berlin
D
hanna.hartman@snafu.de

Shawn Hatfield
USA
us@schematic.net

Dag-Are Haugan
Kamrergatan 25
211 56 Malmö
S
dhaugan@hotmail.com

Jens Hedman
EMS
Sodermalarstrand 61
118 25 Stockholm
S
jens.hedman@composer.e
ms.srk.se

Robin-Julian Heifetz
14543 Burbank Blvd.
201
Van Nuys, CA 91411-4314
USA
heifo1@aol.com

Douglas Henderson
Schwartzkopffstr. 4
10115 Berlin
D
dougfender@hotmail.com

Robert Henke
Monolake/imbalance
Computer Music
Knaackstr. 18
10405 Berlin
D
rob@monolake.de

Scott Herren
USA
us@schematic.net

Donna Hewitt
University Of Western
Sydney, Nepean
2nd Ave
Kingswood 2747
AUS
donnahewitt@one.net.au

Johann Heyss
Caixa Postal 100153
24001-970 Niterói-Rio De
Janeiro
BR
heyss@yahoo.com

Richard Hoadley
Anglia Polytechnic
University
East Rd
Cambridge CB1 1PT
GB
r.j.hoadley@anglia.ac.GB

Elizabeth Hoffman
NYU Dept. of Music
24 Waverly Place, Rm.
268
New York, NY 10003
eh37@is.nyu.edu

Mark Hofschneider
127 Berkeley Pl # 2
Brooklyn, NY 11217
USA
markhof@bway.net

Klaus Hollinetz
Steinhumergutstr. 1
4050 Traun
A
klaus.hollinetz@servus.at

Earl Howard
39-39 45th St.
Sunnyside, NY 11104-2103
USA
sculpsound@unidial.com

Tim Howle
Oxford Brookes
University
Headington
Oxford OX3 0BP
GB
tjhowle@brookes.ac.GB

Peter Hulen
816 Cleo St.
Lansing, MI 48915-1323
USA
hulenpet@msu.edu

Oeivind Idsoe
Stormyrveien 11b
0672 Oslo
N
plateaux@frisurf.no

Alejandro Iglesias Rossi
Arenales 3837 6to
1425 Capital Federal
RA
telecentrosantafe@arnet.
com.ar

Ryoji Ikeda
13 Osward Rd
London SW17 7SS
GB
touch@touch.demon.
co.uk

Matt Ingalls
545 Valle Vista # 4
Oakland, CA 94610
USA
mingalls@concentric.net

James Ingram
York University
Honeystone, Rectory Rd
Bristol BS20 0QB
GB
jai_uk@hotmail.com

Angelo Iotti
Istituto Superiore
Str. Fraore N. 20
43010 Fraore (PA)
I
angelo.iott@libero.it

Haruo Ishii
Aichi Prefectural Art
University
30-1 Ishibata Narumi-cho
Midori-ku Nagoya-shi
458-0801
J
mxc00275@nifty.ne.jp

Szigeti Istvan
Hear Studio
Brody S. U. 5-7
1800 Budapest
H
hear@www.radio.hu

Stanley Jackson
Darwinarts
200 Buena Vista Ave
Boulder Creek, CA 95006
USA
wayne@darwinarts.com

Pierre Alain Jaffrennou
9 Rue du Garet - Bp 1185
69202 Lyon Cedex 01
F
jaffrennou@grame.fr

Juri Jansen
Herler Str. 41
51067 Köln
D
fabia@t-online.de

Robert Jelinek
Burgg. 100/4-5
1070 Wien
A
you@sabotage.at

Promonium Jesters
74 Marietta St
Uxbridge, ON L9P 1J5
CDN
david@daymen.com

Reinsel Joseph
Rensselaer Polytechnic
Institute
126 Second St. R-6
Troy, NY 12180
USA
reinsj@rpi.edu

Arsenije Jovanovic
Adriatic Sound Factory
Santa Croce 57a
52210 Rovinj
HR
arsenije.jovanovic@pu.tel.
hr

Sergio Kafejian
Rutilia, N. 56
01432-050 São Paulo
BR
kafejian@uol.com.br

Frederic Kahn
36, Grande Rue de Vaise
69009 Lyon
F

Jeff Kaiser
PO Box 1653
Ventura, CA 93002-1653
USA
pfmentum@jetlink.net

Bernard Karawatzki
Classensgade 17 B. St.tv.
2100 Ø København
DK
bernard@vip.cybercity.dk

Paras Kaul
Paras West Productions
PO Box 5275
Ms State, MS 39762
USA
paras@ra.msstate.edu

Hideko Kawamoto
University Of North Texas
200 Hann # 8
Denton, TX 76201
USA
hk0008@jove.acs.unt.edu

Kohske Kawase
2-22-14 Sakuradai
176-0002 Nerima-ku
Tokyo
J
neocube@intacc.ne.jp

Joshua Kay
USA
us@schematic.net

Edward Kelly
Music Centre
University Of East Anglia
Norwich NR5 9BW
GB
the_lone_shark@hotmail.
com

Peter Kelly
7 Park View Terrace
Leeds LS 196ES
GB
pjpkz@GBgateway.net

Cyril Kestellikian
CMNT
11 Ave des Coccinelles
13012 Marseille
F
ckestell@aol.com

SGB Jun Kim
2 Carling St. # 9
Hamilton, ON L8S 1M8
CDN
kjun@nas.net

Mark Kirschenmann
1741 Broadview Ln. #501
Ann Arbor, MI 48105
USA
sonikman@umich.edu

Kazuyuki Kishino
(aka Kk.null)
3-690-47
228-0003 Hibarigaoka,
Zama, Kanagawa
J
kknull@fsinet.or.jp

Josef Klammer
Neuholdaug. 51
8010 Graz
A
remmalk@styria.com

Roger Kleier
301 East 12th St. # 3d
New York, NY 10003
USA
rkleier@aol.com

Judy Klein
130 West 17th St #7-s
New York, NY 10011
USA
jak@maestro.com

Georg Klein
Arndtstr.1
10965 Berlin
D
georg.klein@gk-m.de

Petra Klusmeyer
2458 N. Seminary,
Garden Apt.
Chicago, IL 60614
USA
petravonk@aol.com

Paul Koonce
110 North Stanworth Dr.
Princeton, NJ 08540
USA
koonce@princeton.edu

Stevan Kovacs Tickmayer
31, Fbg. Madeleine
45000 Orléans
F
tick@cybercable.fr

Johannes Kretz
Universität für Musik Und
Darstellende Kunst
Wohllebeng. 11/10
1040 Wien
A
johkretz@eunet.at

Schematic Label
USA
us@schematic.net

Brandon Labelle
Errant Bodies
PO Box 931124
Los Angeles, CA 90093
USA
otic@earthlink.net

Francis Larvor
49, Rue Vitry
93100 Montreuil-sous-
Bois
F
flarvor@freesurf.fr

Fabio Lauria
Timecode Records
Postlagernd
4018 Basel 18
CH
timecode@gmx.ch

Steve Law
Solitary Sound
C1-2/120 Arden St.
North Melbourne 3051
AUS
steve@solitary-sound.com

Simon Lazar
Ibn Gvikol Str.
1111 Tel Aviv
IL
lazarjuroz@hotmail.com

Sonia Leber
Wax Sound Media
11 Teak St. Caulfield St.
Victoria, 3162
AUS
wax@waxsm.com.au

Andreas List
Ttmproduction
Thalhauser Fußweg 11
85354 Freising
D
ttmstudio@aol.com

David Little
Amsterdam Conservatory
Cornelis Springerstr. 14-2
1073 LJ Amsterdam
NL
d.little@cva.ahk.nl

Yeeon Lo
California State University
At San Jose
One Washington Square
San Jose, CA 95192-0103
USA
acoustic@netcom.com

Bernhard Loibner
Yppeng. 5/14
1160 Wien
A
bernhard@allquiet.org

Jarryd Lowder
41 E. 28th St. # 3a
New York, NY 10016
USA
jarryd@sva.edu

Michelangelo Lupone
CRM – Centro Ricerche
Musicali
Via Lamarmora 18
00185 Roma
I
michelangelo.lupone@usa
.net

Eric Lyon
Dartmouth College
Music Dept, HB6187
Hanover, NH 03755
USA
eric.lyon@dartmouth.edu

Akitsugu Maebayashi
Nagoya University Of Arts
Nishiharu
Nishikasugai 481-8535
J·
m8@gol.com

Wayne Magruder
441 Lorimer
Brooklyn, NY 11206
USA
waynebmagruder@usa.ne
t

Elio Martusciello
Via Montiano, 8b
00127 Roma
I
emartus@tin.it

Neal Matherne
5656 Heather Lane
Birmingham, AL 35235
USA
nmathbass@hotmail.com

Stephan Mathieu
Scheidter Str. 76
66123 Saarbrücken
D
sssst@musork.com

Kaffe Matthews
Annetteworks
PO Box 14077
London N16 5WF
GB
annettetours@stalk.net

Bohdan Mazurek
Polskie Radio S.a.
Woronicza 17
00-977 Warszawa
PL
bohdan_pl@yahoo.com

Daniel Antonio Miraglia
Universidad De Morón
Petkovic 5379
1678 Caseros
RA
d_miraglia@hotmail.com

Norikazu Mitani
Keio Univ. Graduate
Schoool Of Media And
Governance
Kanagawa Daimachi
2-5602
221-0834 Yokohama
Kanagawa
J
norikazu@sfc.keio.ac.jp

Gareth Mitchell
38 Chiswick Lane
London W4 2JQ
GB
philophersstone@apex
mail.com

Fabrice Mogini
Middlesex University
London
64, Moresby Rd
London E5 9LF
GB
fm101@mdx.ac.GB

Yura Moorush
Moscow
RUS
moorush@mu.ru

Brian Moran
343 E 22 # 7
New York, NY 10010
USA
opsysbug@bway.net

Yoichi Nagashima
Art & Science Laboratory
10-12-301, Sumiyoshi-5
430-0906 Hamamatsu,
Shizuoka
J
nagasm@kobe-
yamate.ac.jp

Sabine Nagy
Hauzenbergerstr. 20
80687 München
D
max.koenig@t-online.de

Fumitaka Nakamura
Kobe Yamate College
Chuo-ku Nakayamate
Dohri
650-0004 Kobe
J
fuming@kobe-
yamate.ac.jp

Marko Neurauter
Grünau 7, Top 5 (elson)
6850 Dornbirn
A
marko.neurauter@student
s.fh-vorarlberg.ac.at

Carsten Nicolai
Raster-noton
Limbacher Str. 270
09116 Chemnitz
D
office@raster-noton.de

Frank Niehusmann
Nelkenweg 33
42549 Velbert
D
niehusmann@planet-
interkom.de

Ken Niibori
Kitamachi-18-12-102
185-0001 Kokubunji City,
Tokyo
J
nibo@annie.ne.jp

Benny Nilsen
Hazard
Branningevagen 17-19
120 54 Arsta
S
hazard_@hotmail.com

Timothy Nohe
University Of Maryland
Baltimore County
Visual Arts, Fa 111,
1000 Hilltop Circle
Baltimore, MD 21250
USA
nohe@umbc.edu

Adolfo Núñez
Aniceto Marinas 2 4f
28008 Madrid
E
adolfo.nunez@cdmc.inae
m.es

Norbert Oldani
709 Buchanan Rd
Utica, NY 13502
USA
noldani@mvcc.edu

Jan Oleszkowicz
Jagiellonska 6 M 21
03-721 Warszawa
PL
kompart@poczta.fm

Brian O'Reilly
CCMIX
62 Rue Morcelin
Berthelot
94140 Alfatville
F
briore@excite.com

Ed Osborn
PO Box 9121
Oakland, CA 94613
USA
edo@curve.to

Claude Pailliot
Dat Politics
175 Rue Colbert
59800 Lille
F
datpol@nordnet.fr

MGBi Pakesch
MGBivision
Glaserg. 4/7
1090 Wien
A
mpakesch@to.or.at

Vicky Paniale
60 Siege House,
Sidney St.
London E1 2HQ
GB
immedia@hotmail.com

Maggi Payne
1324 Acton
Berkeley, CA 94706
USA
maggi@mills.edu

Geoffrey Perrin
Dun Loaghaire Institute
Of Art And Design
39 Corke Abbey
Bray, Co Wicklow
IRL
gperrin@perrin.buyandsell
.ie

Angelo Petronella
Via Trognano 7
27010 Bascape (PV)
I

Jane Philbrick
103 Peaceable St.
Redding, CT 06896
USA
jphilbrick@snet.net

Emanuel Pimenta
Rua Tiervo Galvan,
4 #11b
1070-274 Lisbon
P
edmp@asa-art.com

Andreas Pischläger
Ledererg. 77
4020 Linz
A

Lionel Polard-Cohen
8 Rue Robert Houdin
75011 Paris
F
lionel.polardcohen@free.fr

Gerhard Prandstätter
Dr.-Stumpf-Str. 5
6020 Innsbruck
A
pra@lion.cc

Bert Praxenthaler
Epfenhausen
Bahnhofstr. 12
D-86929 Penzing
D
praxenthaler@t-online.de

Rostislav Prochovnik
Zednicka 953
Ostrava-poruba 708 00
CZ

Richard (dj Q-bert)
Quitevis
Invisibl Skratch Piklz
1441 Rollings Rd
Burlingame, CA 94010
USA
salaryman@pacbell.net

Radian Radian
Ruckerg. 10/21–22
1120 Wien
A
info@mego.at

Dodi Reifenberg
Hochstädter Str. 11
13347 Berlin
D
noel@p-soft.de

Alistair Riddell
Five Degrees
12 Napier St
Fitzroy, Victoria 3065
AUS
amr@alphalink.com.au

Patrick Riley
504 Pinehurst Cove
Poinciana,
FL 34758-3640
USA
priley7601@aol.com

Stuart Rimell
14 Beck Rd, Madeley,
Nr Crewe
Cheshire CW3 9JF
GB
p9e84@keele.ac.GB

Don Ritter
204 15 St.
Brooklyn, NY 11215
USA
ritter@interport.net

Dean Roberts
Formacentric
PO Box 46 251 Auckland
1002 Auckland
NZ
forma@iconz.co.nz

Steve Roden
Box 50261
Pasadena, CA 91115
USA
sroden@deltanet.com

Hans-Joachim Rödelius
Smart-artists
Schimmerg. 35a
2500 Baden
A
roedelius@mycity.at

Abel Rogantini
Rivadavia 9013
1407 Buenos Aires
RA
rogantiniabel@hotmail.
com

H-ed Roland
H.e.r. Kryptic Visions
Marchlewski Str. 31
10243 Berlin
D
ddslash@compuserve.com

Mario Rosivatz
Röckelbrunnstr. 20-3301
5020 Salzburg
A
rosivatz@gmx.de

Julien Roy
2051 Valois
Montreal, QC H1W 3M6
CDN
juroy@yahoo.com

Mark Rudolph
Lucid Actual
3600 Ave Du Parc A2610
Montreal, QC H2X 3R2
CDN
mrudolph@total.net

Alexis Ruiz
Luis G. Vieyra 17
11850 Mexico, D.F.
MEX
lexisrve@prodigy.net.mx

Antti Saario
Merikannotie 7b 33
00260 Helsinki
SF
a.s.saario@bham.ac.GB

Kiawasch Saheb-Nassagh
Triester Str. 128a/11
8020 Graz
A
kiawasch@yahoo.com

Diane I. Samuels
330/400 Sampsonia Way
Pittsburgh, PA 15212
USA
samuels@andrew.cmu.edu

Luz Maria Sanchez
Radio Udg
Ave San Francisco 2170-a
44500 Guadalajara
MEX
mariasc@radio.udg.mx

Riccardo Santoboni
Via Eudo Giulioli 3
00173 Roma
I
md9077@mclink.it

Massimo Sapienza
Microwave
PO Box 11453
1001 GL Amsterdam
NL
microwave@staalplaat.
com

Ricardo Sasaki
Calle 13 Esq. Av.
Hernando Siles No. 5908,
2nd Floor
La Paz
BOL
rgsasaki@hotmail.com

Louis Schebeck
Atm International
PO Box 1259 Strawberry
Hills
Sydney NSW 2012
AUS
atmint@smartchat.net.au

Christof Schläger
Elisestr. 3
44628 Herne
D
schlaeger.smit@wanadoo.
nl

Marcus Schmickler
Vogteistr.18
50670 Köln
D
schmickler@a-musik.com

Gue Schmidt
Rüdigerg. 10/5
1050 Wien
A
gue.s@xpoint.at

Lucas Schwabeneder
Kaiser-Josef-Platz 2
4600 Wels
A
schwabeneder@utanet.at

Elliott Sharp
206 E. 7 St. # 14
New York, NA 10009
USA
esharp@panix.com

Takao Shimizu
982-4 Kosugi
Toyama 939-8063
J
pehoo200@po.incl.ne.jp

Alexei Shulgin
Easylife.org
Polotskaya 29-1-39
121351 Moscow
RUS
alexei@easylife.org

Gary Singh
Cadre Institute
56 S. 2nd St., # 202
San Jose, CA 95113
USA
gsingh@email.sjsu.edu

Pekka Siren
PO Box 13
00024 Helsinki
SF
agaps@nettilinja.fi

Randall Smith
15 Spencer Ave
Toronto, ON M6K 2J4
CDN
peacock@interlog.com

Nelson Soares
O Grivo
Rua Prof. Miguel De
Souza 305/201
30.570-150 Belo
Horizonte / Minas Gerais
BR
nelson.bh@zaz.com.br

David Solursh
Skwerely Studios
115 First St. # 347
Collingwood ON L9Y 4W3
CDN
dsolursh@georgian.net

Tim Song Jones
1044a Royal Oaks Dr.
Monrovia, CA 91016-3788
USA
fodrjonz@ni.net

Bent Sørensen
Novator
Østre Alle 43d
3250 Gilleleje
DK
novator@danbbs.dk

Rogelio Sosa
75, Rue de Wattignies
75012 Paris
F
sosa.ro@wanadoo.fr

Arthur Stammet
29, Rue Léon Metz
4238 Esch-sur-Alzette
L
arthur.stammet@ci.edu
c.lu

Gianni Stiletto
FH Salzburg
Schillerstr.30
5020 Salzburg
A
gianni.stiletto@fh-
sbg.ac.at

Tony Stoufer
Absolute Obscurity
Records
1616 Brandee Ln
Santa Rosa, CA 95403
USA
ptfinch@juno.com

Woody Sullender
2333 N Spaulding # 1
Chicago, IL 60647
USA
rsullend@email.unc.edu

Wolfgang Suppan
Burgg. 67/20b
1070 Wien
A
w-suppan@1012surfnet.at

Hans Sydon
Enghavevej 40,3
1674 Kopenhagen
DK
hearfilm@post9.tele.dk

Nobukazu Takemura
Powerbox
Higashi2-17-10
Shibuya-ku, Tokyo
J
fukutake@powerbox.co.jp

Yasuhiro Takenaka
4-9-8, Ichiban-cho
190-0033 Tachikawa-shi,
Tokyo
J
vyuoo430@nifty.ne.jp

Jeff Talman
338 Berry St., No. 4ne
Brooklyn, NY 11211
USA
jefftalman@mindspring.
com

Jacob Ter Veldhuis
Studio Toonbeeld
Drift 40
3941 DC Doorn
NL
jtv1@xs4all.nl

Aaron Thieme
1200 19th St.
San Francisco, CA 94107
USA
ghede@well.com

Dave Tipper
us@schematic.net

Todor Todoroff
273 Rue du Progrès
1030 Bruxelles
B
todor.todoroff@skynet.be

Takuji Tokiwa
Graduata School Of
Media And Governance,
Keio University
Shonan Fujisawa Campus
6-41-9-102, Shonanndai,
Fujisawashi
Kanagawaken 252-0804
J
tokiwa@sfc.keio.ac.jp

Yasunao Tone
307 West Broadway
New York, NY 10013
USA
yasunaot@worldnet.att.
net

Maciej Toporowicz
97 Crosby St.
New York, NY 10012
USA
noqontrol@aol.com

Germán Toro-Pérez
Leopold-Figl-G. 69
3040 Neulengbach
A
toro@eunet.at

Pierre Alexandre
Tremblay
[iks]
5550 Snowdon St. # 13
Montréal H3X 1Y9
CDN
ora@cam.org

Uli Troyer
Friends Of Carluccio
Schellhammerg. 16-9
1160 Wien
A
uli@monochrom.at

Alejandro Viñao
27 Coolhurst Rd
London N8 8ET
GB
alej@vinao.u-net.com

Klaus Voltmer
Epy
Franzensg. 6/1
1050 Wien
A
klaus@epy.co.at

Otto von Schirach
us@schematic.net

Paulo R. F. Von Zuben
Rua Lisboa, 225, Apt. 33
05413-000 São Paulo-Sp
BR
pvzuben@dialdata.com.br

Frans De Waard
Fagelstraat 40
6524 ce Nijmegen
NL
frans@staalplaat.com

Robert Wacha
Bka-arts-collectiv
Gartenstadtstr. 18
4040 Linz
A
robert_wacha@hotmail.
com

Craig Thomas Walsh
University Of North
Carolina At Greensboro
Corner Of Market And
Mciver St.s
Greensboro, NC 27455
USA
ctwalsh@uncg.edu

Peter Wandmose
Echo Film
Mikkel Bryggers Gade 11
1460 K Copenhagen
DK
danwan@teliamail.dk

Gary Weisberg
23 South Broadway
Red Hook, NY 12571
USA
gr2w@meterpool.com

Andreas Weixler
Atelier Avant
Schaftalbergweg 33
8044 Graz
A
aweixler@sime.com

Holger Wendland
Prießnitzstr. 21
01099 Dresden
D
edition-raute@t-online.de

Frances White
105 Linden Ave
Princeton, NJ 08540-8535
USA
jfwp@earthlink.net

James Whitehead
Jliat
13 Wells Rd
Walsingham Norfolk
NR22 6DL
GB
james@jliat.demon.co.GB

Erik Wiegand
Schwedter Str. 28
10119 Berlin
D
me@errorsmith.de

Garnet Willis
Organoise
401 Richmond St. W.
Suite 361
Toronto, ON M5V 3A8
CDN
garnet_willis@goodmedia.
com

Ryszard Wolny
Zai
Zielonogórska 39a/2
66-016 Czerwieńsk
PL
wolny1@kki.net.pl

Bart Woodstrup
639 Lincoln Hwy. # 6w
Dekalb, IL 60115
USA
bwood@nicemusic2.music
.niu.edu

Shahrokh Yadegari
University Of California,
San Diego
7486 La Jolla Blvd. # 550
La Jolla, CA 92037
USA
sdy@ucsd.edu

John Young
School Of Music, Victoria
University
PO Box 600
6001 Wellington
NZ
john.young@vuw.ac.nz

Michl Achleitner
Niederkalmberg 10
4352 Klam
admin@eurogym.asn-
linz.ac.at

Stephan Adelsberger
Wallrissstr. 59
1180 Wien
stephan@adelsberger.
com

Bernhard Adleff
Wiener Str. 181
4020 Linz

Lisa Aichhorn
Markt-Süd 16
4363 Pabneukirchen
sisley_at@yahoo.com

Christian Aichinger
Baumgartenberg 1
4342 Baumgartenberg
admin@eurogymasn-
linz.ac.at

Astrid Aigner
Markt 5
4352 Klam
bgbaaigner@yahoo.com

Elisabeth Aigner
Machlandstr. 39
4310 Mauthausen
eli_aigner@yahoo.de

Andreas Aigner
Baumgartenberg 1
4342 Baumgartenberg
admin@eurogym.asn-
linz.ac.at

Leo Aistleitner
Baumgartenberg 1
4243 Baumgartenberg
admin@eurogym.asn-
linz.ac.at

Akinwale Olaitan Ajewole
Matzelsdorf 11
8411 Hengsberg

Tolga Akkas
Franz-Grubbauerstr. 1/3/2
3350 Haag

Regina Albinger
C/o HS Orth/Donau,
Schlossplatz 4
2304 Orth/Donau
hauptschule.orth@aon.at

Alexander Aletzdorfer
Nelkenstr. 9
4072 Alkoven

Sabine Alphasamer
Römer Str. 40
2514 Traiskirchen
sumsie12@gmx.at

Jakob Altzinger
Baumgartenberg 1
4342 Baumgartenberg
jakob_altzi@yahoo.de

Sandra Angerer
Stöcklerstr. 12
4320 Perg
admin@eurogym.asn-
linz.ac.at

Ronald Anzenberger
Pestalozzistr. 46/3
4840 Vöcklabruck

Gülsüm Arslan
HS Traiskirchen
2514 Traiskirchen

Latschesar Atanassov
Baumgarten 1
4342 Baumgartenberg
admin@eurogym.asn-
linz.ac.at

Thomas Auderer
Johannesfeldstr. 12
6111 Volders

Thomas Aumayr
Corneliusg. 22
4020 Linz
tommy_a65@hotmail.com

Klaus Bachler
Baumgartenberg 1
4342 Baumgartenberg
admin@eurogym.asn-
linz.ac.at

Christian Bahner
Johann Schuster-Str.13
2514 Traiskirchen
christian_bahner@gmx.at

Melek Barlak
HS Traiskirchen
2514 Traiskirchen

Christian Bauer
Michael-Hofer-G.19
3950 Gmünd
gmuend.hs1@aon.at

Cornelia Bauer
Frankenberg 103
4312 Ried / Riedmark
admin@eurogym.asn-
linz.ac.at

Paul Baumann
Baumgartenberg 1
4342 Baumgartenberg
admin@eurogym.asnt-
linz.ac.at

Norbert Baumgartner
Roisenberg 1
4341 Arbing
noribaum@gmx.at

Florian Beer
Wagnerstr. 21
2371 Hinterbrühl
etpt@ycom.at

Michael Beim
Eibenstr. 26
4600 Wels
iftmikeb@gmx.net

Attila Berei
Baumgartenberg 1
4243 Baumgartenberg
admin@eurogym.asn-
linz.ac.at

Christine Berger
Schumpeterstr.15
4020 Linz

Monika Berghold
Egelsdorf 46
8261 Sinabelkirchen

Harald Bernhard
Hauptstr. 184
3871 Alt-Nagelberg
gmuend.hsi@aon.at

Kastler Bernhard
Schneckenreitsberg 8
4352 Klam
admin@eurogym.asn-
linz.ac.at

Stefanie Besenbäck
Baumgartenberg1
4342 Baumgartenberg
admin@eurogym.asn-
linz.ac.at

Julia Biberauer
Baumgartenberg 1
4243 Baumgartenberg
admine@europym.asn-
linzalat

Wolfgang Biebl
Schlossplatz 4
2304 Orth/Donau
hauptschule.orth@aon.at

Simeon Biedner
Steinkloftstr. 36
5340 St. Gilgen
genesis29@gmx.de

Leopold Bigler
C/o HS Orth/Donau,
Schlossplatz 4
2304 Orth/Donau
hauptschule.orth@aon.at

Michaela Binder
Pürbach 10
3944 Pürbach
gmuend.hs1@aon.at

BHAK Gmünd
Otto-Glöckel-Str. 6
3950 Gmünd
s309418@intra.asn-
noe.ac.at

Martin Bladerer
Hauslehen 109
3342 Opponitz

Peter Bliem
Rothenwand 6
5584 Zederhaus

Angelique Böhm
Birand 27
3873 Birand
gmuend.hs1@aon.at

Patrick Böhm
Kerngraben 49
4320 Perg
admin@eurogym.asn-
linz.ac.at

Jan Böhm
Taferlweg 13
4323 Münzbach
langeder@yline.at

Sanja Bonic
Ehrenbrunng. 6
2320 Schwechat
icys@gmx.at

Konrad Brandl
Baumgartenberg 1
4243 Baumgartenberg
atmin@eurogym.asn-
linz.ac.at

Barbara Brandner
Untergaisberg 13
4352 Klam
admin@eurogym.asn-
linz.ac.at

Katharina Brandner
Baumgartenberg 1
4342 Baumgartenberg
admin@eurogym.asn-
linz.ac.at

Karin Brandstetter
C/o HS Orth/Donau,
Schlossplatz 4
2304 Orth/Donau
hauptschule.orth@aon.at

Michael Brandstetter
Sperken 45
4352 Klam
bgbmbranstetter@yahoo.
com

Georg Brandstetter
Kaiser-Friedrich-Str. 23
4360 Grein
admin@eurogym.asn-
linz.ac.at

Christina Breier
Römerstr. 49
6912 Hörbranz
bg.gallus3@schulen.vol.at

Melanie Breinesberger
Haid 8
4343 Mitterkirchen
breinesberger@yahoo.de

Martin Breithofer
Nestelbach 52
8262 Ilz

Jacquline Brunner
Waldweg 3
3950 Gmünd
gmuend.hs1@aon.at

Julia Buchmayr
Baumgartenberg 1
4342 Baumgartenberg
admin@eurogym.asn-
linz.ac.at

Georg Burdicek
Unt. Augartenstr. 5/4/13
1020 Wien
geebee@lion.cc

Cornelia Buresch
C/o Hs Orth/Donau,
Schlossplatz 4
2304 Orth/Donau
hauptschule.orth@aon.at

Rüstü Caliskan
C/o HS Orth/Donau,
Schlossplatz 4
2304 Orth/Donau
hauptschule.orth@aon.at

Lisa Cibej
Gartenstr. 37
4311 Schwertberg
stella.x@usa.net

Clemens Cizek
Lagerstr.19
3950 Gmünd
gmuen.hs1@aon.at

Bianca Craiovan
Baumgartenberg
Kreuznerstr. 50b
4360 Grein
admin@eurogym.asn-
linz.ac.at

Wolfgang Csacsinovits
Blumau 54
8283 Blumau

Julia Csapo
Ringofeng. 38
2513 Möllersdorf

Andreas Datzreiter
C/o VS Ferschnitz,
Schulstr. 102
3325 Ferschnitz

Alexandra Deibl
Klingerg.24
4210 Gallneukirchen
sonvilla@.ufg.ac.at

Sabrina Deinhofer
C/o VS Ferschnitz,
Schulstr. 102
3325 Ferschnitz

Manuel Dekors
Hölbinglstr. 15
3350 Haag

Philip Deutsch
Graben 31
4221 Steyregg

Andreas Diesner
Bauernzeile 38
3942 Hirschbach
gmuend.hsi@aon.at

Patrick Digby
Edramsbergerstr. 34
4073 Wilhering
digby@aon.at

Fritz Dimmel
Reucklstr. 9
2020 Hollabrunn
gymradio@bghollabrunn.
ac.at

Elisabeth Dirneder
Hafnerplatz.1
4320 Perg
e_dirneder@yahoo.de

Florian Dober
Mayg. 43
1130 Wien
dober@chello.at

Peter Donko
Robert-Stolz-Str. 23/15
4020 Linz
peter.donko@aon.at

Marcel Drabits
C/o Hs Orth/Donau,
Schlossplatz 4
2304 Orth/Donau
hauptschule.orth@aon.at

Claudia Dragosits
Mayg. 43
1130 Wien
claudia.dragosits@chello.
at

Alexander Duggleby
Pensionatsstr. 76
4810 Gmunden
aduggleby@gmx.net

Sarah Dunzinger
Hausleiten 16
4490 St. Florian

Silvia Eberl
Hüttengraben 36
8611 St. Katharein/
Laming
silvia.eberl@gmx.at

Josef Ebner
Schulstr. 102
3325 Ferschnitz
305141@asn.netway.at

Julius Ecker
Schloss Wildberg
4202 Kirchschlag
schloss.wildberg@aon.at

Robert Ecker
Baumgartenberg 1
4342 Baumgartenberg
robert_ecker@yahoo.de

Gerald Eder
Baumgartenberg 1
4243 Baumgartenberg
admin@eurogym asn-
linz.ac.at

Heidemarie Edinger
C/o HS Orth/Donau,
Schlossplatz 4
2304 Orth/Donau
hauptschule.orth@aon.at

Simone Eichinger
Rosenweg 6
4222 Langenstein
admin@eurogym.asn-
linz.ac.at

Gerhard Eichner
Kartitsch 38
9941 Kartitsch
g.eichner@tivoliplan.at

Sebastian Eigner
Hans-Groß-Siedlung 2
8572 Bärnbach
eise@itr.or.at

Julia Eisschiel
Mühlkreisbahnstr.16
4111 Walding
sonvilla@ufg.ac.at

Michaela Beham,
Daniela Fuchs, Elisabeth
Steinkress, Michael Polli
Ramsauer Str. 88–89
4020 Linz
schule@bg-rams.ac.at

Andrea Emhofer
Baumgartenberg 1
4243 Baumgartenberg
admin@eurogym.asn-
linz.ac.at

Karina Emhofer
Baumgartenberg 1
4342 Baumgartenberg
admin@eurogym.asn-
linz.ac.at

Sebastian Endt
Wallackstr. 12
4600 Wels
s.endt@datapool.at

Stephan Enzinger
Lerchenfelderstr. 8
5202 Neumarkt
steve@gmx.at

Maria Ertl
Weidasiedlung 7
6600 Lechaschau
mary.ertl@reflex.at

Viktoria Esch
Franz-Lisztg. 8
2603 Felixdorf
beatrice.esch@chello.at

Viktoria Fabsits
Schenkendorfg. 11/20
1210 Wien
vicky.fabsits@surfeu.at

Stefan Fahrngruber
Erlaaerstr. 105
1230 Wien
stefan.fahrngruber@
gmx.at

Rene Fasching
Egelsdorf 76
8261 Sinabelkirchen

David Feiler
Leopoldsg. 16/20
1020 Wien
davif.feiler@aon.at

Robert Felbermayer
Gutshofweg 4
4343 Mauthausen
glifberg2000@uboot.com

Steffanie Feodorow
C/o Berufsschule 9,
Wiener Str. 181
4020 Linz

Tom Ferstl
Lacken 5
5662 Gries im Pinzgau

Lukas Fichtinger
Neuhaus 35
4943 Geinberg
liderc@gmx.at

Christiane Fichtl
Stummerstr. 5
3350 Haag

Maria Fidesser
Neug. 30
2020 Hollabrunn
mirlifidesser@hotmail.
com

Jürgen Filler
Hans-Reither-G. 12
3950 Gmünd
gmuend.hs1@aon.at

Klaus Fischer
Langenloiserstr. 22
3500 Krems
kfischer@hlfkrems.ac.at

Alexander Fischl
Voltag. 45/3/4
1210 Wien
newtron@yline.com

Eva-Maria Fladerer
Sinabelkirchen 160
8261 Sinabelkirchen

Elisabeth Forstenlechner
Weinzierl 9
4320 Perg
e_forstenlechner1@yahoo.
de

Lukas Fraser
Friesach Dorf 8
8114 Stübing
lukas@aon.at

Roman Freud
C/o Vs Ferschnitz,
Schulstr. 102
3325 Ferschnitz

Verena Friedl
Unterrettenbach
8261 Sinabelkirchen

Katrin Friedl
El.-Hummelstr. 5
3293 Lunz am See
katrin.friedl@aon.at

Bernhard Fröhler
Hochpointstr. 13a
4600 Wels
wolfgang.froehler@memb
er.alpenverein

Carmen Froschauer
Baumgartenberg 1
4342 Baumgartenberg
admin@eurogym.asn-
linz.ac.at

Christina Fröschl
Baumgartenberg 1
4342 Baumgartenberg
messino272@hotmail.com

Michael Frühmann
Puchenstuben 25
3214 Puchenstuben
mezzoforte@utanet.at

Ute Furchheim
Polgarstr. 24
1220 Wien
rhaboz@to.or.at

Alexander Fürmötz
Deich 8
4040 Auberg

Johannes Fürnhammer
Baumgartenberg 1
4342 Baumgartenberg
admin@eurogym.asn-
linz.ac.at

Ursula Gabriel
Schulg. 10
6714 Nüziders
vsn@cable.vol.at

Gregor Gahbauer
Baumgartenberg 1
4342 Baumgartenberg
kecks@uboot.com

Siegfried Ganaus
Kranzbichlerstr. 67
3100 St. Pölten
raptor@kstp.at

Markus Gaszo
C/o Hs Orth/Donau,
Schlossplatz 4
2304 Orth/Donau
hauptschule.orth@aon.at

Nadia Gattringer
Baumgartenberg 1
4342 Baumgartenberg
admin@eurogym.asn-
linz.ac.at

Gregor Gerstorfer
M. v. Aichholz Str. 34b
4810 Gmunden

Barbara Glanz
Ehrenhöbarten 16
3943 Schrems
gmuend.hs1@aon.at

Martin Glasner
Passail 237
8162 Passail
mglasn9@6gweiz.asn-
graz.ac.at

Carina Gleichweit
C/o VS Staudach,
Staudach 126
8230 Hartberg

Jürgen Glück
Gymnasiumstr. 2
4710 Grieskirchen
borggries.dir@mail.asn-
linz.ac.at

Matthias Götz
C/o Hs Orth/Donau,
Schlossplatz 4
2304 Orth/Donau
hauptschule.orth@aon.at

Julia Grabner
Fünfing 36
8261 Sinabelkirchen

Walter Gramberger
Ehrendorferwald 21
4694 Ohlsdorf
walter_gramberger@usa.
net

Stefan Grasser
Machlandstr. 11
4310 Mauthausen
stefang@gmx.at

Edith Greindl
Baumgartenberg 1
4342 Baumgartenberg
admin@eurogym.asn-
linz.ac.at

Thomas Grimschitz
C/o Berufsschule 9,
Wiener Str. 181
4020 Linz

Vera Gröbner
Rainerg. 39
1050 Wien
office@grg5.asn-
wien.ac.at

Thomas Groppenberger
Schulg. 61
3424 Muckendorf
tommisoft@yline.com

Lukas Grossinger
Stift Zwettl 12
3910 Zwettl
ladmin@hs-
stiftzwettl.ac.at

Korrdula Grasser
Baumgartenberg 1
4342 Baumgartenberg
admin@eurogym.asn-
linz.ac.at

Ralph Gruber
Ardning Nr. 169
8904 Ardning
ralphmc@gmx.at

Vela Gusenbauer
Baumgartenberg 1
4342 Baumgartenberg
vgusenbauer@hotmail.
com

Michael Gusenbauer
Mettensdorf 32
4342 Baumgartenberg
michael_gusenb@yahoo.
de

Dominik Gusenbauer
Baumgartenberg 42
4342 Baumgartenberg
karl.gusenbauer@yline.
com

Gerhard Gutlederer
Linden 4
3371 Neumarkt
g_gutlederer@hotmail.
com

Harald Guttmann
Conrathstr. 29 2/1/3
3950 Gmünd
gmuend.hs1@aon.at

Isabella Haas
C/o VS Staudach,
Staudach 126
8230 Hartberg

Thomas Haberkorn
Steinböckstr. 10
3340 Waidhofen / Ybbs

Kathrin Habernek
C/o HS Orth/Donau,
Schlossplatz 4
2304 Orth/Donau
hauptschule.orth@aon.at

Valentin Hablig
Hinterberg 6
4083 Haibach/Donau

Wolfgang Hachl
Höhenstr. 21
3340 Waidhofen/Ybbs
m.zoechmann@aon.at

Rene Hackl
Zeitling 56
4320 Perg
hackl.r@utanet.at

Johanna Hader
Markt 74
4371 Dimbach
bgb.jhader@yahoo.com

Melanie Haderer
Schreineredt 12
4363 Pabneukirchen
admin@eurpgym.asn-
linz.ac.at

Christina Margaretha
Haderer
Baumgartenberg 1
4342 Baumgartenberg
chrissy@eurogym.asn-
linz.ac.at

Samuel Hadeyer
Brevenhuberg. 2
3350 Haag

Robert Hafner
Radau 100
5351 Aigen-Voglhub
reinhold.hafner@schule.at

Doris Hahn
Baumgartenberg 1
4342 Baumgartenberg
admin@eurogym.asn-
linz.ac.at

Franz Haider
Schwindg. 9/4
1040 Wien
franzhai@yahoo.com

Gottfried Haider
Schwindg. 9/4
1040 Wien
gohai@yahoo.com

Harald Haider
Kleinerlau 9
4371 Dimbach
harry_haider@yahoo.de

Regina Hammelmüller
Hollengruberstr. 8
3350 Stadt Haag

Spohiemarie Hammer
Baumgartenberg 1
4342 Baumgartenberg
sophiehammer@eurogym.
asn-linz.ac.at

Michael Hammer
Köhlerweg 74
4222 Luftenberg
k.hammer@mail.asn-
linz.ac.at

Michael Hancianu
C/o VS Staudach,
Staudach 126
8230 Hartberg

Harald Hansl
Zanaschkag. 12/23/1/14
1120 Wien
nexidus@animalhome.net

Michael Harb
Heimweg 75
8053 Graz
harb.mi@gmx.at

Helene Lene Haring
Baumgartenberg 1
4342 Baumgartenberg 1
helene.h@pi-linz.ac.at

Christopher Harter
Höbarten 32
3961 Waldenstein
gmuend.hs1@aon.at

Gerd Haselsteiner
Hauptplatz 20
3362 Mauer
gerd.haselsteiner@aon.at

Michael Haslmayr
Im Neubruch 24
4040 Linz
j.haslmayr@utanet.at

Sabine Haupt
Obergroßau 61
8261 Sinabelkirchen

Philipp Hayder
Baumgartenberg 1
4342 Baumgartenberg
slimfastshady@hotmail.
com

Bernadette Heghi
Baumgartenberg 1
4342 Baumgartenberg
admin@eurogym.
asn-linz.ac.at

Albert Heiligenbrunner
Baumgartenberg 1
4342 Baumgartenberg
admin@eurogym.asn-
linz.ac.at

Barbara Heimel
Baumgartenberg 1
4342 Baumgartenberg
admin@eurogym.asn-
linz.ac.at

Raphael Heiß
Baumgartenberg1
4342 Baumgartenberg
admin@eurogym.asn-
linz.ac.at

Sophie Helbich-
Poschacher
Richterhof
4320 Perg
s_helbich@yahoo.com

Katharina Henninger
Amalienstr. 75/4/18
1130 Wien
katrin.henninger@gmx.at

Christian Herzog
Grubweg 16
8580 Köflach
noeffred@gmx.net

Dominik Hierner
Baumgartenberg 1
4342 Baumgartenberg
admin@eurogym.asn-
linz.ac.at

Marlene Hierzer
C/o VS Staudach,
Staudach 126
8230 Hartberg

Lucia Hiesl
Baumgartenberg 1
4342 Baumgartenberg
admin@eurogym.asn-
linz.ac.at

Elke Hinterdorfer
Baumgartenberg 1
4342 Baumgartenberg
alkelk@hotmail.com

Gregor H. Hinterdorfer
Baumgartenberg 1
4342 Baumgartenberg
h_greg_2000@yahoo.de

Regina H. Hinterreiter
Baumgartenberg 1
4342 Baumgartenberg
franz.hinterreiter@membe
rs.gpa.at

Petra Hinterreiter
Baumgartenberg 1
4243 Baumgartenberg
admin@eurogym.asn-
linz.ac.at

Julia Hinterreither
Großerlau 13
4371 Dimbach Dimbach
bgbjhinte@yahoo.com

Mario Hirschböck
Sinabelkirchen 166
8261 Sinabelkirchen

Thomas Hirtenfelder
Stremayrg. 6/46
8010 Graz
thegrafx@usa.net

Julia Höbarth
Herrenstr.10
4320 Perg
juliahobarth@usa.net

Thomas Hochgatterer
Baumgartenberg 1
4243 Baumgartenberg
t.hochgatterer@aon.at

Patrick Hochholzer
C/o VS Ferschnitz,
Schulstr. 102
3325 Ferschnitz

Oliver Hödl
Dr.-Hans-Ertl-G. 6
8680 Mürzzuschlag
oliver.hoedl@gmx.at

Vanessa Hödlmayr
Adalbert Stifter Str.10
4311 Schwertberg
v_hoedlmayr@yahoo.de

Florian Hofer
Millöckerg. 4
2540 Bad Vöslau
florian_hofer@hotmail.
com

Markus Hofer
Sonnenhang 5
8763 St. Oswald
fhofer@ctc.at

Marion Hofer
Handlerweg 11
8261 Sinabelkirchen

Patrick Höfler
Haring 69
8183 Floing
patman@gmx.at

Mario Höflinger
Stadtplatz 18
4221 Steyregg

Patrick Hofmann
Buckkogelg. 11
8020 Graz
eva.hofmann@utanet.at

Lisa Hofstadler
C/o VS Staudach,
Staudach 126
8230 Hartberg

Gudrun Hofstadler
C/o VS Staudach,
Staudach 126
8230 Hartberg

Bernhard Hoisl
Sh. Milesstr. 10
9100 Völkermarkt
hoisl@gym1.at

Markus Höller
C/o VS Ferschnitz,
Schulstr. 102
3325 Ferschnitz

Marisa Höllrigl
Waldrandstr. 26
3950 Gmünd
gmuend.hs1@aon.at

Patrick Holpfer
Finkeng. 5
2514 Traiskirchen

Sophia Holzer
Baumgartenberg 1
4342 Baumgartenberg
admin@eurogym.asn-
linz.ac.at

Fabian Holzer
Baumgartenberg 1
4243 Baumgartenberg
admin@eurogym asn-
linz.ac.at

Manuel Holzmüller
Gmünder Str. 254
3945 Hoheneich
gmuend.hs1@aon.at

Caroline Hoos
Ried 195
4312 Ried/Riedmark
c_hoos@yahoo.de

Petra Hopfgartner
Reichhartstr. 6
3754 Irnfritz

Stefan Hörschläger
Parkfeld 40
4203 Altenberg

Heinz Hösch
Unterweissenbach 174
8330 Feldbach
heinzh@gmx.at

Benjamin Hotter
Johannes-Messner-Weg 14
6130 Schwaz
b.hotter@aon.at

Carina Howegger
Grießbühlstr. 12
3950 Ehrendorf
gmuend.hs1@aon.at

Julian Hruza
Enderg. 22
1120 Wien
julianhruza@hotmail.com

Robert Huber
Hofau 2
4650 Lambach

Markus Huber
Georg-v.-Trapp-Str.16
5026 Salzburg
markusph2@gmx.de

Stefan Huber
Jeging 36
5222 Munderfing
shuber2@gmx.at

Gerhard Huber
Mettensdorf 7
4342 Baumgartenberg
gerhard_hupsi@yahoo.de

Martin Huemer
C/o BG XIX,
Gymnasiumstr. 83
1190 Wien
ps@to.or.at

Roman Humpelstetter
Wieshöfstr. 33
3121 Karlstetten
rhump@gmx.at

Theresa Hunstorfer
Wienerweg 12
4360 Grein
bgbthunstorfer@yahoo.
com

Sarah Jaksch
Baumgartenberg 1
4342 Baumgartenberg
admin@eurogym.asn-
linz.ac.at

Stefan Jank
C/o HS Orth/Donau,
Schlossplatz 4
2304 Orth/Donau
hauptschule.orth@aon.at

Bernhard Jantscher
Ulmg. 14
8570 Voitsberg
berni_jantscher@hotmail.
com

Simon Jantscher
Hub 116
8046 Stattegg
simon.jantscher@styria.
com

Viktoria Jindra
Bachg. 9
3950 Dietmanns
gmuend.hs1@aon.at

Tomas Jiskra
Mayg. 43
1130 Wien
tomas_jiskra@web.de

Sandra Jöbstl
C/o VS Staudach 126
8230 Hartberg
607391@asn.netway.at

Kamil Jozwiak
Griesplatz 8/23
8020 Graz

Peter Kail
Schottenring 19
1010 Wien
scotpeter@gmx.net

Gernot Kainig
Kleinrojach 20
9431 St. Stefan

Christiane Kainrath
2041 Aschendorf 29
hoermann@asn.netway.at

Stefan Kalchmair
Bergerndorf 21
4600 Wels
akwcsg@hotmail.com

Sabrina Kaltenböck
Schubertplatz 15
3950 Gmünd
gmuend.hs1@aon.at

Thomas Kaltofen
Griesstr. 23
4502 St.
Marien/Nöstlbach
thomas@kaltofen.com

Lukas Kamarad
Bahnstr. 7
2020 Hollabrunn
lukas.kamarad@bgholla
brunn.ac.at

Thomas Kamleitner
Baumgartenberg 1
4342 Baumgartenberg
t_kamleitner@eurogym.as
n-linz.ac.at

Cornelia Kamleitner
Baumgartenberg 1
4342 Baumgartenberg
admin@eurogym.asn-
linz.ac.at

Sabrina Kammla
Eduard-Meoler-G. 14
2514 Traiskirchen

Daniel Kapeller
Graben 21
4240 Freistadt
studiokapeller@magnet.at

Bettina Kapfer
Baumgartenberg 1
4342 Baumgartenberg
kapfi@eurogym.asn-
linz.ac.at

Marina Kapplmüller
Baumgartenberg 1
4342 Baumgartenberg
admin@eurogym.asn-
linz.ac.at

Bernhard Kastner
Pottenhofen 159
2163 Ottenthal
kastner@aon.at

Kathrin Kastner
Baumgartenberg 1
4342 Baumgartenberg
admin@eurogym.asn-
linz.ac.at

Carina Kaufmann
Hummelberg 33
4341 Arbing
carina.kaufmann@eunet.
at

Martin Kerschberger
Alte Weinstr. 5
4522 Sierning
kerschberger.martin@gmx
.at

Thomas Keusch
Kindergartenstr. 34
3363 Neufurth
thomas.keusch@pgv.at

Andrea Kirchhofer
Baumgartenberg 1
4342 Baumgartenberg
admin@eurogym.asn-
linz.ac.at

Justine Kirchner
4243 Baumgartenberg
admin@eurogym.asn-
linz.ac.at

Kerstin Kirchsteiger
C/o VS Staudach,
Staudach 126
8230 Hartberg

Melanie Kitzler
Eichberg 20
3950 Gmünd

Peter Klar
Haupstr. 16
2381 Laab Im Walde
pklar@ibhs-perch.ac.at

Helmut Klauninger
Pius-Parsch-Platz 3
A-1210 Wien
hklauninger@xpoint.at

Nadine Klausner
Gnies 84
8261 Sinabelkirchen

Michaela Kleinbruckner
Baumgartenberg 1
4342 Baumgartenberg
admin@eurogym.asn-
linz.ac.at

Stefan Kleinhagauer
Breitenfurt 7
4452 Ternberg
bigkleini@gmx.at

Christoph Kleinsasser
Stelzhamerstr. 20
4600 Wels
ferris34@hotmail.com

Christian Kletzl
Irrsdorf 90
5204 Straßwalchen
ckletzl@gmx.at

Paul Klingelhuber
Baumgartenberg 1
4342 Baumgartenberg
paul.klingelhuber.@gmx.
at

Doris Knebelreiter
Egelsdorf 43
8261 Sinabelkirchen

Katrin Kober
Egelsdorf 118
8261 Sinabelkirchen

Sabrina Koch
Martinsdorf Nr. 171
2223 Martinsdorf

Georg Kochmann
Lüfteneggerstr. 13
4020 Linz

Berufsschule 9
C/o Berufsschule 9,
Wiener Str. 181
4020 Linz

Katharina Koller
C/o HS Orth/Donau,
Schlossplatz 4
2304 Orth/Donau
hauptschule.orth@aon.at

Kerstin König
Obergroßau 15
8261 Sinabelkirchen

Hannes Kornberger
C/o Vs Staudach,
Staudach 126
8230 Hartberg

Markus Kothbauer
Wielands
3950 Gmünd
gmuend.hs1@aon.at

Sonja Krainer
Griesg. 7
6112 Wattens
skrainer@infowerk.co.at

Alexandra Krämmer
Sinabelkirchen 46a
8261 Sinabelkirchen

Sebastian Kranzlmüller
Edisonstr. 2
4020 Linz

Daniel Krenn
Hintstein 38
4463 Großraming
daniel.krenn@gmx.at

Nicole Krippner
C/o HS Orth/Donau,
Schlossplatz 4
2304 Orth/Donau
hauptschule.orth@aon.at

Sylvia Kroisleitner
Rettenegg 55
8674 Rettenegg

Daniel Kropacek
Heidenreichsteinerstr. 21
3943 Schrems
kresch@lion.cc

Sindek Krzysztof
Sportplatzg. 248/2
3945 Hohenreich
gmuend.hsi@aon.at

Kevin Ku
Larocheg. 20
1130 Wien
spook@aon.at

Konstantin Kugler
Baumgartenberg 1
4342 Baumgartenberg
admin@eurogym.asn-
linz.ac.at

Johannes Kuhn
Elslerg. 9
1130 Wien
jkuhn.usnavycapt@aon.at

Katharina Kulmer
Obergroßau 56
8261 Sinabelkirchen

Stefan Lammer
Untergroßau 6
8261 Sinabelkirchen

Robert Lamprecht
Dirnberg 9
4550 Kremsmuenster

Michael Landl
Baumgartenberg 1
4342 Baumgartenberg
admin@eurogym.asn-
linz.ac.at

Eva Langeder
Obergaisberg 25
4323 Münzbach
eva_langeder_one@a-
topmail.at

Susanne Langer
Baumgartenberg 1
4342 Baumgartenberg
admin@eurogym.asn-
linz.ac.at

Patrick Langwieser
Baumgartenberg 1
43442 Baumgartenberg
admin@eurogym.osn-
linz.ac.at

Dragos Lauric
Sinabelkirchen Nr. 51
8261 Sinabelkirchen

Leonie Lawniczak
Johann-Strauß-Str. 29
4020 Linz

Daniela Lehner
Baumgartenberg 1
4342 Baumgartenberg
bgb.dlehner@yahoo.com

Roland Lehner
Baumgartenberg1
4342 Baumgartenberg
admin@eurogym.asn-
linz.ac.at

Mario Leichtfried
Birkeng. 66/3
3100 St. Pölten
schubfaktor@kstp.at

Roman Leitner
Schörgenhubstr. 18
4020 Linz

Thomas Leitner
Unterlembach 63
3962 Unterlembach
gmuend.hs1@aon.at

Ingrid Leitner
Eichberg 105
3950 Gmünd

Philip Leitner
Lassenberg
8521 Wettmannstätten
philisto@gmx.net

Manuel Leitner
Marktplatz 1
4381 St. Nikola
admin@eurogym.asn-
linz.ac.at

Christiana Leitner
Baumgartenerg 1
4342 Baumgartenberg
admin@eurogym.asn-
linz.ac.at

Mario Leonhartsberger
Einsiedelstr. 25
4323 Münzbach
admin@eurogym.asn-
linz.ac.at

Dominik Lepizh
Mayg. 43
A-1130 Wien
lepizh@chello.at

Martin Lettner
Baumgartenberg 1
4342 Baumgartenberg
duzismehlbox@gmx.at

Daniela Lettner
Baumgartenberg 1
4342 Baumgartenberg
admin@eurogym.asn-
linz.ac.at

Laurenz Lindenberger
Schwemmplatzstr. 36
4320 Perg
admin@eurogym.asn-
linz.ac.at

Sandra Litschauer
Hamerlingg. 39/2/2
3950 Gmünd
gmuend.hs1@aon.at

Lukas Loidl
Margaritenweg 2
4223 Katsdorf

Franzi Loisl
Baumgartenberg 1
4342 Baumgartenberg
f.loisl.jr@aon.at

Stefan Luftensteiner
Baumgartenberg 1
4342 Baumgartenberg
admin@eurogym.asn-
linz.ac.at

Sarah Luger
Baumgartenberg 1
4342 Baumgartenberg
admin@eurogym.asn-
linz.ac.at

Barbara Lilani
Lumesberger
Baumgartenberg 1
4342 Baumgartenberg
b.lumesberger@eurogym.
asn-linz.ac.at

Sandra Lumetsberger
Baumgartenberg 1
4342 Baumgartenberg
admin@eurogym.asn-
linz.ac.at

Carrina Machtinger
Baumgartenberg 1
4342 Baumgartenberg
admin@eurogym.asn-
linz.ac.at

Antonia Mahrl
C/o HS Orth/Donau,
Schlossplatz 4
2304 Orth/Donau
hauptschule.orth@aon.at

Marlene Maier
Feldweg 1
4040 Linz

Clemens Maier
Baumgartenberg 1
4243 Baumgartenberg
admin@eurogym.asn-
linz.ac.at

Matthias Maierhofer
Dechant-Hauer-Str. 16
3950 Dietmanns
m.maierhofer@utanet.at

Klaus Maislinger
Kaiser-Max-Str. 13
6060 Hall In Tirol
karin.peschel@hak-
hall.asn-ibk.ac.at

Florin Manaila
Obergroßau 47
8261 Sinabelkirchen

Hans-Peter Manzenreiter
Billingerstr. 9/1/4
4240 Freistadt
unga@epnet.at

Adrian Marte
Schwedenweg 13
6714 Nüziders
marteadrian@cable.vol.at

Simon Martin
Irnharting 29
4623 Gunskirchen

Christian Mastnak/
Karl Rechberger
Fürstenbergstr.1
8130 Frohnleiten
brg2k@gmx.at

Michael Matt
Gartenstr. 18
6840 Götzis
m_matt@matt.vol.at

Petra Matzhold
Untergroßau 71
8261 Sinabelkirchen

Thomas Mauerhofer
C/o VS Staudach,
Staudach 126
8230 Hartberg

Thomas Maureder
Kleinfeld 6
4210 Gallneukirchen

Christian Mayer
Bahnhofstr. 168
3942 Hirschbach
gmuend.hs1@aon.at

Kornelia Silvia Mayer
Baumgartenberg 1
4342 Baumgartenberg
kornelia_mayer@hotmail.
com

Bernhard Mayer
Fünfing 2
8261 Sinabelkirchen

Christian Mayer
C/o HS Orth/Donau,
Schlossplatz 4
2304 Orth/Donau
hauptschule.orth@aon.at

Ernstl Mayr
Baumgartenberg
4342 Baumgartenberg
mayr_ernst@hotmail.com

Verena Mayrhofer
Baumgartenberg 1
4342 Baumgartenberg
admin@eurogym.asn-
linz.ac.at

Mario Meir-Huber
Spreitzenberg 4
5222 Munderfing
mario_mh@yahoo.de

Georg Meisenberger
Itzingerstrasse 48
4910 Ried Im Innkreis
meisi@rocketmail.com

Tamara Melcher
Obergroßau 71
8261 Sinabelkirchen

Verena Michalitsch
Mayg. 43
1130 Wien
verenam.@gmx.net

Peter Micheuz
Pestalozzistr. 1
9100 Völkermarkt
peter.micheuz@aon.at

Dominik Millinger
Filzen 102
6391 St. Jakob / Tirol
dominik.mil@netway.ar

Peter Mohr
St. Peter Hauptstr. 33c
8042 Graz
anthropos@gmx.net

Stefanie Mohrenberger
Elisabethstr.
2380 Perchtoldsdorf
smohrenberger@ibhs-
perch.ac.at

Anna Mörtenhuber
Diepersdorf 67
4552 Wartberg An Der
Krems

Kathi Moser
Baumgartenberg 1
4342 Baumgartenberg
ratzi_katzi@yahoo.de

Mario Mrkonjic
Promenadeg. 41 - 45/4/21
2391 Kaltenleutgeben
mmrkonjic@ibhs-
perch.ac.at

Gregor Muck
C/o HS Orth/Donau,
Schlossplatz 4
2304 Orth/Donau
hauptschule.orth@aon.at

Martin Muck
C/o HS Orth/Donau,
Schlossplatz 4
2304 Orth/Donau
hauptschule.orth@aon.at

Daniel Mühlbachler
Leonfeldner Str. 68
4040 Linz

Andrea Mühlbachler
Karlingberg 59
4320 Perg
andrea_muehl@yahoo.co
m

Roland Mungenast
Balzerle 148a
6571 Strengen
roland.mungenast@gmx.
at

Lukas Naderer
Baumgartenberg1
4342 Baumgartenberg
nadal@netway.at

Astrid Nagl
C/o Bg XIX,
Gymnasiumstr. 83
1190 Wien
ps@t0.or.at

Matthias Nefischer
Baumgartenberg 1
4342 Baumgartenberg
mathnef@gmx.at

Tina Nefischer
Dirnbergerstr.42
4320 Perg
tina_nefischer@yahoo.de

Elisa Nefischer
Baumgartenberg 1
4342 Baumgartenberg
admin@eurogym.asn-
linz.ac.at

Nadine Nenning
C/o VS Ferschnitz,
Schulstr. 102
3325 Ferschnitz

Andreas Neuhold
Kainbach 122
8047 Graz
a_neuhold@yahoo.de

Michaela Johanna
Neulinger
Baumgartenberg 1
4342 Baumgartenberg
bgb2bmichineu@yahoo.
com

Simon Neumüller
Ried Nr. 47
4312 Ried/Riedmark
admin@eurogym.asn-
linz.ac.at

Thomas Nichelmayer
Voralpensiedlung 22
3350 Haag

Michi Niederer
Wimpassingerstr. 57
4600 Wels
michael_niederer@hotmai
l.com

Jakob Niederhauser
Lärchenauer str. 42a
4020 Linz

Ela Nussbaumer
Baumgartenberg 1
4342 Baumgartenberg
m.nussbaumer@eurogym.
asn-linz.ac.at

Tobias Oberascher
Nelkeng. 6
5204 Strasswalchen
rumblecat@hotmail.com

Thomas Oberhauser
Mitteldorf 30
6886 Schoppernau
toemme@i-one.at

Renate Oblak
Leitnerg. 7/5
8010 Graz
remi@algo.mur.at

Jürgen Oman
(conspirat).
Grillparzer Str. 15
4614 Marchtrenk
j.oman@netway.at

Simon Opelt
Kirchenstr. 2
4864 Attersee
s.opelt@gmx.at

Manuel Orlinger
Pfarrfeld 5
4210 Gallneukirchen

Gernot Ostermann
C/o HS Orth/Donau,
Schlossplatz 4
2304 Orth/Donau
hauptschule.orth@aon.at

Markus Oswald
Prof.-Franz-Spath-Ring
13/17
8042 Graz
m_oswald@gmx.net

Sonja Patzl
C/o HS Orth/Donau,
Schlossplatz 4
2304 Orth/Donau
hauptschule.orth@aon.at

Helmut Pauer
Im Hoffeld 54
8046 Graz
helmut_pauer@hotmail.
com

Irene Pechböck
Baumgartenberg 1
4342 Baumgartenberg
admin@eurogym.asn-ac.at

Mathias Pechböck
Baumgartenberg 1
4342 Baumgartenberg
admin@eurogym.asn-
linz.ac.at

Sandra Pernat
Untergroßau 138
8261 Sinabelkirchen

Daniela Pfeifer
Unterrettenbach
8261 Sinabelkirchen

Benjamin Pflügl
Sacharowweg 28
4030 Linz

Daniel Pfordte
C/o BRG Anton-Bruckner-
Str. 16
4600 Wels
brucknergym.wels@asn-
linz.ac.at

Kerstin Pichler
Rosenhügel 4
3950 Ehrendorf
gmuend.hs1@aon.at

Gernot Pichler
Dorfgraben 266
2873 Feistritz
gernot.pichler@usa.net

Elisa-Maria Pichler
Baumgartenberg 1
4342 Baumgartenberg
admin@eurogym.asn-
linz.ac.at

Markus Piereder
Stelzhamer Str. 11
5280 Braunau
mpeireder@hotmail.com

Christoffer Piereder
Stelzhamer Str. 11
5280 Braunau
cpiereder@hotmail.com

Michaela Pierer
Brahmsstr. 30
4020 Linz

Joachim Pierer
Brahmsstr. 30
4020 Linz

Lukas Pilat
Wilbrandtg. 23
1180 Wien
lukas@workmail.com

Martina Pirklbauer
Baumgartenberg 1
4342 Baumgartenberg
admin@eurogym.asn-
linz.ac.at

Philipp Pirklbauer
Baumgartenberg 1
4342 Baumgartenberg
admin@eurogym.asn-
linz.ac.at

Julia + Sarah Pleiner
Grasbach 69
4210 Gallneukirchen
sonvilla@ufg.ac.at

Simone Podlesnic
Lerchenweg 6
4331 Naarn
simone_podlesnic@yahoo
.de

Marlena Polec
Markt 66/1
4363 Pabneukirchen
bgbmpolec@yahoo.com

Sara Posch
C/o Berufsschule 9,
Wiener Str. 181
4020 Linz

Beate Prantner
Hörmannserstr. 23
3950 Dietmanns
gmuend.hs1@aon.at

Tanja Praxmarer
C/o VS Mötz
Kirchplatz 3
6423 Mötz
moetz01@asn.netway.at

Ulrike Prem
C/o HS Orth/Donau,
Schlossplatz 4
2304 Orth/Donau
hauptschule.orth@aon.at

Philipp Princic
C/o VS Ferschnitz,
Schulstr. 102
3325 Ferschnitz

Wolfgang Prinz
Grazerstr. 34g
8045 Graz

Robert Prückl
Markt 4
4291 Lasberg
veliant@yahoo.com

Elke Pühringer
Fuchsenweg 10
4320 Perg
admin@eurogym.asn-
linz.ac.at

Stefan Pühringer
Columbusstr. 7
4600 Wels
pstefan@liwest.at

Desiree Purkarthofer
Obergroßau 11
8261 Sinabelkirchen

Manuel Purkarthofer
Frösau 27
8261 Sinabelkirchen

Beate Purkarthofer
Obergroßau 9
8261 Sinabelkirchen

Martin Puschner
C/o HS Orth/Donau,
Schlossplatz 4
2304 Orth/Donau
hauptschule.orth@aon.at

Michael Raab
Schulg. 3
4081 Hartkirchen
hs.hartkirchen@eduhi.at

Melanie Rab
C/o VS Ferschnitz,
Schulstr. 102
3325 Ferschnitz

Johanna Melitta
Raffetseder
Baumgartenberg 1
4342 Baumgartenberg
j.raffetseder@eurogym.as
n-linz.ac.at

Iris Raffetseder
Obergassolding 18
4342 Baumgartenberg
admin@eurogym.asn_linz.
ac.at

Lisa Ratzenböck
Purwörth 13
4111 Walding
sonvilla@ufg.ac.at

Veronika Ratzinger
Baumgartenberg 1
4342 Baumgartenberg
ratzi_katzi@yahoo.de

Peter Rauschmeier
Steinberg-Frankg. 8
2380 Perchtoldsdorf
gravedigger@kabsi.at

Bauernfeind Regina
Baumgartenberg 1
4342 Baumgartenberg
admin@eurogym.asn-
linz.ac.at

Gernot Reiber
Frühlingstr. 33
8053 Graz
gernot_paradise@yahoo.
de

Thomas Reiberger
Westbahnstr. 24
3385 Markersdorf
reibsi@mail.gymmelk.
ac.at

Thomas Reichhart
Elz 47
4292 Kefermarkt
t.reichhart@edumail.at

Lukas Reindl
Baumgartenberg 1
4342 Baumgartenberg
admin@eurogym.asn-
linz.ac.at

Kilian Reisiner
Baumgartenberg 1
4342 Baumgartenberg
admin@europagym.asn-
linz.ac.ac

Ingrid Reisinger
Kalmberg 14
4352 Klam/Bad Kreuzen
bgbireisinger@yahoo.com

Ilona Reisinger
Josefstal 16
4311 Schwertberg
ilona_reisinger@yahoo.de

Thomas Reiter
Ried 142
4312 Ried/Riedmark
cmcriedrdm@eduhi.at

Stephan Reiter
Libellenweg 4
4030 Linz
st_reiter@aon.at

Harald Reiter
Hofkirchen 48
4351 Saxen
admin@eurogym.asn-
linz.ac.at

Thomas Reitinger
Stelzhamerstr. 20
4600 Wels
treitinger@hotmail.com

Verena Riedl
C/o BG Waidhofen
Gymnasiumstr. 1
3830 Waidhofen/Thaya

Sabine Riegler
Haidershofen 43
4431 Haidershofen

Michael Riegler
Baumgartenberg 1
4342 Baumgartenberg
admin@eurogym.asn-
linz.ac.at

Maria Ringhofer
Markt 12
2842 Edlitz
bal.hs@aon.at

Eva Nicola Rinner
Kammern 169
6863 Egg
eva.nicola@rinner.at

Robin Ristl
Kaiser-Ebersdorferstr.
90/11/53
1110 Wien
robinristl@gmx.at

Lukas Rittberger
Blindendorf 131
4312 Ried/Riedmark

Julia Rohn
Santner str. 519
5071 Wals

Hanna Rohn
Santnerstr. 519
5071 Wals

Helene Rohrauer
C/o Europagymnasium
Parkstr.
4311 Schwertberg
admin@eurogym.asn-
linz.ac.at

Sophie Rohrauer
Baumgartenberg 1
4342 Baumgartenberg
admin@eurogym.asn-
linz.ac.at

Viktoria Roidinger
Blindendorf
4312 Ried / Riedmark
admin@eurogym.asn-
linz.ac.at

Elisabeth Romero
C/o Helmut Klauninger,
Pius-Parsch-Platz 3
1210 Wien
hklauninger@xpoint.atl
(lehrer)

Wolfgang Rominger
Untergroßau 50
8261 Sinabelkirchen

Martina Rominger
Frösau 5
8261 Sinabelkirchen

Sabrina Rösel
Alois-Ullrich-G. 3
3950 Gmünd

Katrina Rosenberger
C/o VS Ferschnitz,
Schulstr. 102
3325 Ferschnitz

Christoph Rosenlechner
Monsbergerg. 16
8010 Graz
haller@borg-graz.ac.at

Bianca Roßmann
Theodor Körnerstr.
173/4/10
8010 Graz
b_rossmann@yahoo.at

Sinan Saral
Hegerg. 15/15
1030 Wien
ahmet.saral@teleweb.at

Chriestina Satler
Baumgartenberg 1
4342 Baumgartenberg
admin@eurogymasn-
linz.ac.at

Roman Schacherl
Kakteenstr. 6
4481 Asten
roman.schacherl@lion.cc

Benedikt Schalk
Gladbeckstr. 1/12b/2
2320 Schwechat
bschalk@gmx.net

Stefan Schallerl
Wolfgruben 86
8200 Gleisdorf
mrnasty@gmx.net

Eva-Maria Schandl
Schloßg. 7
3950 Gmünd
gmuend.hs1@aon.at

Daniela Schatz
C/o Schulstr. 102
3325 Ferschnitz

Markus Schausbeger
Baumgartenberg 1
4342 Baumgartenberg
m_schausberger@yahoo.
de

Wolfgang Schedl
Maroltingerg. 9a/1
1160 Wien
wolfgang.schedl@black
box.net

Judith Scheibelberger
Baumgartenberg 1
4342 Baumgartenberg
judith_sch@yahoo.de

Claudia Scheidl
Baumgartenberg 1
4342 Baumgartenberg
claudschy@hotmail.com

Julia Schellner
Gugitzg. 15/1
1190 Wien

Jakob Scherzenlehner
C/o Schulstr. 102
3325 Ferschnitz

Kristin Scherzer
Dr.-Th.-Weippelg. 15
3424 Zeiselmauer

Petra-Anna
Scheuchenegger
Baumgartenberg 1
4342 Baumgartenberg
blind212@hotmail.com

Patricia Schiefer
Kindergartenstr. 6
4310 Linz
p_schiefer@a-topmail.at

Rainer Schildberger
Brunnental 12
4571 Steyrling
r.schildberger@edumail.at

Karoline Schillinger
C/o Bg XIX,
Gymnasiumstr. 83
1190 Wien
ps@to.or.at (fr.suko)

Simone Schimpf
Diesterwegg. 39/30
1140 Wien
jschimpf@bhakwien13.at

Eva Schindling
Grazerg. 6
8430 Leibnitz
evaschindling@hotmail.
com

Gerald Schinwald
Baumgartenberg 1
4243 Baumgartenberg
admin@eurogym.asn-
linz.ac.at

Rupert Schlager
Baumgartenberg1
4342 Baumgartenberg
admin@eurogym.asn-
linz.ac.at

Lisa Maria Schmidtberger
Baumgartenberg 1
4342 Baumgartenberg
bgblschmidtberger@
yahoo.com

Andrea Schmidtberger
Baumgartenberg 1
4342 Baumgartenberg
andreaschmidtberger@hot
mail.com

Wolfgang Schmied
Spöttlstr. 10/12
4600 Wels
euprojekt@crosswinds.net

Susanne Schmied
Litschauer Str. 35
3950 Gmünd
gmuend.hs1@aon.at

Jörg Schneider
Neue Heimat 14
4360 Grein
admin@eurogym.asn-
linz.ac.at

Stefan Schober
Baumgartenberg 1
4342 Baumgartenberg
admin@eurogym.asn-
linz.ac.at

Raphael Schön
Baumgartenberg 1
4342 Baumgartenberg
schoen.raph@aon.at

Josef Schreiner
C/o HS Orth/Donau,
Schlossplatz 4
2304 Orth/Donau
hauptschule.orth@aon.at

Christian Schreiner
C/o HS Orth/Donau,
Schlossplatz 4
2304 Orth/Donau
hauptschule.orth@aon.at

Matthias Schreiner
Hinterebenau 27
5323 Ebenau
srem111@hotmail.com

Paul Schreitl
Nr. 162
2223 Martinsdorf
stromkabl@gmx.at

Julia Schreyer
Stadtplatz 4
4360 Grein
admin@eurogym.asn-
linz.ac.at

Philipp Schreyer
Baumgartenberg 1
4243 Baumgartenberg
admin@eurogym.asn-
linz.ac.at

Dieter Schuh
Schulg. 1
3950 Gmünd
d.schuh@i-plus.at

Sabrina Schuhaber
Panoramastr. 6
4311 Schwertberg
admin@eurogym.asn-
linz.ac.at

Stefanie Schulmeister
Primelweg 7
2384 Breitenfurt Bei Wien

Katrin Schuster
Retzerweg 4
3741 Pulkau
310062@asn.netway.at

Klaus Schustereder
Pötting 5
4720 Pötting
schusti@utanet.at

Georg Schusterschitz
Grillparzerstr. 27
8010 Graz
georgschusterschitz@web
.de

Philipp Schweiger
Neumarkt/ Raab 255
8380 Neumarkt/Raab
ph_schweiger@hotmail.
com

Sabrina Schwingenschlögl
HS 1 Gmünd
Schulg.
3950 Gmünd
gmuend.hs1@aon.at

Gerhard Schwoiger
Heudörfelg. 5
A-1230 Wien
gerhard@v-a-m-p.com

Verena Seidl
C/o VS Staudach 126
8230 Hartberg
607391@asn.netway.at

Martin Semmernegg
Messendorferstr. 140
8042 Graz

Manuel Simlik
Friedhofstr. 425
3945 Hoheneich
gmuend.hsi@aon.at

Grigory Sipachev
Möllwaldplatz 6a
1040 Wien
greg.sipachev@hotmail.
com

Katharina Soder
Walzwerkstr. 16
4050 Traun

Roman Sollböck
Amonstr. 36a
3293 Lunz Am See
big_m@websms.at

Felicitas Sonvilla-Weiss
Novarag.38a/17
1020 Wien
feli@yline.com

Claudia Sottner
Gilmstrasse 51a
6130 Schwaz
css@firemail.de

Michaela Spettel
Arlbergstr. 88
6900 Bregenz
oeller@roemer.vol.at

Michaela Spirk
Königsbergstr.
8261 Sinabelkirchen

Katharina Stadlbauer
Untergassolding 1
4342 Baumgartenberg
admin@eurogym.asn-
linz.ac.at

Clemens Stadler
C/o Bg XIX,
Gymnasiumstr. 83
1190 Wien
ps@to.or.at (fr.suko)

Daniel Stallinger
Baumgartenberg 1
4342 Baumgartenberg
admin@eurogym.asn-
linz.ac.at

Ewald Stangl
Grillparzerstr. 1
3860 Heidenreichstein
ewald.stangl@wvnet.at

Benjamin Stangl
Adlerg. 10
4020 Linz
benjaminstangl@geocities
.com

Lukas Staudinger
Agnesg. 2
1190 Wien
7terzwerg@maerchenland.
net

Silvia Steffan
C/o VS Mölz
Kirchplatz 3
6423 Mötz
moetzo1@asn.netway.at

Thomas Steidl
Prof.-Ploner-Str. 11a
9900 Lienz
steidl.seeber@tirol.com

Sara Steinberger
Baumgartenberg 1
4342 Baumgartenberg
bgbssteinberger@yahoo.
com

Markus Steinbichler
Katzgraben 27
4203 Altenberg/oö.

Markus Steiner
Weisching
4343 Mitterkirchen
admin@eurogym.asn-
linz.ac.at

Christina Steininger
Baumgartenberg 1
4342 Baumgartenberg
bgbcsteininger@yahoo.de

Sandra Steininger
Egelsdorfberg 6
8261 Sinabelkirchen

Lisa Steinkelllner
Baumgartenberg 1
4342 Baumgartenberg
admin@eurogym.asn-
linz.ac.at

Thomas Massimo
Stermole
Brucknerg. 6
8753 Fohnsdorf

Manfred Steyer
Mitterfladnitz 124
8322 Studenzen
m.steyer@inode.at

Christina Stieger
Windhaag 144
4322 Windhaag
sweet_chris_at@yahoo.de

Arnold Stocker
Bach 83
9546 Bad Kleinkirchheim
jo.stocker@aon.at

Leopold Stockinger
Rotte Nöchling 27
3332 Rosenau/S.
lstockinger@hotmail.com

Phillip Strahl
Sobothg. 10
8054 Graz
joesoftstudios@yahoo.
com

Gerhard Strassegger
Fünfing 35
8261 Sinabelkirchen

Martin Strasser
Adam-Mölk-Str. 20
6341 Ebbs
strasser@bg-kufstein.asn-
ibk.ac.at

Silvia Strasser
C/o VS Ferschnitz
Schulstr.102
3325 Ferschnitz

Kristina Strasser
Lainsitzweg 8a
3950 Gmünd

Johanna Strasser
Thurnhof 1
4320 Perg
johanna_strasser@yahoo.
de

Isabella Strasser
Hochtor 31
4322 Windhaag
admin@eurogym.asn-
linz.ac.at

Ines Streifert
Baumgartenberg 1
4342 Baumgartenberg
i.streifert@mail.asn-
linz.ac.at